Environmental Science and Engineering
Subseries: Environmental Science

Series Editors: R. Allan • U. Förstner • W. Salomons

L. Håkanson · A.C. Bryhn

Tools and Criteria for Sustainable Coastal Ecosystem Management

Examples from the Baltic Sea and Other Aquatic Systems

Springer

Prof. Dr. Lars Håkanson
Uppsala University
Dept. Earth Sciences
Villavägen 16
SE-752 36 Uppsala
Sweden
lars.hakanson@geo.uu.se

Dr. Andreas C. Bryhn
Uppsala University
Dept. Earth Sciences
Villavägen 16
SE-752 36 Uppsala
Sweden
Andreas.Bryhn@geo.uu.se

ISBN: 978-3-540-78361-9 e-ISBN: 978-3-540-78363-3

Environmental Science and Engineering ISSN: 1863-5520

Library of Congress Control Number: 2008924331

Cover Design: deblik, Berlin

Printed on acid-free paper

9 8 7 6 5 4 3 2 1

springer.com

Contents

Prologue

This book addresses sustainable coastal management with a focus on practically useful (= operational) tools, methods, models and criteria and how targets variables for monitoring and management vary among and within coastal ecosystems (at the ecosystem scale). The ecosystem scale is, we would argue, an important scale in aquatic management, e.g., in contexts of impact assessment, when remedial measures are discussed and when basic questions are asked, e.g.: What is the status of this ecosystem? What can be done to improve the conditions? Which positive and negative consequences could be linked to a given or suggested remedial measure? Few people would be interested in the content of a sampling bottle; most people in science and management are interested in what this content may actually represent. That is, the interest is on a larger entity, the ecosystem. But there is no contradiction between work at this larger ecosystem scale and sampling and work at smaller scales, since the mean or median values characterizing ecosystem conditions and the standard deviations in empirical data characterizing the variability around such mean values of necessity must emanate from sampling at individual sites.

The aim of this book is to address important questions related to the role a number of operational bioindicators (such as the Secchi depth, concentrations of chlorophyll-a and cyanobacteria, the oxygen concentration in the deep-water zone and the macrophyte cover) and how the variability within and among coastal areas of these bioindicators is related to differences in coastal morphometry, salinity, temperature and nutrient loading. We discuss an index of coastal area sensitivity to nutrient loading (eutrophication) and an index of the "biological value" of coastal areas. These are central aspects of coastal eutrophication studies and also key questions related to the European Water Framework Directive. There is a long tradition in lake studies to carry out comparative investigations and look for factors causing variations in functional characteristics among lakes. This work follows that tradition for marine systems.

This book is mainly based on results from the Thresholds-project, an EU-project on threats to European coastal ecosystems, coordinated by Carlos Duarte, CSIC University, Mallorca. The aim has been to try to write a state-of-the-art book for coastal scientists and managers with different educational backgrounds (biology, ecology, geosciences, oceanography, physics, chemistry, economics, etc.). Both simple and more advanced models are used, but this volume is not a textbook in modeling, and there are, for example, no differential equations in the running text. It is enough to know addition, subtraction, multiplication and division to follow this text. But it is not possible to understand how coastal systems function without a basic knowledge of the transport processes (inflow, outflow, sedimentation, resuspension, biouptake, etc.) that exist in all aquatic systems and apply to all substances.

Most of the data discussed in this book emanate from comprehensive "data-mining" of public sources available via the Internet. Several persons in our group at Uppsala University have participated in the data-mining and the work to develop and test the models and tools discussed in this work, especially Dan Lindgren, Jenny Eklund, Julia Hytteborn, Thorsten Blenckner and Maria Stenström-Khalili. This book is also a compilation and review of the results from our group during the last 4 years related to coastal studies.

Chapter 1
Introduction and Aim

Due to the importance and complex nature of coastal areas, it is easy to understand why so much interest and research concern this zone. The coast is a zone of conflicts where many users, such as professional and leisure-time fishermen, people responsible for recreation, shipping and environmental management and research, place different demands on the coastal waters and apply different criteria to set the value of coastal waters and to define desired conditions (Wallin et al. 1992; Lundin 1999, 2000a, b; Wulff et al. 2001a). The coastal zone is also a "recipient" of many types of pollutants, such as organic matter, radionuclides, nutrients and organic toxins (Pearson and Rosenberg 1976; Ambio 1990, 2007; Meeuwig et al. 2000; Aertbjerg 2001). The coast may be regarded as "a pantry and a nursery" for the sea. It has been demonstrated that shallow coastal areas can have a bioproduction many times higher than the most productive areas on land (see Chap. 5 and Rosenberg 1985). All three functional groups of primary producers – phytoplankton, benthic algae and macrophytes – are present in coastal areas (but not in open water areas). Where there is a high primary production, there is also a high secondary production of zooplankton, zoobenthos and fish (Mann 1982; Sandberg et al. 2000; Håkanson and Boulion 2002).

One purpose of this book is to discuss and review a set of operational bioindicators for coastal management (see Moldan and Billharz 1997; Livingston 2001; Bortone 2005) mainly in terms of eutrophication and to illustrate how these bioindicators relate to one another and express fundamental structural and functional properties of coastal ecosystems. From the perspective of the ecosystem scale, relatively little research has been devoted to the important but complex problem of developing practically useful bioindicators of ecosystem status, thresholds and regime shifts (see Carpenter 2003). Since 1987, many countries have accepted "sustainable management" as a goal for environmental and economic policy. The term was introduced in the final report of the Commission for Environment and Development (the Brundtland Commission). However, this phrase is empty unless it is defined in terms of operationally measurable properties, desired goals and relevant data (Bailey et al. 1985). There are alternatives to choosing the ecosystem as the basis for the environmental typology (see, e.g., O'Neill et al. 1982; Cairns and Pratt

1987). There is, however, a clear international trend towards consideration of the "health" of the different ecosystems.

Practically useful bioindicators should be (see Håkanson 1999):

- representative for the given ecosystem,
- simple and inexpensive to measure,
- clearly interpretable and predictable by validated quantitative models,
- internationally applicable,
- relevant for the given environmental threat.

Ideally, operational bioindicators for water management should be comprehensible without expert knowledge. In fact, one reason to develop such measures is so that politicians and the general public can understand the present condition and reasons for changes in the environment. The complicated nature of ecosystems makes it very difficult indeed to carry out causal, mechanistic analyses concerning the quantitative linkages between a given threat or pressure (like increased nutrient loading) and variables expressing ecosystem effects. As a background to the forthcoming discussions about bioindicators, Table 1.1 provides information on how different European countries (Finland, Sweden and the Netherlands) have classified a number of standard bioindicators in the context of the European Water Framework Directive.

The classes, high, good, moderate, poor and bad are generally used as descriptive terms. By reviewing the different reference guidelines for coastal management and the European Water Framework Directive, it is obvious that different reference levels are used in different countries, which is an evident problem. In addition to the information compiled in Table 1.1, there are many more criteria existing in the guidelines, in particular concerning biological variables for the assessment of

Table 1.1 Water quality criteria from different countries for mainly coastal areas for chlorophyll-a concentrations (Chl), concentrations of total phosphorus (TP), total nitrogen (TN) and water transparency (Secchi depth) from different sources: www.environment.fi/waterquality, www.environ.se

	Classes				
	I high	II good	III moderate	IV poor	V bad
Finland					
Chl lakes (μg/l)	<4	<10	<20	20–50	>50
Chl sea (μg/l)	<2	2–4	4–12	12–30	>30
TP lakes (μg/l)	<12	<30	<50	50–100	>100
TP sea (μg/l)	<12	12–20	20–40	40–80	>80
Secchi (m)	>2.5	1–2.5	<1		
Sweden					
Chl (μg/l)	<1.5	1.5–2.2	2.2–3.2	3.2–5	>5
Secchi (m)	>5.4	4–5.4	3.4–4	2.5–3	>2.6
TN (μg/l)	<266	266–350	350–490	490–756	>756
TP (μg/l)	<22.6	22.6–28	28–34	34–40	>40
Netherlands					
Chl (μg/l)	<9.3	9.3–14	14–28	28–56	>56

species composition of, for example, macrophytes, phytoplankton, benthic fauna, fish, etc. Such biological variables may be very important indicators of water quality, but they are often descriptive and often also coupled to large areal, spatial and temporal variabilities. They generally have very high coefficients of variation, which means that many samples are required to obtain mean values with small uncertainty bands (see Chap. 4) and if they can be operationally modeled, it can generally be done only with a high uncertainty and low predictive power. Therefore, the concept of "operational" bioindicators (or effect variables) is very important for coastal management.

Many of the examples and data used in this book emanate from extensive "data-mining", and Table 1.2 gives a list of websites which have been used in this work. The public accessibility of data from databases has been of paramount importance for this work, and some of these websites are excellent in terms of overview and accessibility of data.

The "best of the best" would, we would argue, be the database for Swedish lakes and the database for Chesapeake Bay (see Cooper and Brush 1993; Boesch et al. 2000). People responsible for other databases should study the structure of those databases. There are many databases around and some of them are so poorly structured that it is virtually impossible to use them and since these databases exist because of governmental funding, the data should not just be assessable in theory but also in practice.

Looking at the examples and case-studies in this book, the reader might get the impression that this book mainly concerns the conditions in the Baltic Sea. That was never our intention. The aim has been to address general tools, methods and criteria, which would apply to most coastal areas, but it is difficult to find better data than

Table 1.2 Compilation of main references for data used in this work

Area	Country	Main references
Chesapeake Bay	U.S.A.	http://www.chesapeakebay.net/data/wq_query1.cfm?db =CBP_WQDB
European coastal zone		http://dataservice.eea.europa.eu/dataservice/ http://www.loicz.org/public/loicz/typology/typodoc.pdf http://www.bgs.ac.uk/products/digbath250/ http://www.esri.com/data/download/basemap/index.html
Italian coast	Italy	http://dataservice.eea.europa.eu/dataservice/metadetails.asp?id=836
Bothnian Bay	Sweden	http://dataservice.eea.europa.eu/dataservice/metadetails.asp?id=836
West coast	Sweden, Norway	http://dataservice.eea.europa.eu/dataservice/metadetails.asp?id=836
Swedish lakes	Sweden	http://info1.ma.slu.se/db.html
Baltic Sea		Wallin et al. (1992); ICES (2006a, b, 2007); SMHI (2007); www2.ecology.su.se/dbHFJ/index.htm, 2007-09-04
Ringkobing Fjord	Denmark	Petersen et al. (2006)

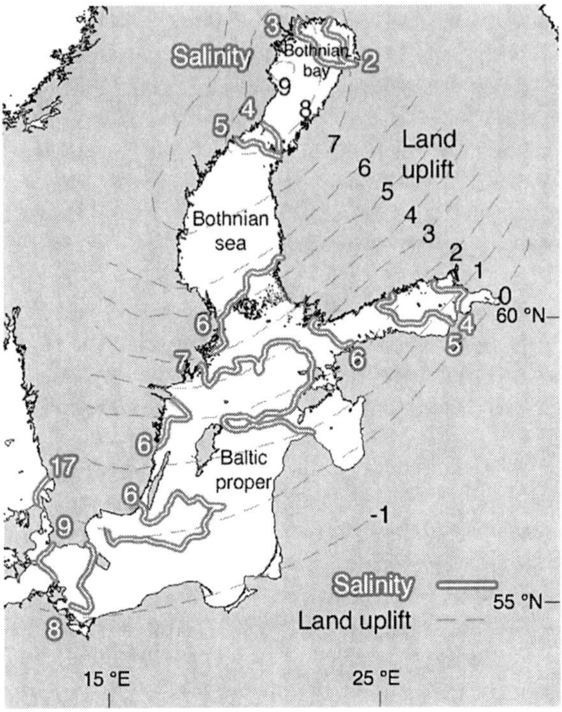

Fig. 1.1 Geographical overview of the Baltic Sea (from Håkanson and Gyllenhammar 2005). Land uplift in mm/yr. Salinity in psu. The scales give latitudes and longitudes

those from the Baltic Sea and this explains why so many examples and case-studies use data from this part of the world. So, for persons not familiar with this system, Fig. 1.1 gives a geographical overview and the names of the three main basins (the Bothnian Bay, the Bothnian Sea and the Baltic Proper) in this system. This figure also shows the salinity gradient in this brackish system. One can note that the salinity decreases from about 17 psu at the Danish Straits to about 3 psu in the northern part of the Bothnian Bay. It is easy to imagine the enormous water dynamics of the system which is responsible for the inflow of salt water from the south (Kattegat and Skagerack), the freshwater outflow and the rotation of the earth (the Coriolis force), the variations in winds and air pressures that causes the necessary mixing and water transport revealed by the salinities seen in Fig. 1.1.

These salinities demonstrate that this is a very dynamic system – and so are most coastal systems!

The water fluxes needed to explain the measured salinities can be calculated by means of process-based mass-balances for salt and Fig. 1.2 gives these water fluxes for the Baltic Sea and its three main basins. All these water fluxes are given in km^3/year. This figure also gives water fluxes from rivers, precipitation and evaporation. For these fluxes, data from Omstedt and Axell (2003) for the period 1981–1998 were used and these data also largely agree with data used by Monitor

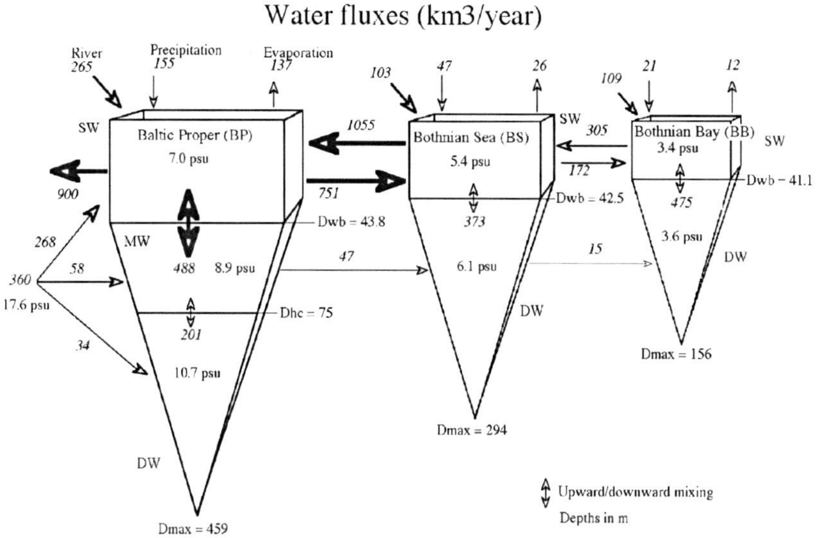

Fig. 1.2 Calculated fluxes of water to, from, between and within the three main basins of the Baltic Sea, the Baltic Proper, the Bothnian Sea and the Bothnian Bay (from Håkanson et al. 2007d)

(1988). The model used for these calculations is the process-based mass-balance model, CoastMab, outlined in Sect. 9.1 and used for the case-studies discussed in Chap. 6.

For coastal management, it is important to:

- Define permissible ranges (lower and upper values), thresholds and points of no return for all bioindicators. This should be done to minimize risks related to changes in ecosystem structure and biodiversity.
- Keep an open dialogue between scientists, policy makers, administrators and the general public based on facts and reason (rather than feelings and emotions, which are ingredients in many "environmental" debates and discussions).

Table 1.3 gives an updated and revised compilation of results from a project (see Håkanson et al. 2000) that had the following goals:

- to develop a system of water quality indices according to specific requirements of different water users;
- to establish normal values (corresponding to natural, reference conditions) of the chosen set of indices;
- to estimate the environmental sensitivity and stability of the ecosystems by applying mathematical models of fluxes for suspended particulate matter and nutrients.

Some of the variables listed in Table 1.3 and many bioindicators based on individual species of, e.g., zoobenthos, plankton or fish, discussed in other contexts (see, e.g., Bortone 2005) do not meet the previously discussed criteria set up for this

Table 1.3 Compilation of general, operational target variables for water management

Basic management objectives	Target bioindicators (y_i)
1. Conservation (water quality)	1. Secchi depth 2. Oxygen saturation in deep-water 3. Chlorophyll-a concentration 4. Concentration of cyanobacteria 5. Macrophyte cover or biomass 6. Number of coliform bacteria
2. Recreation (angling, swimming, etc.)	1. Secchi depth 2. Max. phytoplankton biomass 3. Chlorophyll-a concentration 4. Macrophyte cover or biomass 5. Concentration of cyanobacteria (harmful algae) 6. Number of coliform bacteria
3. Fishery (professional fishing and aquaculture)	1. Fish biomass (predatory fish biomass) 2. Toxic substances in fish e.g., PCB, dioxins, ^{137}Cs, Hg, 3. Biomass/production of target fish species, e.g., cod

work concerning operational bioindicators valid for entire coastal areas (the ecosystem scale). Therefore, we will focus on the following operational bioindicators:

- Secchi depth (a standard measure of water clarity).
- Chlorophyll-a concentrations (a simple, operational, standard measure of algal biomass).
- The concentration of cyanobacteria (as a measure of the biomass of "harmful" algae).
- The oxygen saturation in the deep-water zone (regulating the survival of an important functional group, the zoobenthos).
- The macrophyte cover (or biomass; as a measure of coastal productivity; the "biological value" of the coastal area).

These bioindicators are easily understood by coastal managers and also simple and relatively inexpensive to measure and they can be predicted by general well tested models at the ecosystem scale (see Chap. 5).

Cyanobacteria are selected since they cause the most resented algal blooms in the Baltic Sea and many other systems and because their biomass can be predicted at certain timescales. Other types of harmful algae may be more relevant for coastal management in other parts of the world.

Table 1.3 lists biotic and abiotic target variables for three different categories of water users, i.e., from three management perspectives:

1. Conservation of the aquatic ecosystem at a state allowing sustainable development and maintenance of desired water quality. Evidently, different users might have different demands on "water quality" and different criteria to define "water quality" and setting management targets. An effect variable for this category, which is important but not operational in the sense that it can be predicted today (to the best of our knowledge) at the ecosystem scale with validated (well

tested) general models, is the number of coliform bacteria (which is also of great interest in contexts of recreation, swimming, etc.). Key abiotic variables influencing the operational bioindicators include suspended particulate matter (SPM; see Chap. 5), nutrient concentrations (phosphorus and nitrogen; see Chap. 4), salinity, coastal morphometry (size and form) and the water exchange between the coast and the sea (see Chap. 3) and in estuaries also the characteristics of the catchment areas (see Stålnacke et al. 1999a, b, 2003, 2004). These target bioindicators and the key abiotic variables will be discussed in greater detail in the following chapters.

2. Recreation, with a focus on suitable conditions for angling, swimming, etc. Target bioindicators are, e.g., Secchi depth (water clarity), maximum (rather than mean or median values) phytoplankton biomass, chlorophyll-a concentrations, macrophyte cover (regulating access to the shoreline for recreation), concentration of cyanobacteria (which may cause damage for animals and man) and concentration of bacteria.
3. Fishery, with a focus on biomass and production of "attractive" species of fish with low concentrations of toxic substances (like organics, radionuclides and metals).

In a study of the coast of Britain, Andrle (1996) concluded that coastline complexity varies in a continuous and systematic way with scale. Figure 1.3 illustrates that it is important to take scale into account when analyzing geographical data in order to find enclosed and sensitive areas. An area that does not appear to be enclosed on a regional scale may, in fact, be enclosed and highly sensitive when studied at a local scale.

An overview of the coastal classification system discussed in this work is shown in Fig. 1.4. The classification is done in steps, using the following key attributes: topographical openness (= exposure), salinity regime, trophic regime and ecological sensitivity of a given coastal area. The first attribute, exposure, is determined

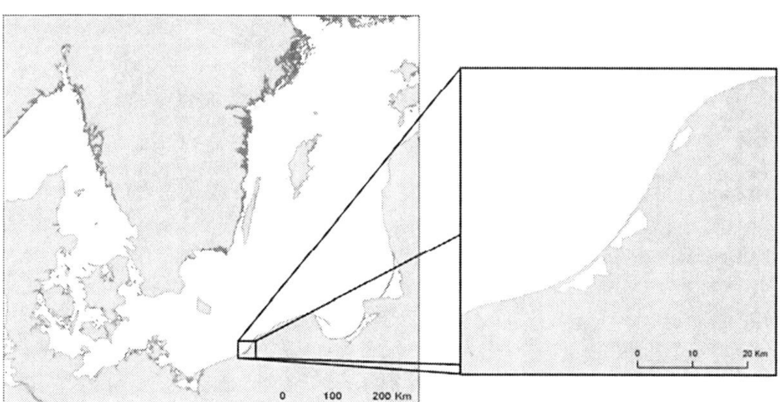

Fig. 1.3 Example illustrating the importance of taking scale into account

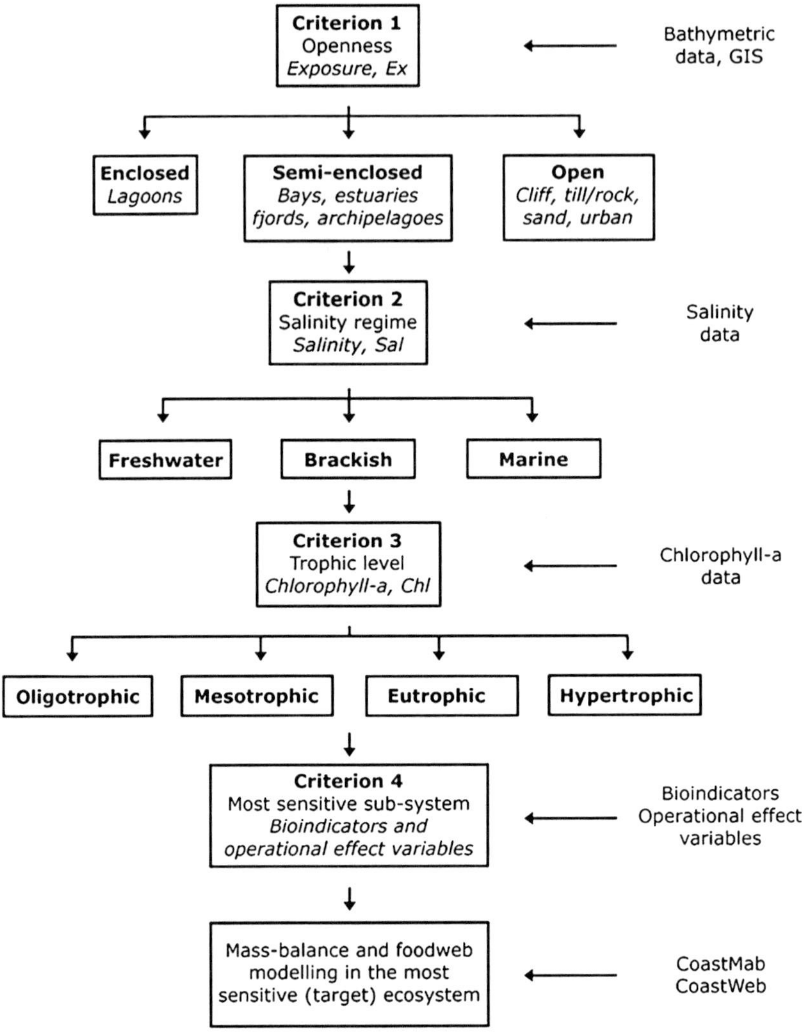

Fig. 1.4 Overview of the coastal classification workflow (from Lindgren and Håkanson 2007)

using bathymetric data. The second attribute is the salinity regime. The salinity is important for a number of reasons, as described in more detail in Chap. 3. The third attribute is the productivity of the system, characterized by the trophic regime using chlorophyll-a concentrations in the surface-water layer for the growing season as the key criteria.

Håkanson et al. (2007a) gives further motives for the different classes and here we will just stress that the three classes for salinity are mainly motivated by the threshold values for the number of species (the number of species reaches a minimum value at a salinity of 5 psu, see Remane 1934).

After having selected an interesting target ecosystem at the local and regional levels with a high loading (pressure) and/or a high sensitivity, mass-balance modeling can be performed for a given coastal area to get a good understanding of which processes are most important in that system so that the appropriate remedial actions may be taken and critical thresholds avoided.

The set-up of this book is as follows:

- One and the same load of nutrients and/or toxins may cause very different effects in coastal areas of different size and form. The effect-load-sensitivity (ELS) analysis is an important approach in this context and the basic elements of ELS-analysis and modeling will be discussed in Chap. 2.
- Chapter 3 discusses and motivates abiotic variables influencing coastal area sensitivity to nutrient loading and mainly factors that are inherent and can not readily be modified by man or remedial actions at local to regional scales.
- Chapter 4 focuses on nutrients, which regulate primary phytoplankton production. The anthropogenic loading of nutrients can be reduced if the proper remedial actions are taken and this chapter discusses fundamental aspects of coastal eutrophication, such as "limiting" nutrient, the "predictive power" of operational models for the target bioindicators based on different nutrients (nitrogen and phosphorus) and different forms of the nutrients (total-N, total-P, dissolved forms and organic forms), statistical aspects related to the variability and representativity of different nutrient forms, patterns in variability and criteria for remedial actions (whether remedial actions should focus on reductions in nitrogen or phosphorus loading or both).
- Chapter 5 sets the focus on the target bioindicators: Secchi depth, the concentration of chlorophyll-a, the concentration of cyanobacteria, the oxygen concentration (or saturation) of the deep-water zone regulating the survival of zoobenthos (an important functional group regulating fish production in coastal areas and also diffusion of phosphorus from the sediments, an important aspect of "internal loading"), and the macrophyte cover regulating fish production and aspects of water clarity. The chapter discusses how these target bioindicators may be predictied from nutrients (see Chap. 4) and the key abiotic factors (see Chap. 3).
- Chapter 6 gives three case-studies exemplifying the practical use of the bioindicators and concepts discussed in Chaps. 2, 3, 4, and 5. The first case-study concerns how point-source emissions of nutrients to coastal areas affect the receiving waters when all other important nutrient fluxes to, within and from, the given area are accounted for – what can be expected of often costly nutrient reductions to individual coastal areas? The second case-study focuses on a very important and much debated question – how to find the proper reference values to set the targets for remedial actions. Coastal systems will never return to pristine conditions unless mankind is obliterated from the planet. What is the natural nutrient loading, the loading from point sources and diffuse sources, and how much of the anthropogenic loading can be reduced? Those are target questions in this case-study. The third case-study concerns reconstruction of eutrophication using process-based mass-balance modeling. If the development during the last 100 years can

be explained and quantified, key prerequisites to turn the development would be at hand and one would know better how to remediate a given system.

- Chapter 9 (the appendix) presents and motivates the mass-balance model for phosphorus used in the case-studies in Chap. 6, the CoastMab-model. This model is based on ordinary differential equations and calculates inflow, outflow and internal fluxes on a monthly basis. There are algorithms for all major internal TP-fluxes (sedimentation, resuspension, diffusion, mixing and burial). Many of the model structures are general and can be used for, e.g., for open coasts, estuaries or tidal coasts and also for other substances than phosphorus. The main reasons why this model is not presented in the main text is that we have tried to minimize the number of equations in the running text, and especially equations based on differential equations, so that people with different educational backgrounds can read the text.

Chapter 2
Effect-Load-Sensitivity Analyses – Basic Concepts

The aim of this chapter is to give a brief overview of concepts related to effect-load-sensitivity analysis. In a way, this chapter may be regarded as a second introduction to the following chapters of the book.

Richard Vollenweider presented the first load models for phosphorus and nitrogen for lakes in the late 1960s (Vollenweider 1968). By means of simple mass-balance calculations and statistical regressions, he calculated critical loadings of total nitrogen (TN) and total phosphorus (TP) to avoid or reverse eutrophication. Since then, many studies have demonstrated where the Vollenweider approach can and cannot be used (Schindler 1977, 1978; Bierman 1980; Chapra 1980; Boynton et al. 1982; Boers et al. 1993; Håkanson 1999). The Vollenweider model (and later versions, such as OECD 1982), and the analysis behind this load model, constitutes a fundamental base for practically all assessments of eutrophication for lakes. The interesting part, however, is not to predict a concentration of a chemical element like a nutrient, but to predict ecological effects related to nutrients (see Fig. 2.1). It is evident that the concentration of a nutrient can be influenced by emissions from many types of sources, like point sources (domestic sewage, industries and fish farms), atmospheric deposition (to the water surface and the catchment area), internal loading (linked to resuspension, diffusion, etc.) and, in estuaries, inflow from the sea and tributary input, where the characteristics of the catchment, like bedrocks, soils, land use, etc., influence the nutrient concentration in the coastal area (see Fig. 2.2).

Differential equations are often used to quantify fluxes (g X/yr), amounts (g X) and concentrations (g X/m^3) of all types of materials (such as toxins and nutrients), but not generally for bioindicators such as the Secchi depth, chlorophyll-a concentrations, concentrations of cyanobacteria and the oxygen saturation in the deep-water zone (O_2Sat) or other types of operational effect variables (Fig. 2.3). Regressions based on empirical data on nutrient concentrations in the coastal water are often necessary to predict the target bioindicators or variables expressing ecosystem effects. In theory, both model approaches (see Fig. 2.1A and B) may be used for the effect-load-sensitivity analyses (ELS; see Håkanson 1999) provided that at least one operationally defined ecological effect variable or bioindicator relevant for the load variables(s) in question is included in the model. Step B in Fig. 2.1 illustrates

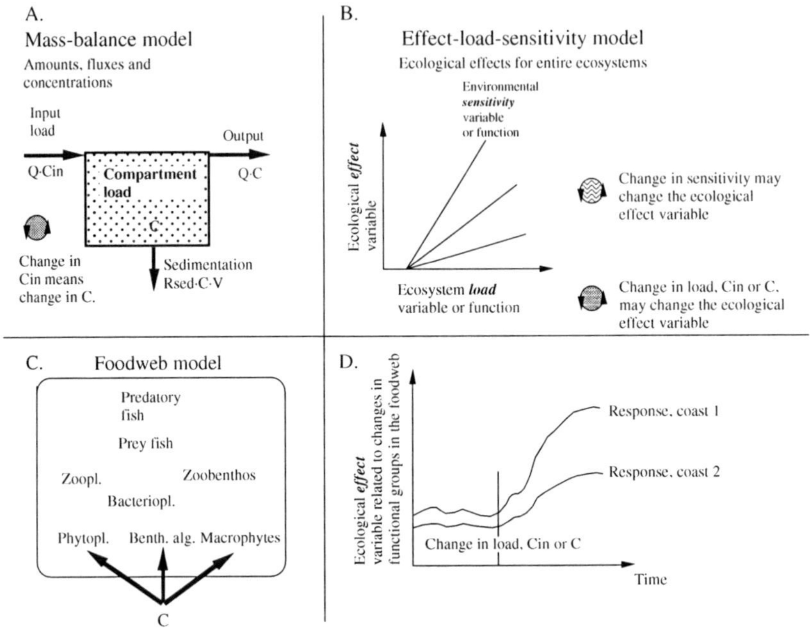

Fig. 2.1 Illustration of the fundamental difference between dynamic, mass-balance models (**A**) and effect-load-sensitivity models (ELS) based on regressions (**B**) and ELS-models related to dynamic foodweb models (**C**) and (**D**) how changes in the load at a given time may cause different responses in the aquatic foodweb in coastal systems of different size and form (coast 1 compared to coast 2). The wheels indicate that by means of remedial measures one may reduce/change the load variable in dynamic models and the load and the sensitivity variables in ELS-models. Q = Water discharge (m^3/time); C_{in} = concentration of substance in inflow (g/m^3); C = concentration of substance in the system (g/m^3); R_{sed} = sedimentation rate (1/time); V = volume (m^3)

a regression and this book will discuss many regressions of that type. Ideally, the effect variable should express the production or biomass of defined functional organisms (preferably fish at the top trophic level, see Figs. 2.1C and 2.2), which characterize a given coastal system. Figure 2.1D illustrates schematically that two coastal areas are likely to react differently to a change in the load of nutrients to the system. The classical approach (from Vollenweider 1968) to carry out ELS-analysis is to use mass-balance models to predict concentrations of nutrients and empirical models (like regressions) to link these concentrations to measured data on the operational bioindicators (see Fig. 2.2). In contexts of coastal management, one must generally for practical and economical reasons seek simpler operational bioindicators than the ideal ones related to production or biomasses of functional groups or species illustrated in Fig. 2.2. The mean concentration of a given toxic substance (or substances) in predatory fish, the Secchi depth, the oxygen saturation/concentration in the deep-water layer and chlorophyll-a concentrations are examples of simple, but relevant operational bioindicators.

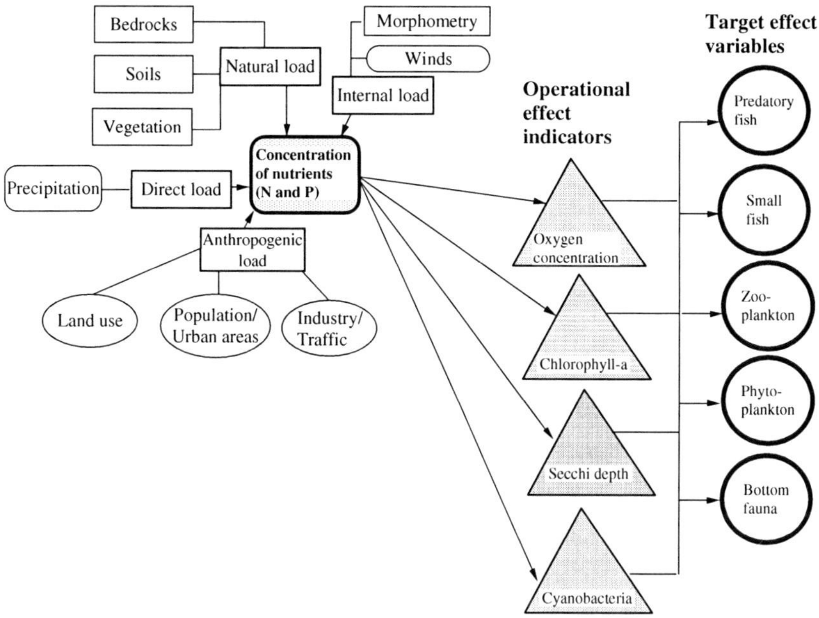

Fig. 2.2 The character of the drainage area influences the water quality in estuaries. This is a statement that is simple to make, but how can it be quantified? It is evident that the geology, hydrology, land use and precipitation influence the inputs of substances to many coastal areas, including the key nutrients, phosphorus and nitrogen. It is an important task to relate drainage area characteristics to coastal area characteristics and to predict variables of primary biological importance, like total-P concentration, which in turn may be related to operational ecological effect variables (like oxygen concentration), and to the target effect variables (like reproduction and abundance of key species)

Figure 2.4 gives the principles of an ELS-model illustrated as a ELS-diagram. Environmental goals (generally set by National Environmental Protection Agencies) should concern the ecological effect variables and not the load variables, since one and the same load may cause different effects in ecosystems of different sensitivities. From this diagram, important concepts like natural background concentration and critical load can be scientifically defined (see Håkanson 1999). When no practically useful validated ELS-models are available, there exists ample room for speculation about cause and effect, and about the best strategies to remediate aquatic systems.

In contexts of ELS-analyses, the primary interest is not on site-specific conditions ("the sampling bottle"), on the individual, organ or cell level, but at the ecosystem level. That perspective should be of main interest from a management point of view where questions are posed concerning the status of larger water bodies (ecosystems), and the remedial actions that could be used in practice to improve the conditions in such systems.

It should be stressed that ELS-analyses are of fundamental importance in water management and that ELS-models are essential tools to examine consequences of remedial measures that may influence a target effect variable. One can reduce

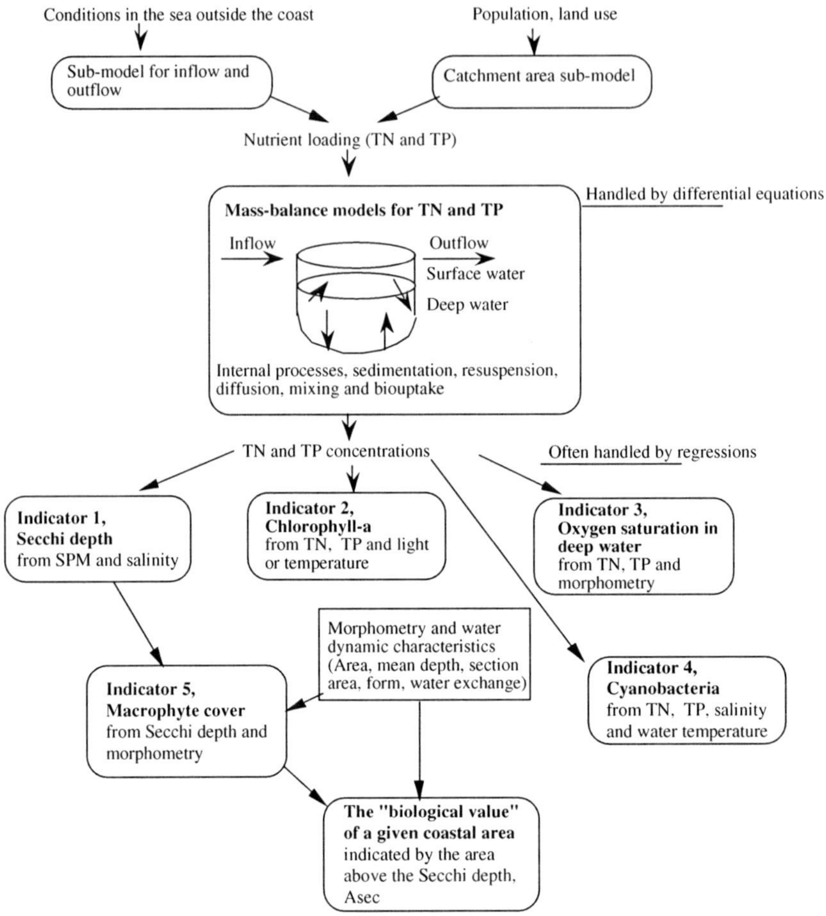

Fig. 2.3 Basic elements in Effect-Load-Sensitivity (ELS) modeling for coastal water eutrophication utilizing mass-balance modeling and regression analyses relating nutrient concentrations to operational bioindicators (Secchi depth, chlorophyll-a concentrations, concentration of cyanobacteria and oxygen saturation in the deep-water zone)

negative ecosystem effects of water pollutants by reducing the load to the system or by changing the sensitivity (for example, by changing the salinity, as in Ringkobing Fjord, where there is a sluice regulating the fluxes of salt water from the sea to the fjord; see Håkanson et al. 2007b).

One cannot generally change the morphometry of the coastal area, but coasts of different size and form will react differently to remedial measures and it is essential to know this so that one can have realistic expectations of the remedial measures for a given coastal area.

It is generally not possible to derive ELS-models, which apply with equal success to all types of ecosystems. Therefore, the operational range, the domain, of the ELS-model must be explicitly given to avoid abuse of the model for ecosystems for which

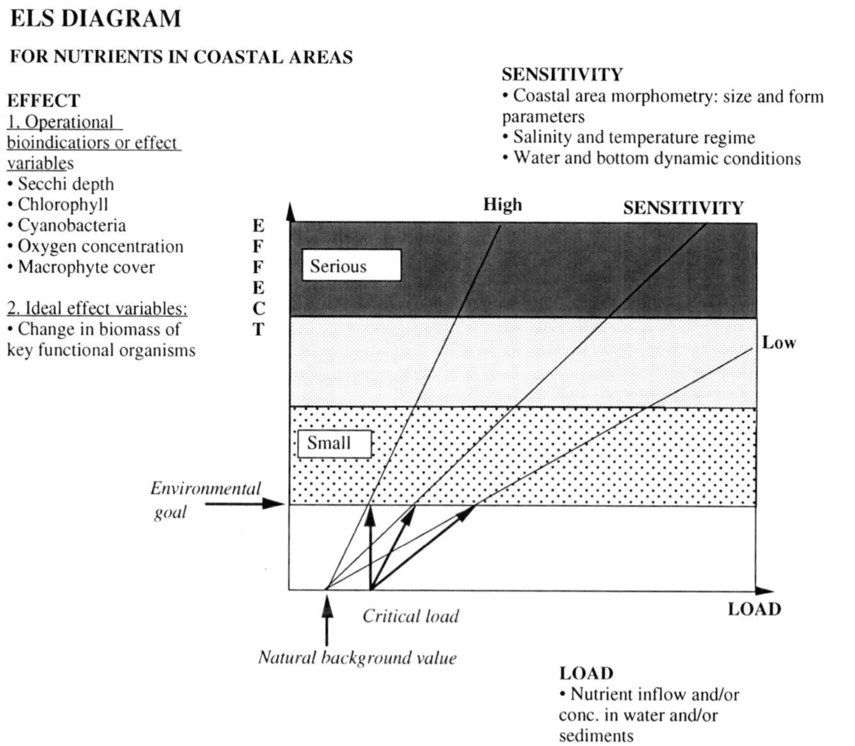

Fig. 2.4 Illustration of the ELS-diagram for coastal eutrophication

it was never intended to be used. If dynamic (time-dependent) ELS-models can meet these requirements, they would generally be preferable to statistical/empirical models because they can provide better understanding of mechanisms and processes.

This book will discuss many ELS-models based on regressions and some of those regressions are included in the more comprehensive process-based mass-balance model for phosphorus (CoastMab) to link dynamically modeled TP-concentrations to the target bioindicators (i.e., the operational effect variables). The CoastMab-model is presented in Sect. 9.1 and applied in Chap. 6.

Chapter 3
Coastal Classifications and Key Abiotic Variables Regulating Target Bioindicators

This chapter will focus on abiotic factors (coastal morphometry, salinity, temperature and water exchange) affecting coastal ecosystem function and structure. These factors cannot generally be influenced by remedial actions. They may, however, be important in contexts of effect-load-sensitivity analyses. Different coastal areas with different temperature regimes, different salinities and different size and form characteristics would react differently to one and the same nutrient loading.

3.1 Coastal Classifications

A given coastal area may be defined and characterized in many ways, e.g., according to territorial boundaries, pollution status, water stratification (thermoclines/haloclines; see Bird 1984, 2000; Dal Cin and Simeoni 1994; de Jonge 2000; Casazza et al. 2003; Irvine 2004), etc. One such system is shown in Fig. 3.1 as a background to the following discussions. Figure 3.1 gives a geographical zonation into the following categories:

1. The drainage area. For example, the drainage area of the Baltic Sea covers 1,700,000 km^2, which is more than four times larger than the entire water area (415,266 km^2).
2. The coastal zone, i.e., the zone inside the outer islands of the archipelago and/or inside barrier islands. This is the zone in focus in this work. The coastal zone is of special importance for recreation, fishing, water planning and shipping and is a zone where different conflicts and demands overlap. The natural processes (water transport, fluxes of material and energy and bioproduction) in this zone are of utmost importance for the entire sea (see Chap. 5 and Håkanson and Rosenberg 1985).
3. The transition zone, i.e., the zone between the coastal zone and the deep-water areas. This is the zone extending down to depths at which episodes of resuspension of fine materials occur in connection with storms and/or current activities

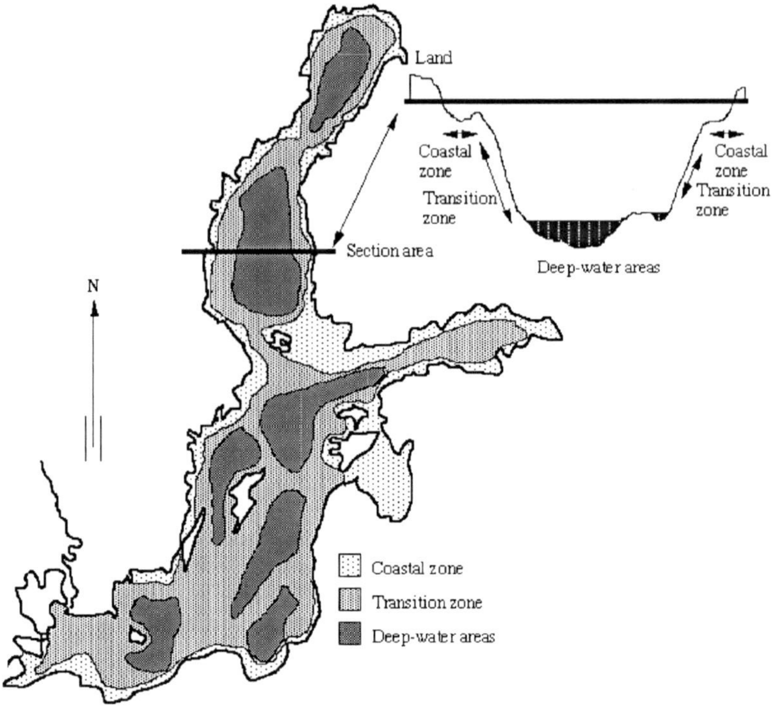

Fig. 3.1 A given coastal system, here the Baltic Sea, may be divided into the following three functional zones: the coastal zone, the transition zone and the deep-water areas. Modified from Håkanson (1991)

(at about 44 m water depth in the Baltic Proper; see Håkanson 1991). The conditions in terms of water dynamics and distribution of pollutants and suspended and dissolved materials in this zone are of great importance for the ecological status of the entire system. This zone geographically dominates the open water areas outside the coastal zone.

4. The deep-water zone, by definition the areas beneath the theoretical wave base (defined in Fig. 3.15, later). In these areas, there is a continuous deposition of fine materials. It is the "end station" for many types of pollutants and these are the areas in which conditions with low oxygen concentrations are most likely to occur.

There are also many other types of classification systems for coastal areas (Inman and Nordström 1971). For example, Johnson (1919); Davies (1964, 1972, 1980) and Davis (1996) have presented coast type classifications based on geological considerations, e.g., according to form-creating processes. Valentin (1952, 1979) based a coastal classification on whether the coast is expanding or retreating.

There are also different regional classification systems (Olsson 1966; Abrahamsen et al. 1977) generally based on or related to Davies's system. Coastal classifications like these are useful to get a broad overview of the general features and they can explain why the coastal geomorphology looks the way it does, but they do not provide the data necessary for ecological modeling or quantifications, e.g., how and why a given coastal area functions as a "nursery and pantry" for the bioproduction or as a receiving system for water pollutants.

The trophic level classification system given in Table 3.1 (from Håkanson et al. 2007a) is based on practically useful, operational effect variables or bioindicators for coastal management. Such bioindicators will be discussed in Chap. 5. Note that in this book, we will not discuss bioindicators related to the abundance or changes in individual species or bioindicators that require in-depth biological knowledge for measurements and interpretations because such bioindicators generally do not meet the mentioned criteria for practical usefulness.

The bioindicators and trophic level classification system discussed here may be useful tools for enhancing communication between natural scientists, water managers, economists, policymakers and/or the general public. A wide variety of indices for coastal areas have emerged during recent years (Swedish EPA 2000; Aertberg et al. 2003; Diaz et al. 2004; Andersen et al. 2006). The index TRIX (TRophic IndeX) has been described by Vollenweider et al. (1998) and is also included in Italian legislation regarding coastal management (Penna et al. 2004). However,

Table 3.1 Characteristic features in (A) freshwater dominated systems, (B) brackish systems and (C) marine coastal systems of different trophic levels (see also OECD 1982; Håkanson and Jansson 1983; Wallin et al. 1992; Håkanson and Boulion 2002; Håkanson et al. 2007a). All data represent characteristic (median) values for the growing season for the surface water layer

Trophic level	Secchi* (m)	Chl-a (μg/l)	Total-N (μg/l)	Total-P (μg/l)	Cyanobacteria** (μg ww/l)
A. Freshwater dominated systems, salinity < 5 psu					
Oligotrophic	>5	<2	<60	<8	<2.2
Mesotrophic	3–5	2–6	60–180	8–25	2.2–250
Eutrophic	1–3	6–20	180–430	25–60	250–1400
Hypertrophic	<1	>20	>430	>60	>1400
B. Brackish systems, salinity 5–20 psu					
Oligotrophic	>8	<2	<70	<10	<9.5
Mesotrophic	4.5–8	2–6	70–220	10–30	9.5–380
Eutrophic	1.5–4.5	6–20	220–650	30–90	380–2500
Hypertrophic	<1.5	>20	>650	>90	>2500
C. Marine systems, salinity >20 psu					
Oligotrophic	>11	<2	<110	<15	<55
Mesotrophic	6–11	2–6	110–290	15–40	55–680
Eutrophic	2–6	6–20	290–940	40–130	680–4040
Hypertrophic	<2	>20	>940	>130	>4040

Relationships between chlorophyll, TP, TN and salinity calculated from Håkanson (2006).
*Secchi depth calculated from Håkanson (2006).
**Concentration of cyanobacteria (CB) calculated using the model from Håkanson et al. (2007c) when TP/TP is set to 15, surface water temperature to 17.5°C and the salinity to 2.5, 12.5 and 36, respectively for fresh water, brackish and marine systems.

TRIX has yet to be critically tested for areas outside the Mediterranean Sea. TRIX is based on chlorophyll-a, oxygen saturation in the deep-water zone, total-P concentration and total-N concentration. It should be noted that the TRIX scale from 0 to 10 gives the impression that one of the ends is good (oligotrophy) and the other is bad (hypertrophy). We argue that a more suitable index for coastal management should instead start from what is regarded as the norm (the normal = reference = target situation = "good" ecological status), which describes a requested situation (e.g., the situation as it was 50 or 100 years ago). The end value of the index should describe an extremely detrimental anthropogenic impact (the maximum possible deviation from the norm) – in both directions. Such indices are also available for lakes (Håkanson et al. 2000), but not, according to our knowledge, for coastal areas. Such an index would account for the fact that artificially low nutrient levels may be just as undesirable as an artificially high nutrient level. The development of a norm-based index for brackish and marine coastal areas would be an apt goal for future scientific work.

The importance of using both morphometric as well as chemical information (salinity, chlorophyll-a and dissolved oxygen) when subdividing and classifying coastal systems was also addressed by Ferreira et al. (2006). They also included loading of nutrients into their methodology.

A Geographical Information System (GIS) is a useful tool for many types of geographical analyses (e.g., Bonham-Carter 1994; Persson et al. 1994a; MEDAR Group 2002; Liu and Jezek 2004; LOICZ 2007) and GIS has also been applied in marine and coastal sciences (Wright and Bartlett 2000; Breman 2002). GIS has also been a component in several coastal classification methods (Cooper and McLaughlin 1997). Also when determining important morphometric features of coastal areas, GIS is a very useful tool. The general maps of the water variables discussed in this work were produced by Lindgren and Håkanson (2007) using the GIS-software ArcGIS 9.1 by Environmental Systems Research Institute, Inc. (ESRI; http://www.esri.com). Having access to the GIS itself with all types of data stored in different layers, evidently gives more analytical opportunities than just the general maps.

3.2 Coastal Morphometry

The morphometry of a coastal area – i.e., the way the coastal area looks, its size and form characteristics – is very important for how sensitive it is to pollution. This is partly because the morphometry influences the sediment and bottom dynamic conditions (Håkanson 2006), but also because the morphometry influences the theoretical water retention time (Håkanson et al. 1986; Rasmussen and Josefson 2002). The latter is important since the in- and outflowing sea water either dilutes ("purifies") or pollutes the coastal area (Le Pape and Menesguen 1997).

An important question concerns the definition of the coastal ecosystem limits, i.e., where to place the boundaries toward the sea and/or adjacent coastal areas. It is crucial to use a technique that provides an ecologically meaningful and practically useful definition of the coastal ecosystem. How should one define this area so that parameters, like the mean depth or the volume, can be relevant as model variables (x) to predict target bioindicators (y)? Arbitrary borderlines can be drawn in many ways and the mean depths of such areas would be devoid of meaning in relation to bioindicators. The approach in this work (from Pilesjö et al. 1991) assumes that the borderlines are drawn at the topographical bottlenecks so that the exposure (Ex) of the coast from winds and waves from the open sea is minimized (Fig. 3.2).

The exposure of the coastal area is the ratio between the section area and the enclosed coastal area. It is easy to use the Ex-value as a tool to test different alternative borderlines and define the coastal ecosystem where the Ex-value is minimal. For more open coastal areas, a significant part of the fine materials suspended in the water can "escape" from the coastal area to the sea or to surrounding coastal areas. This is not the case for more closed coastal lagoons.

Once the coastal area is defined, one can also determine important variables for mass-balance calculations, such as the coastal volume (regulating the concentration of any given substance), and important morphometric parameters for internal fluxes, such as the mean depth and the water surface area, and key variables regulating the water exchange between the coast and the sea, such as the volume and the section area (see Fig. 3.3).

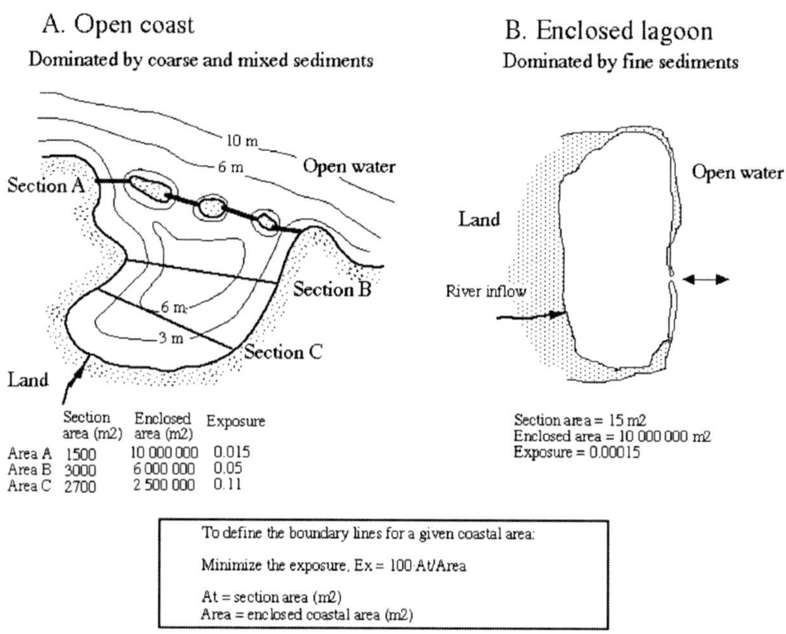

Fig. 3.2 Schematical illustration of the topographical bottleneck approach to determine the boundary lines for open coastal systems. The coastal ecosystem is defined by the borderline marked A, which gives a minimum exposure (Ex). To define the boundary line for, e.g., lagoons is generally quite straightforward

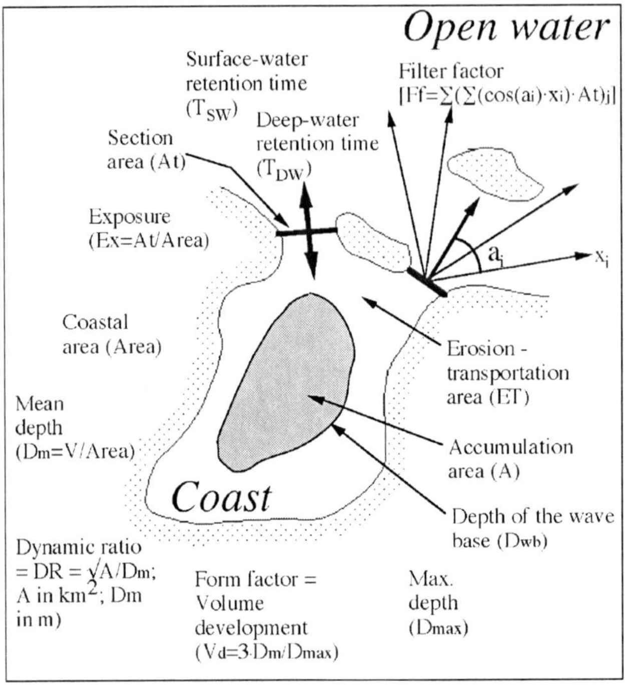

Fig. 3.3 Illustration of key coastal parameters in mass-balance modeling which can be determined quickly from digitized bathymetric maps using GIS-methods (GIS = Geographical Information System). The key issue is to define the boundary lines, i.e., where the coastal area ends and the sea or adjacent coastal area begins. The approach used in this book is to define the boundary lines so that the topographical openness (the exposure, Ex, defined by the ratio between the section area, At, and the enclosed coastal area, Area) attains a minimum value. This is the "topographical bottleneck" method. The filter factor (a measure similar to the fetch) describes how islands and topographical barriers between the defined coastal area and the sea act as energy filter for the wave impact on the coastal area. In this figure, the filter factor is illustrated for one of the two section areas in this coastal area (ai = the angle between two radials sent out from the opening over open water; xi = the length of each radial; i = the number of radials at each opening; j = the number of openings; j = 2 in this example)

This method of defining coastal areas also opens a possibility to use empirical models to estimate, e.g., the theoretical water retention times of the surface water and the deep water, and the bottom dynamic conditions (regulating sedimentation, resuspension and diffusion) from morphometrical parameters (such as area, mean depth, form factor and section area; see Table 3.2).

In large enclosed coastal areas, the theoretical water retention time may be relatively long, whereas in small open coasts the value may be just a few days. In such coastal areas, it is evident that the concentrations of nutrients or pollutants are close to the values in the outside sea. Hence, it is also difficult to maintain a concentration gradient from point source emissions to the coasts if the water mass is exchanged many times each month. So, concentrations of pollutants in coastal areas often depend much on the conditions outside the given coastal area.

Table 3.2 Empirical regressions to estimate theoretical surface and deep-water retention times (T_{SW} and T_{DW} in days) from morphometric parameters

Regression	r^2	n	Reference
$\ln(T_{SW}) = (-4.33 \cdot (\sqrt{Ex}) + 3.49)$	0.95	14	Persson et al. (1994b)
$T_{DW} = (-251 - 138 \cdot \log(At) + 269 \cdot \log(Vd))$	0.79	15	Håkanson and Karlsson (2004)

Model domain: $0.002 < Ex < 1.3$; $0.0006 < At < 0.08$; $0.5 < Vd < 1.5$; note that T_{SW} and T_{DW} are never permitted to be < 1 day and T_{DW} never > 120 days.
$Ex = 100 \cdot At/A$; $Vd = 3 \cdot D_m/D_{max}$; At = section area (km^2); Vd = form factor ($= 3 \cdot D_m/D_{max}$; D_m = mean depth (m), D_{max} = max. depth (m)).

3.2.1 Topographical Openness or Exposure

Open coastal areas are generally not very sensitive to pollution due to the short water retention times that usually characterize such systems (Clark 2001). There are many examples of enclosed water bodies that suffer from pollution including both large systems, such as the Black Sea (Bakan and Büyükgüngör 2000) and the Baltic Sea (HELCOM 1986, 1990, 2003) and smaller enclosed lagoons, e.g., the Oder Lagoon (Glasby et al. 2004; Schernewski and Dolch 2004) and Ringkobing Fjord (Håkanson et al. 2007a, b) at regional and local scales.

The topographical openness or the exposure (Ex) is defined by (3.1) as the ratio between the section area (At, see Fig. 3.2 for illustration) and the enclosed coastal area (Area) (Håkanson et al. 1986):

$$Ex = 100 \cdot At/Area \tag{3.1}$$

The exposure (Ex) can easily be calculated for most coastal areas and at all scales. Even though the exposure was originally defined for the local to regional scales, it can also be calculated for large water bodies as an indication of their connectivity to the world's oceans. Table 3.3 gives a classification system based on the exposure. Coastal areas with Ex-values lower than 0.002 may be called "very closed" systems, and coastal areas with Ex-values higher than 1.3 may be referred to as "open" systems. Figure 3.4 gives a frequency distribution of Ex-values from 540 coastal areas in the Baltic Sea (from Lindgren and Håkanson 2007). One can note that in this region, most coastal areas have Ex-values in the range between 0.002 and 1.3 and should be classified as semi-enclosed systems.

Table 3.3 Classification criteria for topographical openness (exposure) of coastal areas (see Lindgren and Håkanson 2007)

Ex	Topographical openness	Typical systems
0–0.002	Enclosed, very closed systems	Most coastal lagoons
0.002–1.3	Semi-enclosed systems	Bays, fjords, archipelagos
>1.3	Open systems	Open coasts (cliff, sand, rock, man-made, etc.)

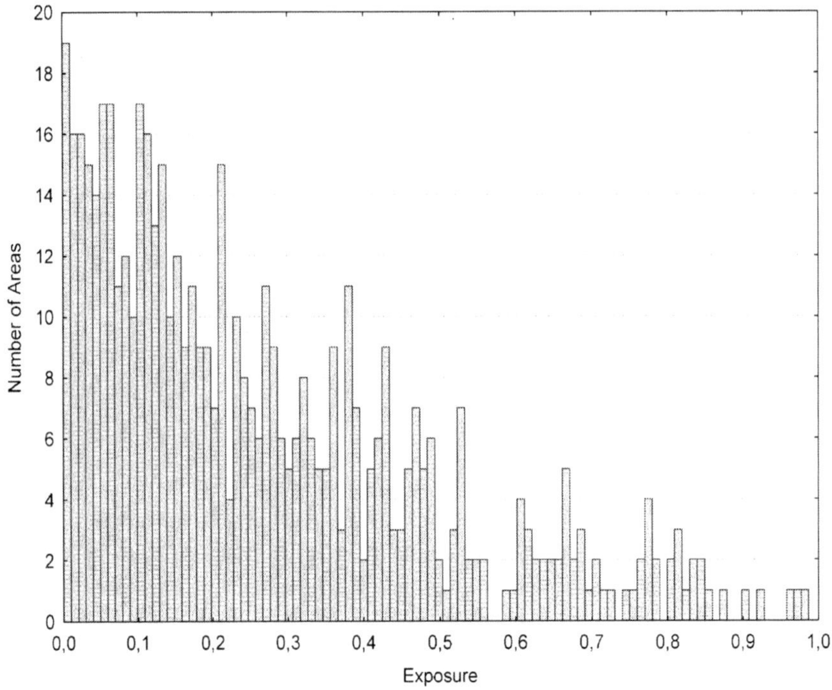

Fig. 3.4 Distribution of exposure values calculated for 540 Swedish coastal areas (from Lindgren and Håkanson 2007)

The exposure has been shown to give good predictions of the theoretical surface-water retention time (T_{SW} in days; see Table 3.2) and the section area and the form factor (Vd; see Sect. 3.2.3) may be used to predict the theoretical deep-water retention time (T_{DW}; see Fig. 3.5 for illustration) for coastal areas defined by the topographical bottleneck method.

The regression for T_{SW} in Table 3.2 yielded an r^2-value (coefficient of determination) of 0.95 when tested using data from 14 coastal areas of different size, form

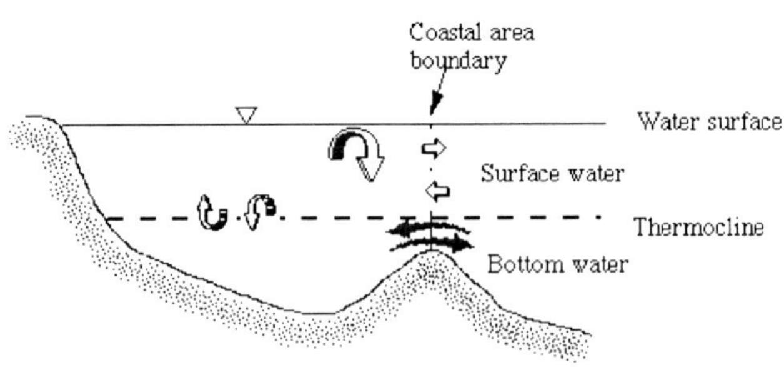

Fig. 3.5 Illustration of surface-water and deep-water layers of a coastal area

and exposure. The empirical data on T_{SW} were measured by the dye method, salt budgets and/or current meters. The models in Table 3.2 may be used for non-tidal coastal areas where the exposure is between 0.002 and 1.3 and these are also the class limits for the exposure in Table 3.3.

There are also simple operational methods to estimate the theoretical turnover times for tidal coasts and open coastal areas dominated by coastal currents (see Sect. 9.1). Figure 3.6 illustrates how the theoretical surface-water retention time depends on coastal area and section area using the equation in Table 3.2. One can note that T_{SW} is highly dependent on the openness of the coast. In enclosed coastal areas, T_{SW} may be as long as 30 days.

It is costly and laborious to empirically determine the theoretical surface or deep-water retention times (T_{SW} and T_{DW}). The factors regulating inflow, outflow and retention of substances in coastal areas depend on many more or less stochastic processes. This makes it difficult to give a reliable prediction of T_{SW} or T_{DW} at a given time. In a coastal area, T_{SW} may be indefinitely long on a calm summer day and very short (say a few hours in a small area) in connection with a storm or a sudden change in air pressure. T_{SW} or T_{DW} always emanate from frequency distributions (see Fig. 3.7). In coastal management, it is often important to get information on the characteristic T_{SW} or T_{DW}-values, e.g., the mean or median value

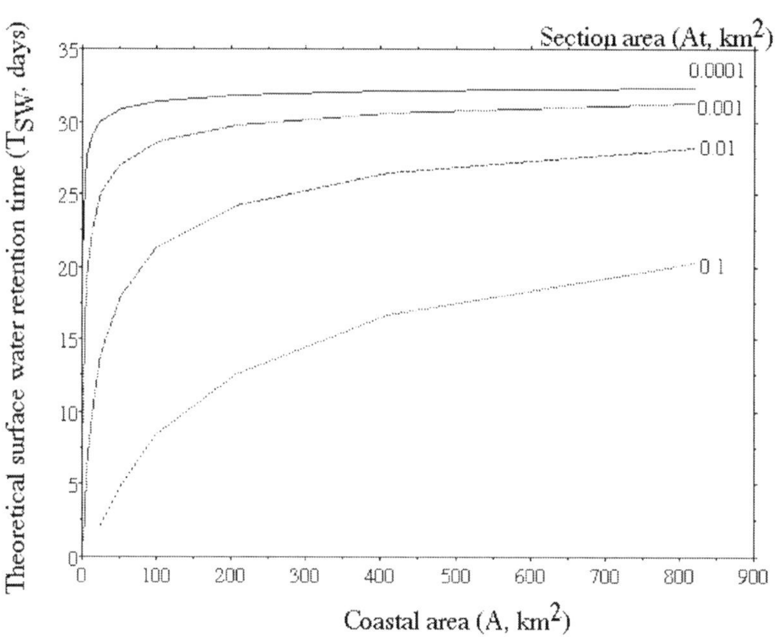

Fig. 3.6 The relationship between the theoretical surface-water retention time (T_{SW}, a key regulator of the water exchange between a given coastal area and the sea), the coastal area (A) and the openness of the coast towards the sea (as given by the section area, At). This nomogram is based on the empirical model in Table 3.2

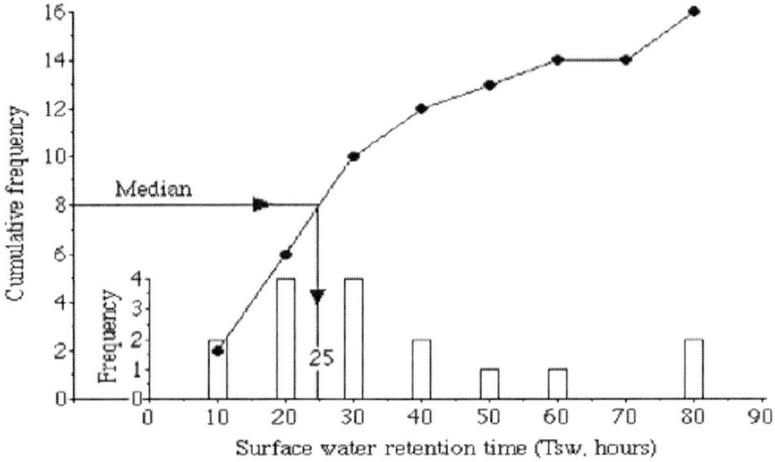

Fig. 3.7 Illustration of empirical data from a dye experiment from a coastal bay to determine the theoretical surface water retention time (T_{SW}) (from Håkanson et al. 1986)

(25 hours as illustrated in Fig. 3.7). As a rule of thumb, one can conclude that the costs of establishing a frequency distribution such as that in Fig. 3.7 from traditional field measurements (using dye, current meters, etc.) is about 25,000 USD for one coastal area. For many coastal areas, it may not be possible to arrive at management decisions if it would be a prerequisite that such expansive field work first must be carried out to determine T_{SW}. This means that it is of major importance that T_{SW} may be predicted easily and inexpensively from one coastal morphometric variable, the exposure (Table 3.2).

Figure 3.8 gives examples of empirical data on the surface and deep-water exchange (T_{SW} and T_{DW}) as empirically determined for Swedish coastal areas (see Håkanson et al. 1984 and Persson and Håkanson 1996). The median T_{SW}-value is 5 days and the median T_{DW}-value 13 days. Evidently, these values vary within and among coastal areas depending on the prevailing meteorological conditions. T_{DW} is rarely longer than 120 days in this part of the world where there is water turnover in the spring and the fall.

- The values for T_{SW} and T_{DW} are of fundamental importance in mass-balance calculations since, by definition, the surface-water flow (Q_{SW} in m^3 per day) between the given coast and the Sea is V_{SW}/T_{SW}, where V_{SW} is the surface-water volume of the coast.
- The water velocity across the section area (At) is $(Q_{SW}+Q_{DW})/(2\cdot At)$ and characteristic values for this velocity generally in the range between 1 to 15 cm/s. The division by 2 comes from the fact that there would be both inflow and outflow to a given coastal areas so the characteristics velocity in one direction may be given by this simple approximation.

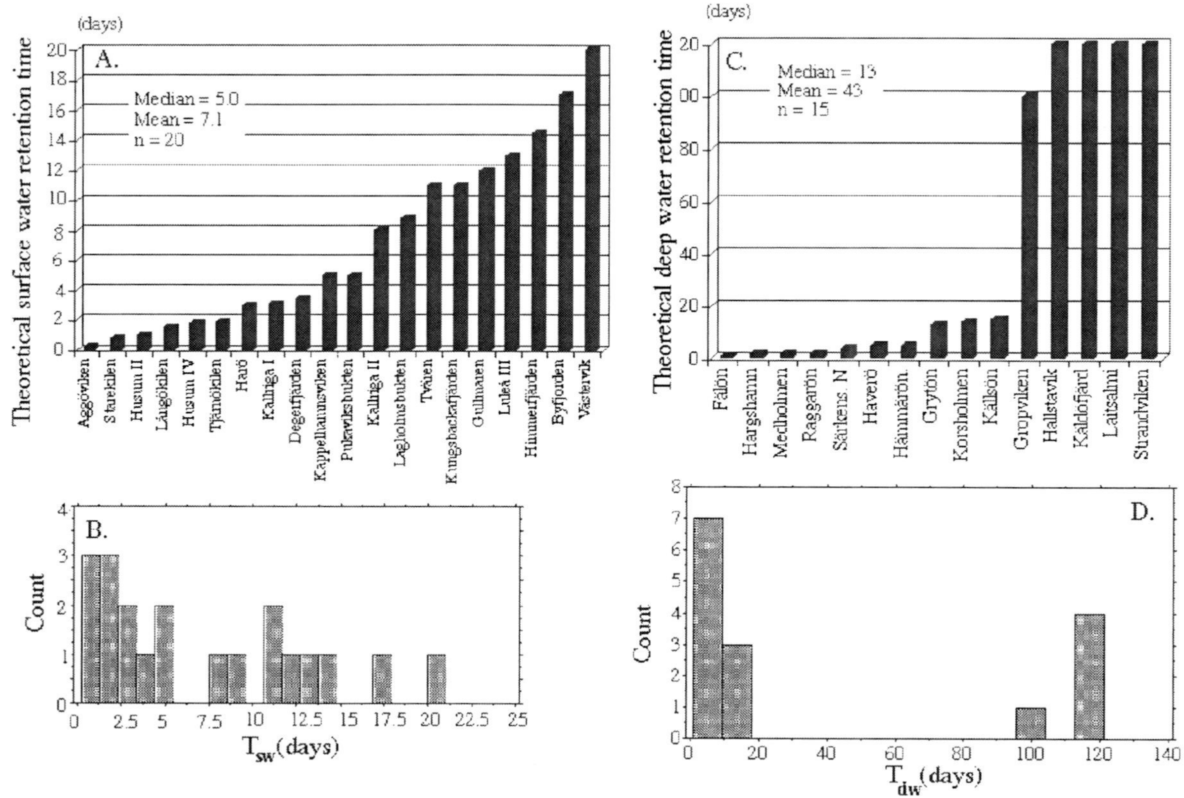

Fig. 3.8 Compilation of empirical data on the surface-water and deep-water retention times in Swedish coastal areas (T_{SW} and T_{DW}; data from Håkanson et al. 1984 and Persson and Håkanson 1996). The upper figures (**A** and **C**) give data from the studied coastal areas, the lower figures (**B** and **D**) give the corresponding frequency distribution. The mean value for T_{SW} is 7.1 days and for T_{DW} 43 days

- A typical value for the velocity of currents outside the coastal area is about 0.2–0.4 knots or 10–20 cm/s (see FRP 1978 and Fig. 3.23, later). Evidently, this value varies with the meteorological conditions.
- If the deep-water volume is not very small, the theoretical deep-water retention time (T_{DW}) is generally longer than that of the surface water (T_{SW}), and the deep water is often exchanged episodically (Persson and Håkanson 1996). Evidently, T_{DW} depends on many factors, e.g., the section area (At, i.e., the openness towards the Sea and/or adjacent coastal areas), and the form of the coast (the form factor Vd; see Table 3.2 which gives a morphometrical model to estimate a characteristic T_{DW}-value).

Table 3.4 exemplifies that the exposure is very important in predicting key bioindicators of coastal eutrophication. The bioindicator in this case is the chlorophyll-a concentration, the load factor is given by the concentration of total nitrogen (TN) in the coastal water and the sensitivity factor by the exposure (Ex). This ELS-model yielded an r^2-value of 0.84 (84% statistical explanation of the variability in the chlorophyll values) and it is based on data from 23 Baltic coastal areas.

These data show that the higher the TN-concentration, the higher the values of the effect variable, and the higher the Ex-value, the lower the value of the effect variable. That is, enclosed coastal areas have a higher sensitivity to nutrient loading. Other examples showing this have been presented by Wallin et al. (1992); Håkanson (1999) and Nordvarg (2001) related to nutrient emissions from point sources to coastal areas.

Figure 3.9 demonstrates the importance of the exposure in a different way. In this case, simulations have been carried out using the CoastMab-model (see Sect. 9.1). In this example, one coastal area has been selected at random (the Gävle Bay area in the Bothnian Sea; area = 17.1 km^2, maximum depth = 17 m, mean depth = 6.4 m, section area = 0.0063 km^2, salinity = 4.2 psu) and the idea has been to make a sensitivity analysis and change the exposure in 10 steps (from the default value of 0.037 to 0.37), keep everything else constant (including the nutrient loading from tributaries and from the open sea, the salinity, etc.) and calculate how this would likely influence four target bioindicators included in the model (the chlorophyll-a concentration in the water, the Secchi depth and the oxygen saturation in the deep-water layer) as well as the key nutrient in this coastal area, the TP-concentration.

One can note from Fig. 3.9 that by increasing the exposure, the TP-concentration in the coastal area would approach the TP-concentration in the outside sea

Table 3.4 Results of stepwise multiple regression for chlorophyll-a concentrations (Chl in µg/l; mean value for the growing season) using data from a Baltic coastal database (from Wallin et al. 1992). n = 23 coastal areas; y = log(Chl). TN = total-N concentration (load factor in µg/l); Ex = exposure (dim. less). r^2 = the coefficient of determination (r = the correlation coefficient)

Step	r^2	x-variable	Model
1	0.78	x_1 = TN	$y = 3.10 \cdot x_1 - 7.43$
2	0.84	x_2 = Ex	$y = 3.00 \cdot x_1 - 0.15 \cdot x_2 - 7.21$

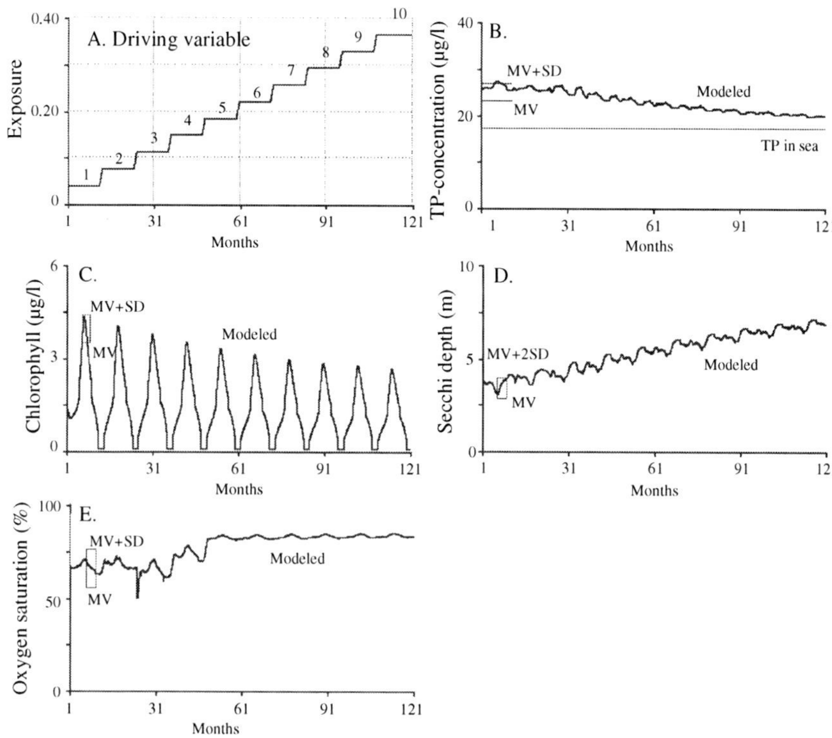

Fig. 3.9 In this scenario, the exposure has been changed in annual steps (**A**; from the default conditions, Ex = 0.037 to 10·0.037) and the consequences have been calculated using data from the Gävle Bay coastal area in the Bothnian Sea for (**B**) the TP-concentrations in water, (**C**) the chlorophyll-a concentration in water, (**D**) the Secchi depth and (**E**) the oxygen concentration in the deep-water zone. The predicted values have also been compared to the empirical data for the given coastal area (initial Ex-value) and one can note the close correspondence between modeled values and empirical data

(Fig. 3.9B), the chlorophyll concentration would go down (Fig. 3.9C), water clarity would be higher (Fig. 3.9D), and the oxygen saturation in the deep-water zone would increase, which would be beneficial for the production of the bottom fauna, a very important functional group or organisms (a staple food for the fish).

Figure 3.10 shows the exposure for the major water bodies of Europe. These values demonstrate quantitatively that some of the major water bodies at the European scale are quite enclosed (e.g., the Baltic Sea and the Black Sea), while others are more open (e.g., the Bay of Biscaya or the Adriatic Sea). Both the Baltic Sea, with a theoretical water retention time of about 20 years, and the Black Sea are facing major pollution problems (Bakan and Büyükgüngör 2000; HELCOM 2003; Neumann and Schernewski 2005; Arslan and Ökmen 2006). One reason for this is related to their enclosed character. So, also at this scale, the Ex-value may be a useful descriptor, even though it is at the regional and local scales that it may have its primary use in contexts of coastal management and modeling.

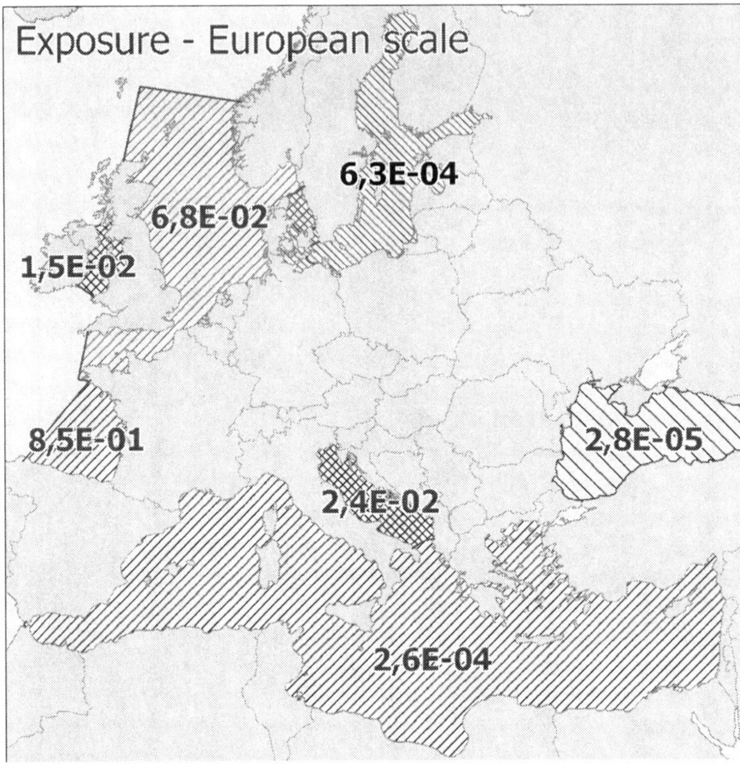

Fig. 3.10 Map of Europe showing exposure values for major water bodies (from Lindgren and Håkanson 2007)

The results of the corresponding calculation for the Baltic Sea are presented in Fig. 3.11, which shows that this area as a whole has a very low exposure, and this is also the case for certain sub-basins within the Baltic Sea, such as the Bothnian Bay and the Gulf of Riga.

3.2.2 Size Parameters

The classical size parameters are the maximum depth (D_{max} in m), coastal area (in km^2 or m^2) and volume (in km^3 or m^3). The size of a coastal area is important for a number of reasons. The water area together with the mean depth defines the water volume, which in turn is essential for calculating the concentration of any substance. Lindgren and Håkanson (2007) suggested that coastal areas may be classified into size classes according to the system given in Table 3.5 (very large, large, intermediate, small and very small). These classes correspond to the A–F categories used in the Swedish Marine Area Registry (SHR; SMHI 1994).

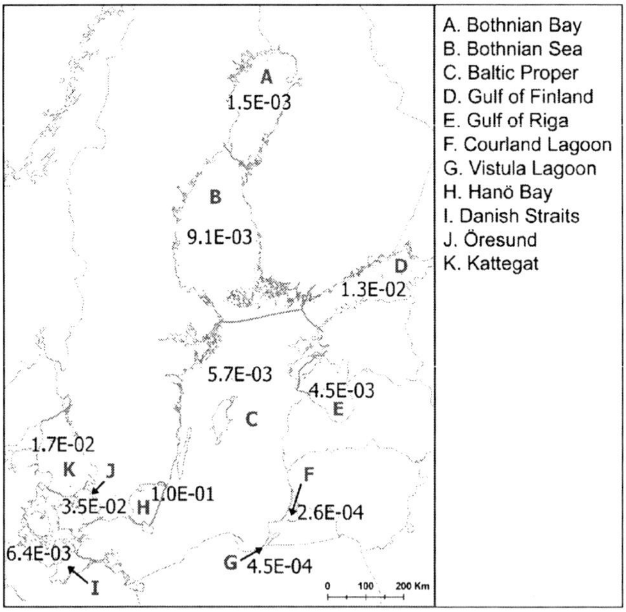

Fig. 3.11 Map of the Baltic Sea with the exposure calculated for several sub-basins (from Lindgren and Håkanson 2007)

3.2.3 Form Parameters

Another important morphometric feature of a coastal area is its form (shape). The form has direct impact on, e.g., how much of the sediment surface that is influenced by wind and wave processes, i.e., the bottom areas where sediment resuspension occurs. This, in turn, is important for the internal loading and hence also for all water concentrations of substances with a particulate phase. The form of the coast is also important, for example, for the growth of macrophytes and benthic algae and hence for the production potential of the coastal system. Figure 3.12 illustrates relative hypsographic curves (depth-area curves) for systems with different form factors (Vd; see (3.2)). Deep, U-formed coasts generally have smaller areas above the Secchi depth. The larger the area above the Secchi depth, the

Table 3.5 Surface area classification

Surface area (km^2)	Class name
>10,000	Very large
1000–10,000	Large
100–1000	Intermediate
10–100	Small
<10	Very small

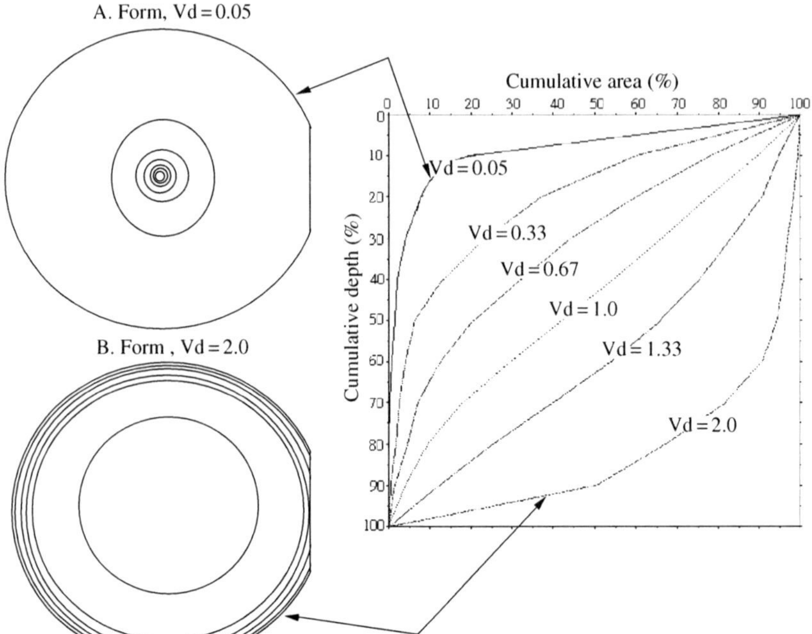

Fig. 3.12 Schematical bathymetrical illustration of two the coastal areas, (**A**) is very convex with an Vd-value (form factor, Vd = 0.05) and (**B**) the other is very concave (Vd = 2.0)

higher the primary production of phytoplankton, benthic algae and macrophytes, the higher the biological value, and the more important it is to protect and preserve such areas. So, these are areas where one should be particularly careful not to not build marinas, harbors or emit contaminants (see Chap. 5 and Håkanson and Rosenberg 1985).

The form factor, or the volume development (Vd), (3.2), is a useful standard measure of the form of any aquatic system (Håkanson 2004).

$$\mathrm{Vd} = 3 \cdot \mathrm{D_m}/\mathrm{D_{max}} \tag{3.2}$$

Håkanson (2004) also gave a classification scheme based on the form factor (Table 3.6). Since Vd is calculated from general features, the mean depth (D_m in m) and the maximum depth (D_{max} in m), it applies to most systems. To illustrate the use of this morphometric form parameter, Lindgren and Håkanson (2007) calculated values of the form factor for all Swedish coastal areas in SVAR (SMHI 2003), where data on D_m, and D_{max} were available and all areas were classified according to the criteria given in Table 3.6 from very convex to concave coastal forms (see Fig. 3.13). The dominating coastal form is slightly convex (SCx).

Table 3.6 Morphometric classification for aquatic systems based on the form factor, Vd (from Håkanson 2004)

Form of lake or coastal area	Class name	Vd
Very convex	VCx	0.05–0.33
Convex	Cx	0.33–0.67
Slightly convex	SCx	0.67–1.00
Linear	L	1.00–1.33
Concave	C	1.33–2.00

3.2.4 Depth Conditions and Coastal Sensitivity

The depth of a coastal area also is evidently an important morphometric feature, e.g., for sedimentation and resuspension (see Håkanson 1999). A useful parameter describing the depth conditions is the dynamic ratio (DR), (3.3).

$$DR = \sqrt{\text{Area}}/D_m \tag{3.3}$$

Area is the enclosed coastal area in km^2 and D_m the mean depth of the coastal area in m. Håkanson (1999) also discussed the close relationship between the

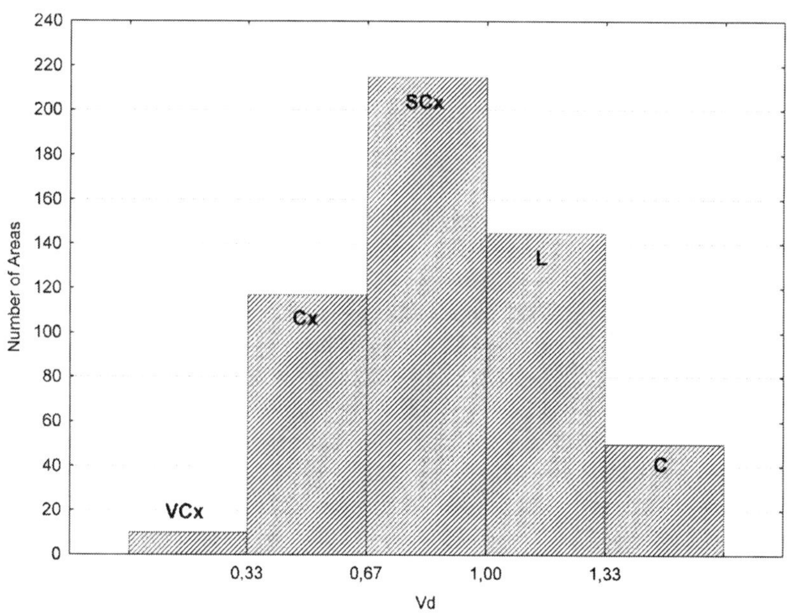

Fig. 3.13 Classification of 538 Swedish coastal areas based on their form factor (Vd) (from Lindgren and Håkanson 2007)

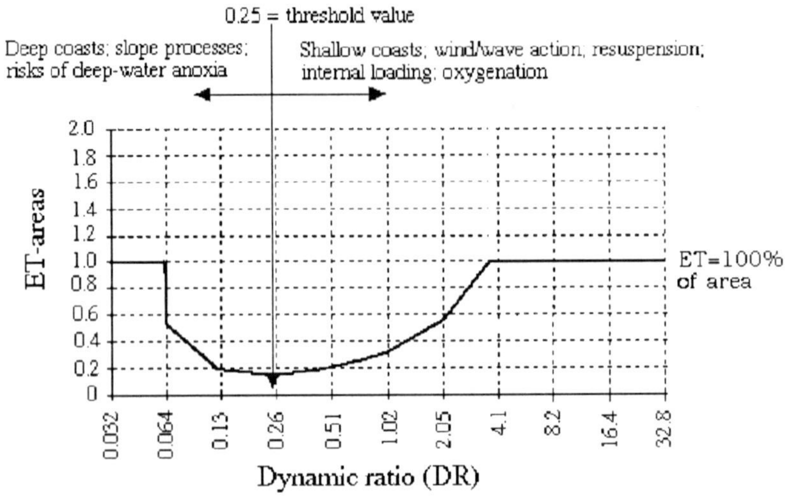

Fig. 3.14 The relation between the dynamic ratio (DR) and the proportion of bottom areas dominated by erosion and transport processes (ET). The threshold value of 0.25 will be used in the following sensitivity index to separate deep areas from shallow coastal areas. Redrawn from Håkanson (2006)

dynamic ratio and the potential bottom dynamic conditions, e.g., the area of the system dominated by processes of fine sediment erosion and transport (ET), see Fig. 3.14.

There is always in all coastal areas a shallow zone where the bottom dynamic conditions are dominated by wind/wave action and erosion and transport processes for fine materials (ET-sediments). From Fig. 3.14, one can note that this zone is generally larger than 15% of the coastal area corresponding to the threshold DR-value of 0.25.

This also implies that the areas of continuous fine sediment accumulation are rarely larger than 85% of the total coastal area. In coastal areas with DR-values higher than 0.25, the conditions are likely to be increasingly influenced by wind/wave action, oxygenation and resuspension processes causing advective internal loading of nutrients. One the other hand, in deeper areas, the risks of getting deep-water anoxia increases with long theoretical deep-water retention times and with decreasing DR-values (see Håkanson 1999). This would threat the survival of the bottom fauna and may trigger diffusive internal loading of phosphorus.

The grain size and/or the composition of the sediments are often used as criteria to distinguish different sediment types (Sly 1978). One can also differentiate between different sediment types by means of functional criteria (like erosion, transportation and accumulation) of coarse sediments (friction material) or fine sediments (cohesive material). In geoecological contexts, it is common to focus on the finer materials most easily set in motion/resuspension and having the highest capacity to bind pollutants (Thomas 1972; Thomas et al. 1972, 1976).

In defining the bottom dynamic conditions (erosion, transportation and accumulation), this work uses the following definitions (from Håkanson 1977):

- Areas of erosion (E) prevail where there is no apparent deposition of fine materials but rather a removal of such materials, e.g., in shallow areas or on slopes; E-areas are generally hard and consist of sand, gravel, consolidated clays and/or rocks.
- Areas of transportation (T) prevail where fine materials are deposited periodically (areas of mixed sediments). This bottom type generally dominates where wind/wave action regulates the bottom dynamic conditions (see the ETA-diagram in Fig. 3.15). It is sometimes difficult in practice to separate areas of erosion from areas of transportation. The effective fetch gives a measure of the free water surface regulating how the winds govern the waves (wave length, wave height, etc.) by accounting for several wind directions (see Håkanson and Jansson 1983).
- Areas of accumulation (A) prevail where the fine materials are deposited continuously (soft bottom areas). These are the areas (the "end stations") where high concentrations of pollutants may appear (see Table 3.7).

The generally hard or sandy sediments within the areas of erosion (E) often have a low water content, low organic content and low concentrations of nutrients and pollutants. The conditions within the T-areas are, for natural reasons, variable, especially for the most mobile substances, like phosphorus, manganese and iron, which react rapidly to alterations in the chemical "micro-climate" (given by the redox potential) of the sediments. Fine materials may be deposited for long periods during stagnant weather conditions. In connection with a storm or a mass movement on a slope, this material may be resuspended and transported up and away, generally in

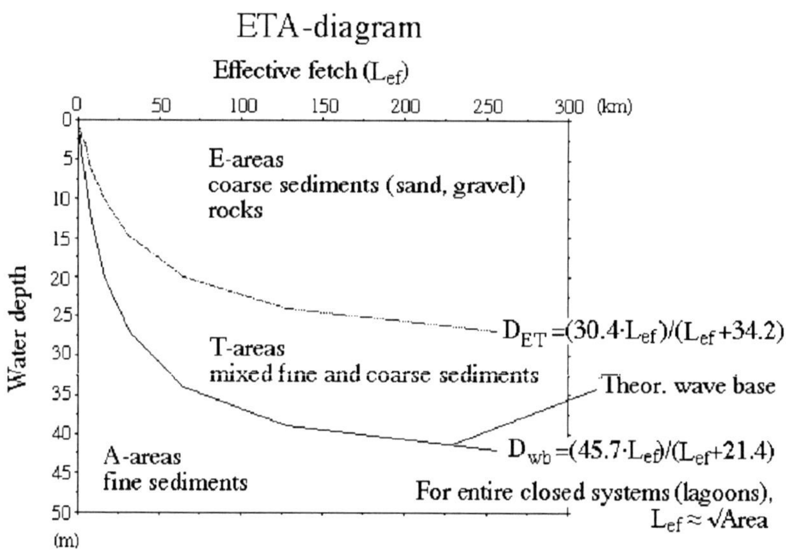

Fig. 3.15 The ETA-diagram (Erosion-Transportation-Accumulation; for more information, see Håkanson 1977) illustrating the relationship between effective fetch, water depth, bottom dynamic conditions and sediment type

Table 3.7 The relationship between bottom dynamic conditions (erosion, transportation and accumulation) and the physical, chemical and biological character of the surficial sediments. The given data represent characteristic values from marine coastal areas based on data from 11 Baltic areas (from Håkanson et al. 1984). ww = wet weight; dw = dry weight

	Erosion	Transportation	Accumulation
PHYSICAL PARAMETERS			
Water content (% ww)	<50	50–75	>75
Organic content (% dw)	<4	4–10	>10
NUTRIENTS (mg/g dw)			
Nitrogen	<2	10–30	>5
Phosphorus	0.3–1	0.3–1.5	>1
Carbon	<20	20–50	>50
METALS			
Iron (mg/g dw)	<10	10–30	>20
Manganese (mg/g dw)	<0.2	0.2–0.7	0.1–0.7
Zinc (μg/g dw)	<50	50–200	>200
Chromium (μg/g dw)	<25	25–50	>50
Lead (μg/g dw)	<20	20–30	>30
Copper (μg/g dw)	<15	15–30	>30
Cadmium (μg/g dw)	<0.5	0.5–11.5	>1.5
Mercury (ng/g dw)	<50	50–250	>250

the direction toward the A-areas in the deeper parts, where continuous deposition occurs. Thus, resuspension is a most natural phenomenon on T-areas.

Lindgren and Håkanson (2007) defined four different DR-classes (very deep, deep, intermediately deep and shallow), see Table 3.8. The dynamic ratio has been calculated for 538 coastal areas in the Swedish coastal database (SMHI 2003) where data on A and D_m were available and all areas have been then classified according to the criteria given Table 3.8. Most of these Baltic coastal areas have DR-values between 0.25 and 4.1 ("intermediately" deep; see Lindgren and Håkanson 2007).

The depth conditions also influence target bioindicators for coastal management. This is exemplified in Fig. 3.16. Here, simulations have been carried out using the

Table 3.8 Classes for the dynamic ratio (DR)

Class	DR	Description
Very deep	<0.064	Areas dominated by slope processes and erosion and transport processes for fine particles
Deep	0.064–0.25	Areas influenced by slope processes were erosion, transport and accumulations for fine particles occur
Intermediate	0.25–4.1	Areas more influenced by wind and wave processes were erosion, transport and accumulations for fine particles occur
Shallow	>4.1	Area dominated by wind and wave processes and erosion and transport processes for fine particles

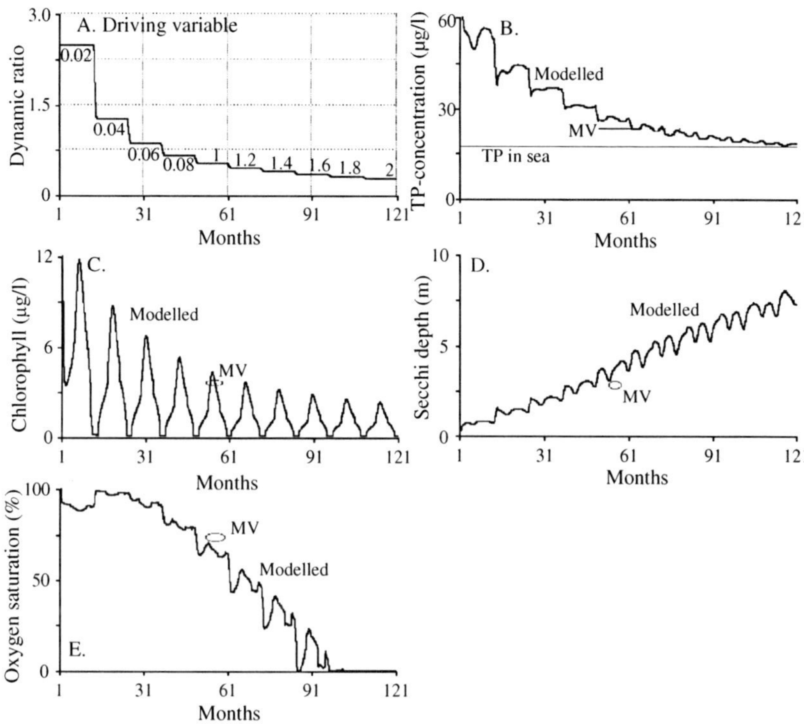

Fig. 3.16 Illustration of the practical use of how changes in the dynamic ratio would likely influence a given coastal area, in this case the Gävle Bay coastal area in the Bothnian Sea. The dynamic ratio has been varied in 10 annual steps by changing the mean depth from 0.02·8.4 = 0.17 m (a very shallow system) to 2·8.4 = 18.8 (a deep system; see **A**) and the consequences of this has been calculated for (**B**) the TP-concentrations in water, (**C**) the chlorophyll-a concentration in water, (**D**) the Secchi depth (as a criteria for water clarity) and (**E**) the oxygen concentration in the deep-water zone. The predicted values have also been compared to the empirical data for the given coastal area (dynamic ratio = 1) and one can note the close correspondence between modeled values and empirical data

same dynamic mass-balance model (CoastMab) as used for Fig. 3.9 for the Gävle Bay coastal area in the Bothnian Sea. In this test, the dynamic ratio was changed in 10 steps by changing the mean depth from 0.02·8.4 to 2·8.4 m (8.4 m is the actual value in this bay), while all else have been kept constant. The default DR-value is 1. How will this likely influence the bioindicators?

One can note from Fig. 3.16 that by increasing the mean depth and decreasing the dynamic ratio, the TP-concentration would approach the TP-concentration in the outside sea (Fig. 3.16B), the chlorophyll concentration would go down (Fig. 3.16C), water clarity would be significantly higher (Fig. 3.16D), and the oxygen saturation in the deep-water zone would approach zero, which is critical for the bottom fauna, which will die of the oxygen concentration is lower than 2 mg/l or the oxygen saturation lower than 20% (see Chap. 5).

So, one can conclude that the exposure and the dynamic ratio are key morphometric parameters regulating coastal area sensitivity to nutrient loading and they will be used to define and index of sensitivity or vulnerability.

3.2.5 A Sensitivity Index (SI) Based on Morphometric Papameters

From this, one can conclude that very enclosed areas, very deep areas and very shallow coastal areas should be most sensitive to nutrient loading. This is expressed in a simple manner in by (3.4), which defines the sensitivity index (SI, dimensionless) from the exposure (Ex, dimensionless) and the dynamic ratio (DR, dimensionless).

$$\begin{aligned} &\text{If DR} \geq 0.25, \quad \text{then SI} = \sqrt{((DR/0.25)/\text{Ex})} \text{ and} \\ &\text{If DR} < 0.25, \quad \text{then SI} = \sqrt{((0.25/DR)/\text{Ex})} \end{aligned} \tag{3.4}$$

The dynamic ratio generally varies between 0.06 and 6 and the exposure between 0.002 and 1. By accounting for the threshold DR of 0.25 (see Fig. 3.14) and by taking the square root of the expression, SI will generally vary between 0 and 100 (extremely sensitive enclosed and deep areas). A typical Baltic Sea coastal area would have a dynamic ratio of 1 and an exposure of 0.05, which would give an SI-value of 8.9 and according to the categories given in Fig. 3.17, this would indicate

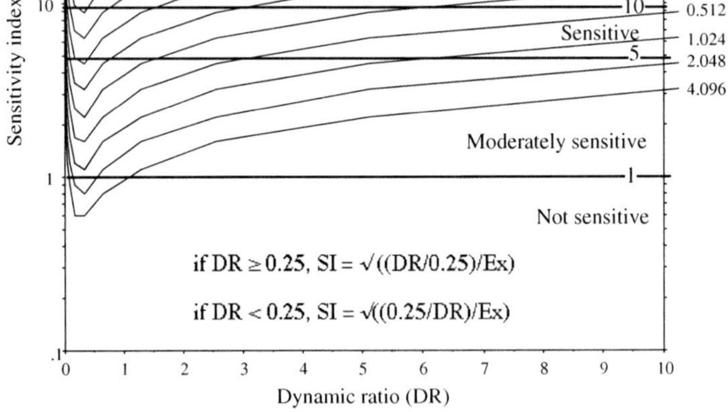

Fig. 3.17 Nomogram illustrating how the sensitivity index (SI) is related to changes in dynamic ratio (DR) and exposure (Ex)

a sensitive coastal area to nutrient loading. The class limits for SI, 1, 5, 10 and 50 and the categories "extremely sensitive, very sensitive, sensitive, moderately sensitive and not sensitive", can, of course, be discussed and this nomenclature is given as a suggestion. However, the definition of the sensitivity index is well motivated by empirical data and process-based dynamic modeling. Since the exposure and the dynamic ratio are easy to define and understand, SI is also easy to apply in practice in coastal management.

Figure 3.18 gives frequency distributions and statistics (mean values, medians, standard deviations, minimum and maximum values) based om data from 478 Baltic Sea coastal areas for (A) coastal area, (B), section area, (C) exposure, (D) mean depth, (E) maximum depth, (F) dynamic ratio, (G) latitude, (H) sensitivity index and (I) the logarithm of SI. One can note that the sensitivity index varies between 0.69 (not sensitive) to 116 (extremely sensitive); the frequency distribution is positively skewed and the mean value is higher (5.7; sensitive) than the median value (3.8; low sensitive) and that the log-transformation of SI provides a more normal frequency distribution (Fig. 3.18E). In this dataset, there are 2 (0.4%) "extremely sensitive" coastal areas (SI > 50), 50 (10.5%) "very sensitive" coastal areas (10 < SI < 50), 121 (25.3%) "sensitive" coastal areas (5 < SI < 10), 301 (63.0%) "low sensitive" coastal areas (1 < SI < 5) and 4 (0.8%) "not sensitive" coastal areas (SI < 1).

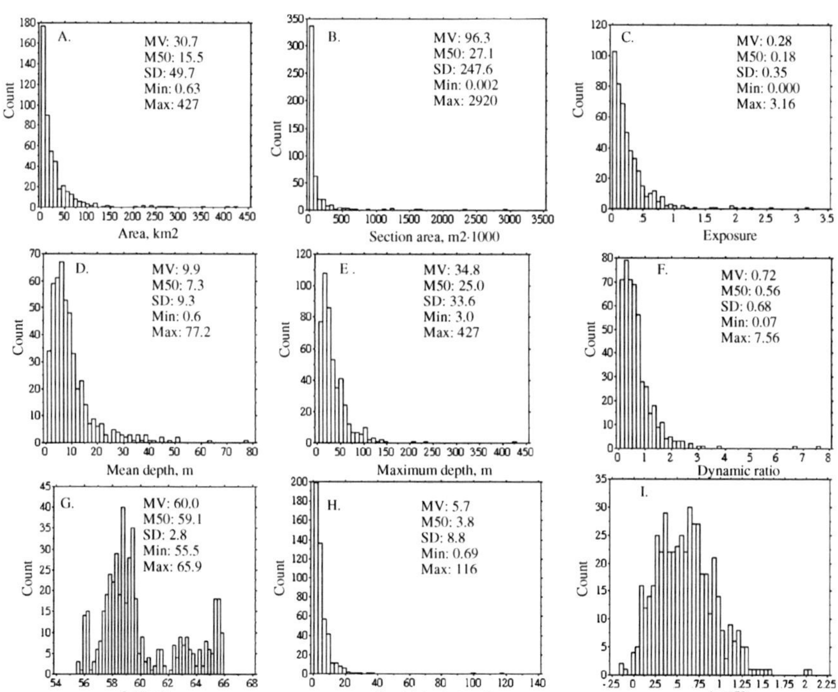

Fig. 3.18 Frequency distributions and statistics (mean values, medians, standard deviations minimum and maximum values) based on data from 478 Baltic coastal areas

Using geographical information systems (GIS) based on digitized bathymetric data, it is easy to apply these concepts and determine SI for any given coastal area.

3.3 Water Exchange

The water exchange is important for several reasons, and one of those concerns the sensitivity to eutrophication (see Le Pape and Menesguen 1997). The sensitivity index discussed in the previous section is based solely on morphometric data derived from bathymetric maps, so it is simple to apply. Nordvarg and Håkanson (2002) discussed another type of sensitivity index based on the concept of the theoretical water retention rate, R_{wat} (dimension 1/time), defined from the ratio between the total water discharge to a given coastal area (Q) and the total volume of the coastal area (V). Q is the sum of the freshwater inflow from tributaries (Q_{trib}), the surface-water inflow from the outside sea (Q_{SW}) and the deep-water inflow from the sea (Q_{DW}). V is the product of the coastal area (Area) and the mean depth (D_m). Evidently, this approach is more difficult to apply. It is more difficult to interpret R_{wat} than SI because the coastal area would be different if the water discharge to the area is dominated by the freshwater inflow rather than the inflow from outside sea, if the inflow from the outside sea is regulated by tidal variations, coastal current or mixing processes between the given coastal area and the outside sea, or if the coastal area is small and deep or large and shallow with the same total volume. Processes regulating water exchange will be discussed in this section.

3.3.1 Surface and Deep-Water Exchange

The hydrodynamical conditions are generally very dynamic in marine coastal areas (a typical surface water retention time is 2–6 days for a Baltic Sea coastal area and about 1 year for a lake; see Håkanson 2000). This has, as already stressed, implications for the conditions in the coastal areas, which are evidently greatly influenced by the conditions in the outside sea and/or adjacent coastal areas in direct contact with the sea, see Muir Wood (1969); Stanley and Swift (1976); Dyer (1972); McCave (1981); Seibold and Berger (1982); Postma (1982) and Kjerfve (1994).

Emissions of pollutants cannot be calculated into concentrations without knowledge of the water retention time of a given coastal area. If concentrations cannot be predicted, it is also practically impossible to predict the related ecological effects. Thus, it is important to highlight some basic concepts concerning the turnover of water in coastal areas. The water exchange can be driven by many processes, which vary in time and space. The importance of the various processes will vary with the topographical characteristics of the coast, which do not vary in time, but vary widely between different coasts. The water exchange sets the framework for the entire biotic spectrum; the prerequisites for life are quite different in coastal waters where the characteristic retention time varies from hours to weeks. Factors influencing the water exchange are listed below (see Fig. 3.19).

- The fresh water discharge (Q_{trib}; in m^3/sec) is the amount of water entering the coast from tributaries per time unit. In estuaries, Q_{trib} may be the most important factor for the water retention time.
- Tides. When the tidal variation is > 50 cm, it is often a dominating factor for the surface water retention time.
- Water level fluctuations always cause a flux of water. These variations may be measured with simple gauges. They vary with the season of the year and are important for the water retention time of shallow coastal areas. Thus, the mean depth is a useful coastal parameter.
- Boundary level fluctuations. Fluctuations in the thermocline and the halocline boundary layers may be very important for surface and deep water retention times, especially in deep and open coasts.
- Local winds may create a water exchange in all coastal areas, especially in small and shallow coasts.
- Thermal effects. Heating and cooling, e.g., during warm summer days and nights, may give rise to water level fluctuations which may cause a water exchange. This is especially true in shallow coasts since water level variations in such areas are more linked to temperature alterations in the air than is the case in open water areas.
- Coastal currents are large, often geographically concentrated, shore-parallel movements in the sea close to the coast. They may have an impact on the water retention time, especially in coasts with a great exposure.

In theory, it may be possible to distinguish driving processes from mixing processes. In practice, however, this is often impossible. Surface water mixing causes a change in boundary conditions, which causes water exchange, and so on. The theoretical surface water retention time (T) for a coastal area is the time it would take to fill a coast of volume V if the water input from rivers is given by Q and the net water input from the sea by (Q_{SW}+Q_{DW}), i.e.: $T = V/(Q_{trib}+Q_{SW}+Q_{DW})$. This definition does not account for the fact that actual water exchange normally varies temporally, areally and vertically, e.g., above and beneath the thermocline.

As mentioned in Sect. 3.2.1, it is demanding to empirically determine the theoretical surface or deep-water retention time (T_{SW} and T_{DW}, respectively) in a given coastal area using field methods. T_{SW} can be predicted very well ($r^2 = 0.95$) with the regression in Table 3.2, which is based on the exposure (Ex), which, in turn is a function of section area (At) and coastal area (Area). The range of this model for T_{SW} is given by the minimum and maximum values for Ex in Table 3.2. The model should not be used without complementary algorithms if the tidal range is > 20 cm/day or for estuaries, where the fresh water discharge must also be accounted for. For open coasts, i.e., when Ex > 1.3, T_{SW} may be calculated not by this equation but from a model based on coastal currents (see Sect. 9.1). The T_{DW}-formula in Table 3.2 is only applicable to stratified coastal areas, which are not influenced by

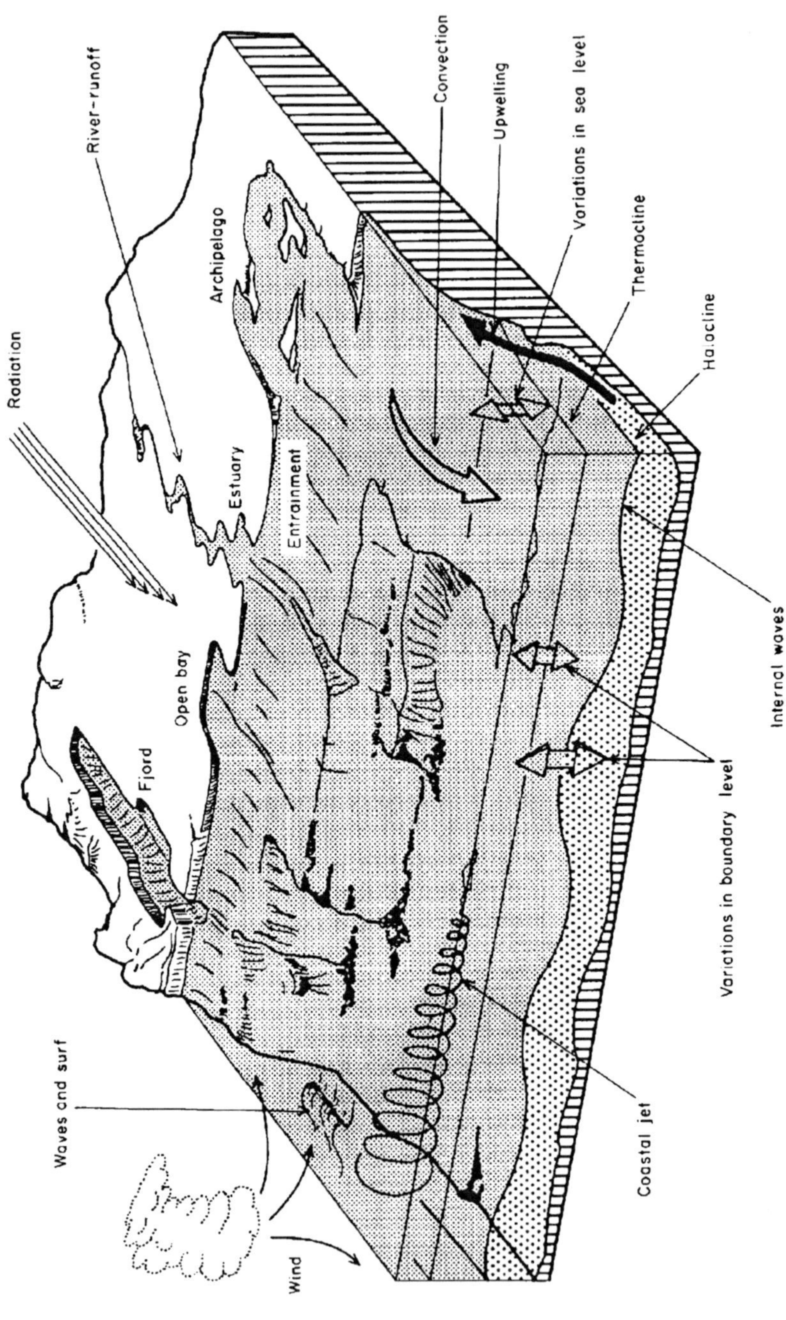

Fig. 3.19 Illustration of key factors regulating water exchange in coastal areas (from Håkanson et al. 1986)

tides since it is based on data from Baltic coastal areas. For open and tidal coasts, there are also simple models available to predict the theoretical water retention time (Håkanson 2000).

3.3.2 Tides

The water exchange between the coast and the sea can affect the conditions in a coastal ecosystem in many profound ways, e.g., the concentrations of pollutants and the recovering ability of coastal areas (Kraufvelin et al. 2001). As already stressed, the morphometry of the coast is a key regulating factor influencing the water exchange. Other important factors are tides and coastal currents. The CoastMab-model (see Sect. 9.1) includes three different modes of calculating the water exchange between the coast and the sea: (1) for tidal areas, (2) for semi-enclosed non-tidal areas and (3) for open non-tidal coastal areas. In the latter case, coastal currents play an important role. This means that information on tides and coastal currents are important in understanding and modeling how a given coastal area will respond to changes in loading of pollutants.

Even though the forces that drive the tides are well-known (e.g., geographical position, the topography of the ocean basin, the Coriolis effect and the form of local basins), the calculation of the tidal effects in a specific coastal area are complicated. Detailed studies of tidal effects may require that between 40 and 100 variables should be taken into account (Davis 1996). The tidal range is also of major importance since it is the main factor in determining the strength of tidal currents (Davies 1980). Much effort has been put into understanding and modeling tides. However, for the purpose of this book, the exact prediction of tides is not the focal issue. The main task here is to determine if the water exchange in a certain coastal area is dominated by tidal processes or not, which is important in mass-balance modeling. A map of the average tidal range for the European coastal zone is given in Fig. 3.20 and this figure gives a general overview. When studying or modeling a specific coastal area, either tidal tables or some tidal prediction software should be used to calculate the tidal range for that specific area, since it may be important to take the local tidal conditions into account.

3.3.3 Coastal Currents

Depending on winds, pressure gradients, tides, etc., water currents vary in space and time, direction and velocity. Nevertheless, when studying the average general circulation, there are clear patterns (Pinet 2003). Longshore coastal currents are evident in many coastal areas and average current velocities can be determined for many of these. In areas where the tidal effects are small, morphometry and coastal currents are the major factors influencing the water exchange (Håkanson 2000). In

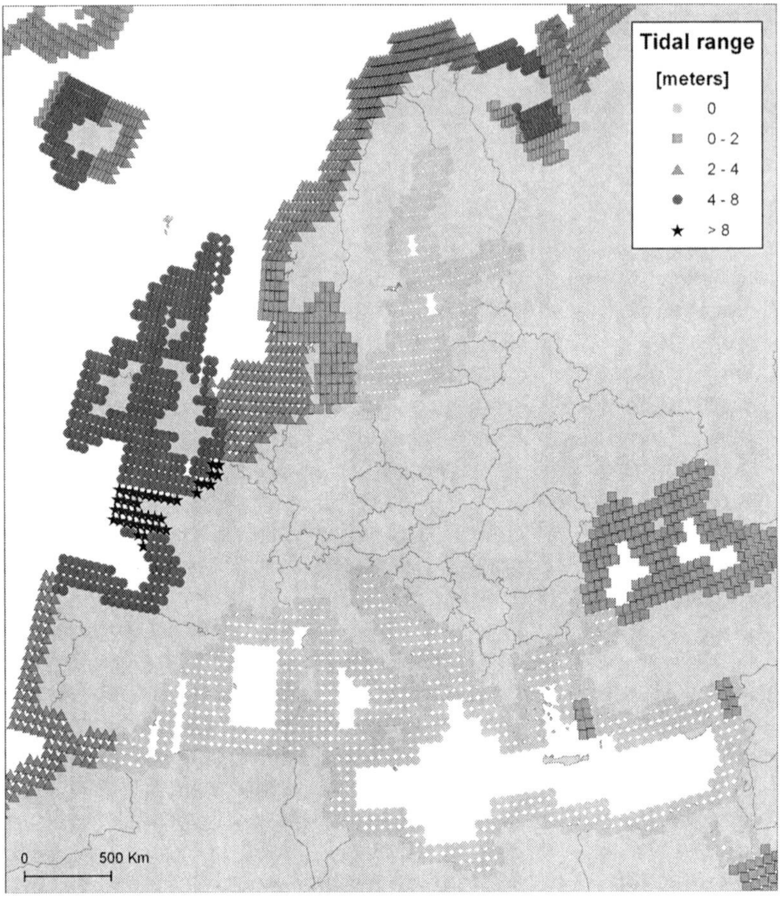

Fig. 3.20 Characteristic tidal ranges along the European coast

shallow waters, the impact of deep-water currents can be considered as negligible compared to the tide and wave-generated currents. For open coastal areas, in non-tidal areas, coastal currents are a major factor regulating the water exchange (FRP 1978). So, knowing the average velocity of coastal currents is important when performing mass-balance modeling of open coastal areas. The focus here is on surface currents, since these often have the greatest effect on coastal areas. This compilation of surface currents aims to give an overview and illustrate basic principles. It should be noted that in many areas there may be local residual currents, closer to land, that move at different velocities and even in opposite directions (Ursella and Gačić 2001) than the general coastal current. Such residual currents that appear on local to regional scales can for obvious reasons not be presented on a map at the international scale. Information on local/regional residual currents may be needed in modeling at the local scale.

In the work presented by Lindgren and Håkanson (2007), it was not possible to find enough data to create a map on coastal currents for the entire European coastal zone. Instead, they gave a map on the average surface-water currents in the Baltic Sea an example. This map was created by digitizing information in FRP (1978) and it is given here as Fig. 3.21.

This figure provides an example of the characteristic circulation pattern for coastal currents (the coastal jet) in the Baltic Sea driven by the rotation of the earth (the Coriolis effect) and the prevailing wind climate of the region. Note that this figure shows the average conditions. During storm events and situations with alternating high and low pressures the surface-water currents can be quite different from the average pattern shown in Fig. 3.21. The Coriolis effect means that water flowing in any marine system on the northern hemisphere will be deflected toward the right and hence the tributary water from Finnish rivers will be deflected to the north and water from Swedish rivers to the south creating the anti-clockwise circulation pattern shown in this figure. This hydrodynamical flow pattern will also regulate the distribution of pollutants from rivers and the net result is that the coastal areas receive a high pollution load (see Håkanson 1999).

Figure 3.22 illustrates the geographical distribution pattern of what may be regarded as a tracer pollutant from point source emissions, namely EOCl (extractable organically bound chlorine; see Håkanson 1999) from paper and pulp mills (PPMs)

Fig. 3.21 Average surface water currents in m/s and hydrodynamical circulation pattern in the Baltic Sea. Redrawn after FRP (1978)

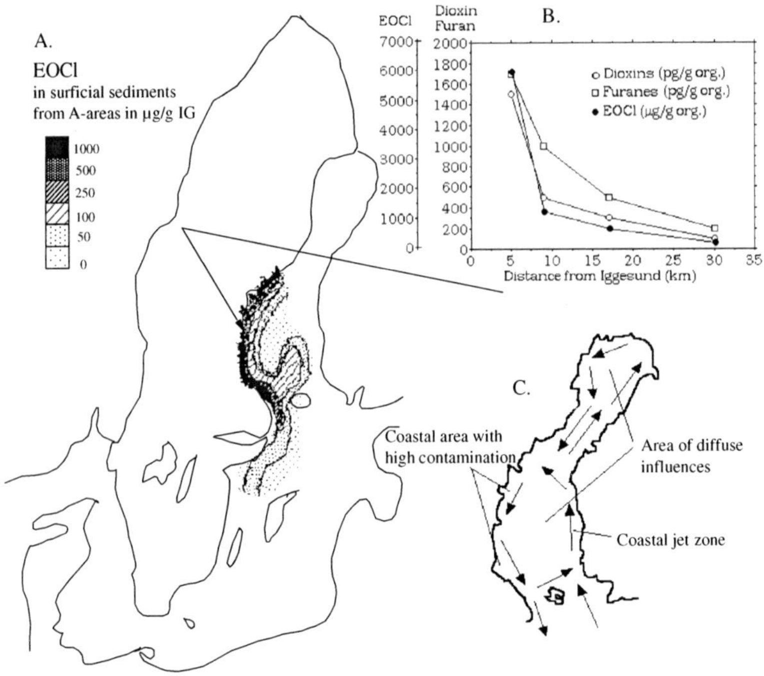

Fig. 3.22 (**A**) Areal distribution of EOCl in surficial accumulation area sediments in the Baltic Sea, (**B**) a diagram illustrating the relationship between EOCl, dioxins and furanes in surficial accumulation area sediments taken at different distances from the Iggesund paper and pulp mill towards the open Bothnian Sea, and (**C**) illustration of the coastal jet-zone and the major hydrological flow pattern in the Bothnian Bay and the Bothnian Sea. Note that the point source pollutants from the coastal-based paper and pulp mills stay close to the coast mainly because of the net hydrodynamical coastal current (Coriolis) illustrated in C (figures from Håkanson et al. 1988)

along the Swedish (western) side of the Baltic Sea. It has been estimated (Håkanson et al. 1988) that about 1/10 of the discharges of EOCl to the southern part of the Bothnian Sea could be found in the sediments of this area; between 5 and 50 tons EOCl were, e.g., found in the coastal sediments outside the paper and pulp mill at Iggesund in the Bothnian Sea (Fig. 3.22).

From Fig. 3.22 one may also note:

1. There is a large-scale spread of EOCl from Swedish PPMs. The larger the emissions, the larger the impact areas in the Baltic, the greater the potential ecological problems.
2. The net coastal current acts as a barrier to the spread of EOCl and related substances and the main direction of transport of the substances is not primarily towards the centre of the Bothnian Sea but in a predominantly southern direction along the Swedish coast.
3. The dominating water circulation in each basin (here the Bothnian Bay, the Bothnian Sea and the Baltic Proper) constitutes an anti-clockwise cell, which

distributes the settling particles, the suspended material and the pollutants in a typical pattern, reflecting the flow of the water. This anti-clockwise cell is created by the rotation of the earth, which deflects any plume of flowing water to the right in relation to the direction of the flow in the northern hemisphere (and to the left in the southern hemisphere). Thus, when a Swedish river enters the Baltic, the water turns to the right and follows the shore.

4. The net hydrological flow is to the south on the west (Swedish) side of the Baltic.
5. The coastal currents are rather strong and stable close to land and weaker towards the center of each basin.
6. The arrows in Fig. 3.21 illustrate only the net component of the flow – this means that the water would also flow in most other directions during the year.
7. The contamination of EOCl in Fig. 3.22 is greater on the Swedish side than on the Finnish side because of greater Swedish emissions of chlorinated organic substances from PPMs.

So, the distribution of point source emissions with a particulate fraction higher than zero (as EOCl in this example) generally shows a characteristic pattern with high concentrations in sediments close to the points of emission, lower concentrations in areas of erosion and transport (out to about 10–15 km from the coast in Fig. 3.22), and comparatively high concentrations again in open water areas within the zones of continuous sediment accumulation. Figure 3.23 illustrates why it is

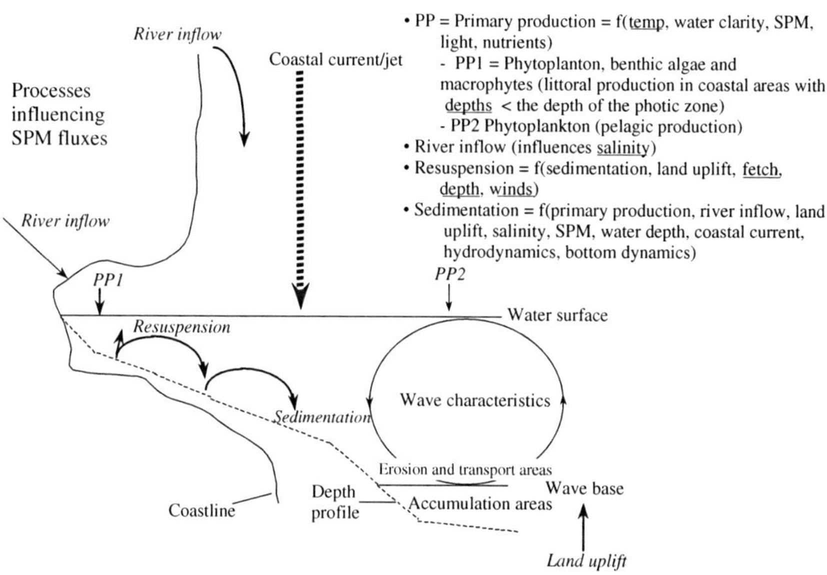

Fig. 3.23 Processes influencing flow patterns in SPM and particulate fractions of the nutrients. The underlined factors are included in the dynamic SPM-model (CoastMab) presented by Håkanson (2006). Note that many of the processes and factors influencing SPM are interrelated in a complex manner. There is also a very dynamic exchange of water, SPM, nutrients and toxins between the Sea and coastal areas

important to know the bottom dynamic conditions in contexts dealing with the spread of pollutants from a point source or from a river mouth. The fine materials (clays, humus, SPM, particulate forms of nutrients, etc.) may be transported via several resuspension cycles from the site of emission to the true accumulation area at water depths greater than the wave base (at about 44 m in the Baltic Proper). Some coastal areas may have well-defined accumulation areas (deep-holes and topographically sheltered areas within coastal areas), and then the concentrations of pollutants emitted from a given point source increase markedly towards the site of emission (see Håkanson 1999).

3.4 Salinity

The salinity is of vital importance for the biology of coastal areas influencing, e.g., the number of species in a system (see Fig. 3.24 from Remane 1934), and also the reproductive success, food intake and growth of fish (Rubio et al. 2005; Nissling et al. 2006). Furthermore, a higher salinity has been shown to increase the flocculation and aggregation of particles (Håkanson 2006) and hence affect the rate of sedimentation, which is of particular interest in understanding variations in water clarity among coastal areas. The saltier the water, the greater the flocculation of suspended

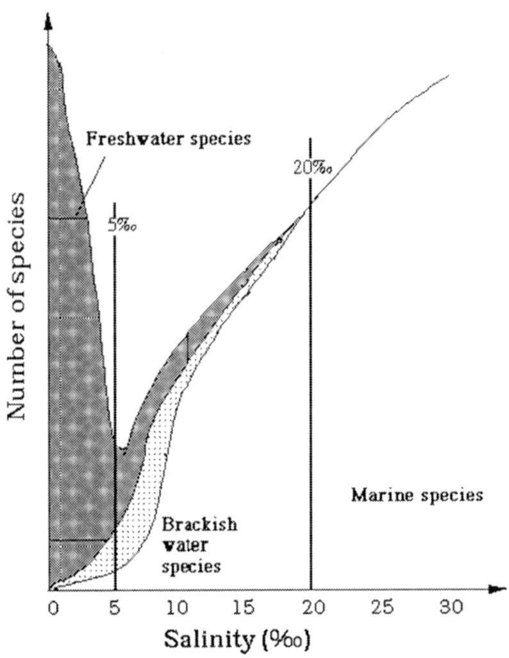

Fig. 3.24 The relationship between salinity and number of species. Redrawn from Remane (1934)

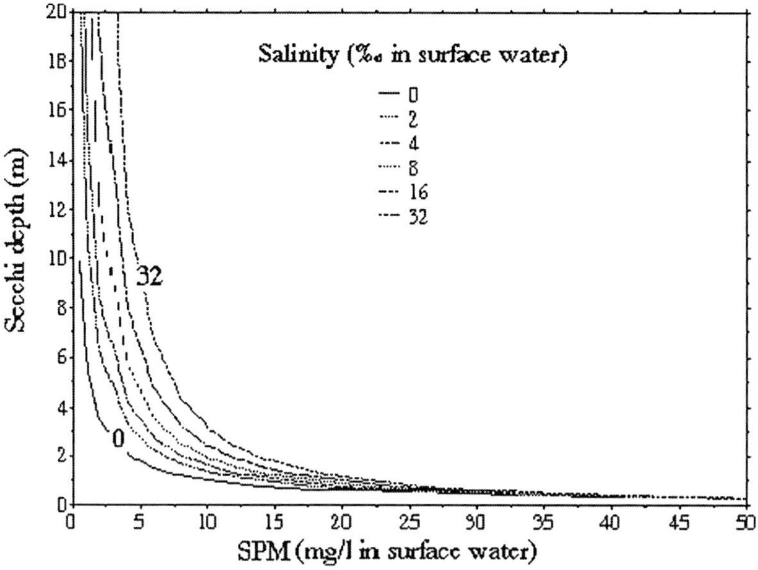

Fig. 3.25 Illustration of the relationship between Secchi depth, SPM in surface water and salinity in surface water using log-data (for further information, see Håkanson 2006)

particles (see Fig. 3.25). This does not only influence the concentration of particulate matter, but also the concentration of any substance with a substantial particulate phase such as phosphorus and many pollutants. The salinity also affects the relationship between total phosphorus and primary production/biomass (chlorophyll-a; Håkanson et al. 2007a). The salinity is easy to measure and the availability of salinity data for the European coastal zone is better than for most other water variables.

Figure 3.26 illustrates the role of salinity on the Chl/TP-ratio (from Håkanson and Eklund 2007a). The figure gives the number of data in each salinity class, the box-and-whisker plots give the medians, quartiles, percentiles and outliers, and the table below the diagram provides information on the median Chl/TP-values, the coefficients of variation (CV) and the number of systems included in each class (n). These results are based on many data. The interesting aspect concerns the pattern shown in the figure. One can note:

- The salinity influences the Chl/TP-ratio very much indeed.
- The median value for lakes is 0.29, which is almost identical to the slope coefficient for the reference model for lakes (0.28 in the OECD-model; see OECD 1982).
- The Chl/TP-ratio changes in a wave-like fashion when the salinity increases. It is evident that there is a minimum in the Chl/TP-ratio in the salinity range between 2 and 5. Then, there is an increase up to the salinity range of 10–15, and then a

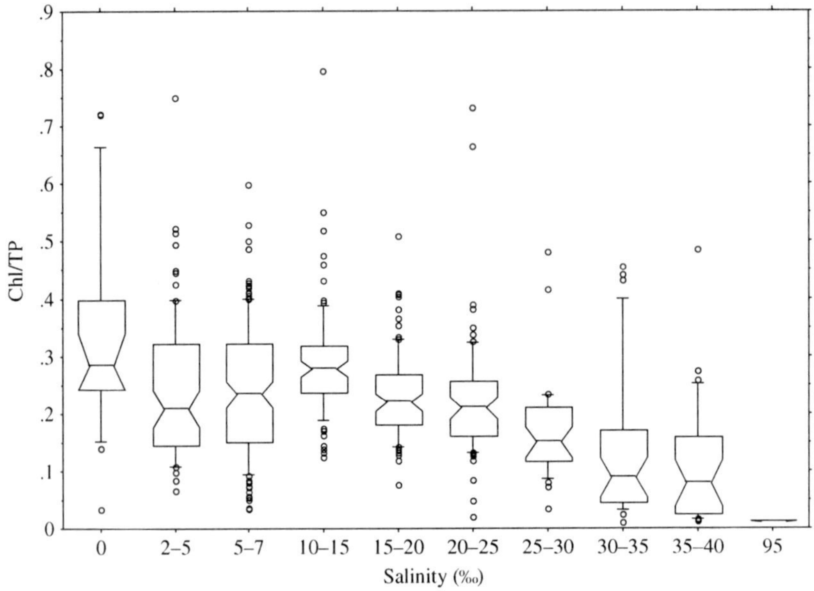

	0	2–5	5–7	10–15	15–20	20–25	25–30	30–35	35–40	95
Median:	0.29	0.21	0.24	0.28	0.22	0.21	0.15	0.11	0.079	0.012
CV:	0.54	0.52	0.48	0.32	0.29	0.42	0.53	0.89	0.98	0.16
n:	20	74	140	82	79	61	30	28	24	3

Fig. 3.26 Box-and-whisker plot (showing medians, quartiles, percentiles and outliers) illustrating the Chl/TP-ratio for 10 salinity classes. The statistics give the median values, the coefficients of variation (CV) and the number of data in each class

continuous decrease in the Chl/TP-range until a minimum value of about 0.012 is reached in the hypersaline Crimean systems.

- The relationship between the Chl/TN-ratio and salinity (from Håkanson and Eklund 2007a) is summarized in Fig. 3.27 in the same manner as was done for the Chl/TP-ratio in Fig. 3.26. One can identify differences and similarities between the results presented in Figs. 3.26 and 3.27.
- At higher salinities than 10–15, there is a steady decrease also in the Chl/TN-ratio (note that there are no data on TN from the hypersaline Crimean lakes).
- The Chl/TN-ratio attains a maximum value for systems in the salinity range between 10 and 15, and significantly lower values in lakes and less saline brackish systems.
- The table in Fig. 3.27 gives the median Chl/TN-values and they vary from 0.0084 (for lakes), to 0.017 for brackish systems in the salinity range between 10 and 15, to very low values (0.0041) for marine coastal systems in the salinity range between 35 and 40.

A salinity map for the entire European coastal zone is shown in Fig. 3.28. The salinity gradient in the Baltic Sea and the transient areas of Kattegat and the Danish straits are clearly seen. A gradient with slightly less saline water closest to the coast can also be observed for almost the whole Norwegian coast, as well as the eastern

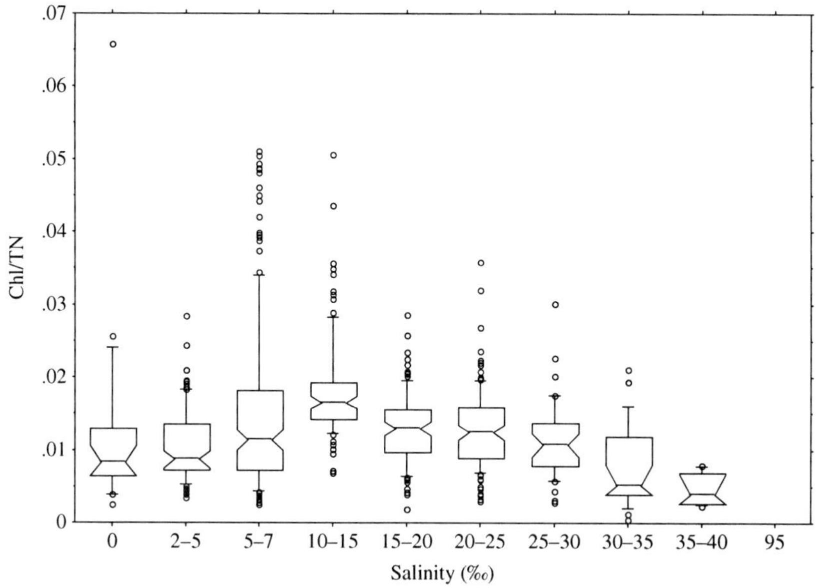

Median:	0.0084	0.0088	0.012	0.017	0.013	0.012	0.011	0.0054	0.0041	-
CV:	1.10	0.47	0.78	0.34	0.38	0.43	0.49	0.79	0.49	-
n:	20	81	162	82	102	118	36	14	6	0

Fig. 3.27 Box-and-whisker plot illustrating relationship between Chl/TP-ratio and salinity. The statistics give the median values, the coefficients of variation (CV) and the number of data in each class

coasts of the North Sea and the Irish Sea. This is due to extensive freshwater input from large drainage areas. In the case of the Norwegian coast, it is also because of the coastal currents transporting brackish water from the Baltic Sea.

Estuaries and fjords are affected by fresh-water discharges from land and such systems often have marked seasonal variations in salinity. The general map presented here only shows the general spatial pattern in surface-water salinity and does not address seasonal variations. Vertical variations in salinity, and also in water temperature, TP and TN-concentrations, are exemplified in Fig. 3.29. This graph shows data from 100 randomly selected verticals from months 5 to 9 for the period 1997 to 2005 at stations with water depths larger than 100 m in the Baltic Proper. The idea is to give an overview of how these variables vary vertically during the summer and to illustrate the relevance of the given depth intervals. The surface-water layer (SW) is defined here by the water depth above the theoretical wave base at 44 m in the Baltic Proper (see the ETA-diagram in Fig. 3.15). Above this water depth, one should expect that wave orbital action may cause resuspension of fine particles which settle out in water by laminar motion, as calculated from Stokes' law (see Håkanson and Jansson 1983).

The middle-water layer (MW) is defined as the depth between the theoretical wave base at 44 m and the average depth of the halocline at 75 m. The deep-water

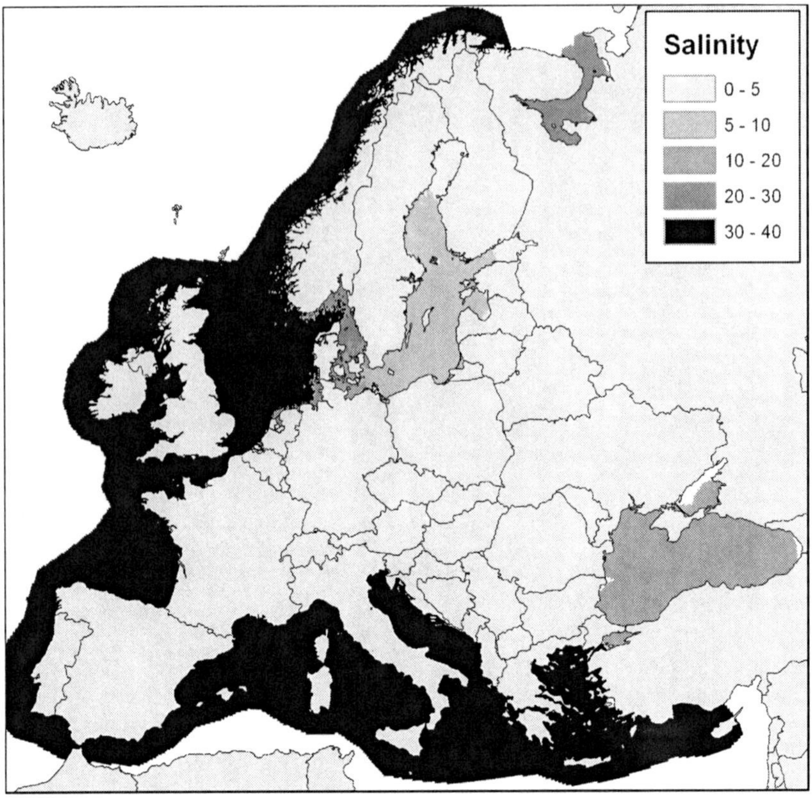

Fig. 3.28 Average annual surface-water salinities in the European coastal zone in the upper 10 m water column for the period from 1990 to 2005

layer (DW) defines the volume of water beneath the average halocline. It should be noted that both the theoretical wave base and the depth of the halocline at 75 m describe average conditions. It is clear from Fig. 3.29C that the halocline varies considerable around 75 m, and the actual wave base also varies around 44 m; during storm events, the wave base will be at greater water depths and during calm periods at shallower depths. The actual wave base also varies spatially within the studied area. From Fig. 3.29 however, it is evident that the two boundary depths relevantly describe the conditions in the Baltic Proper.

3.5 Water Temperature

The water temperature regulates many important processes and functions in aquatic systems, e.g., the bacterial decomposition of organic matter, and hence also the oxygen consumption; water temperatures also regulate the stratification and hence

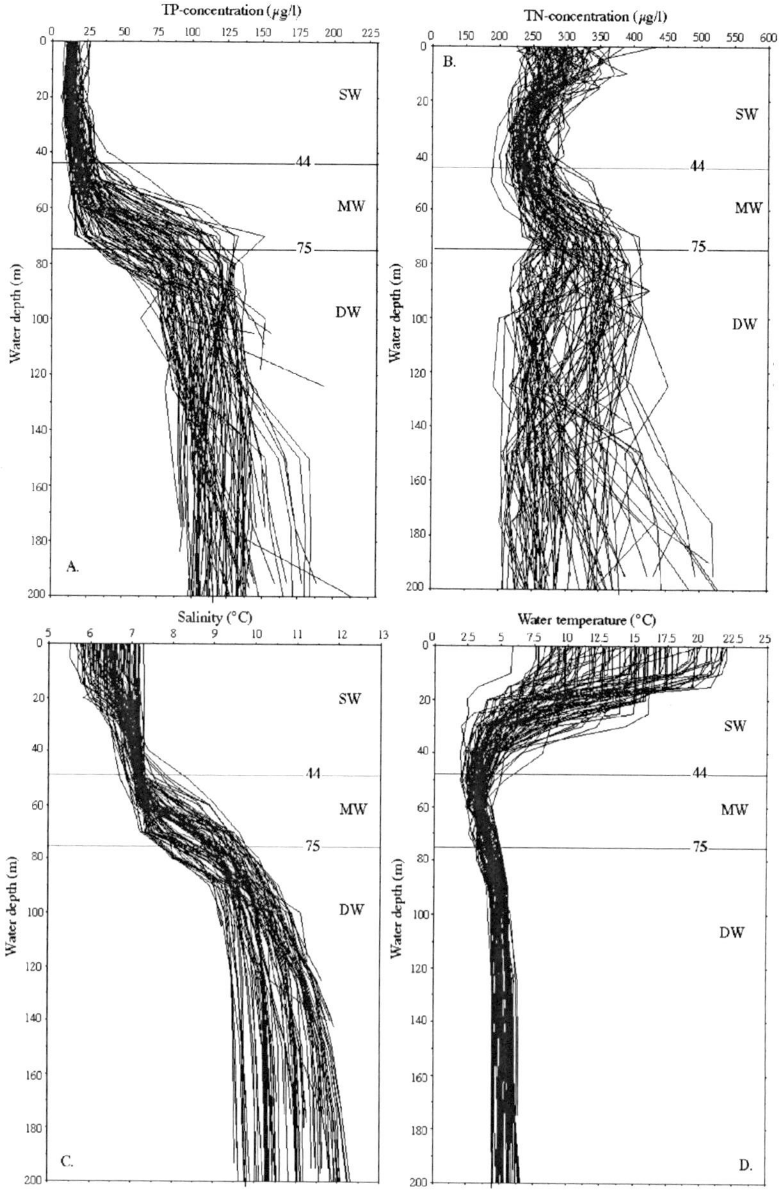

Fig. 3.29 One hundred daily verticals selected at random from stations deeper than 100 m from the Baltic Proper collected months 5–9 between 1997 and 2005: (**A**) TP-concentrations, (**B**) TN-concentrations, (**C**) salinities and (**D**) temperatures; and lines indicating surface-water areas (SW), middle-water areas (MW) and deep-water areas (DW)

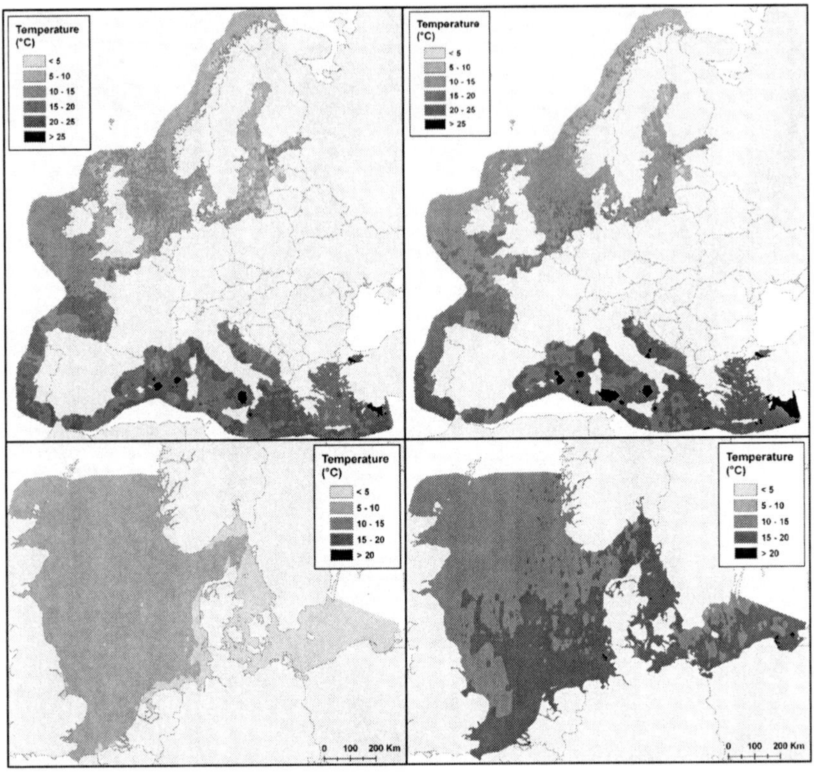

Fig. 3.30 (*Upper left*) Average annual surface-water temperatures, (*Upper right*) median surface-water temperatures for the growing season (May–September) along the European coastal zone, (*Lower left*) surface-water temperatures for January for the North Sea and parts of the Baltic Sea and (*Lower right*) surface-water temperature for July in the upper 10 m water column for the period from 1990 to 2005

also mixing across the themocline. In total, about 410,000 temperature data (ICES 2006a, b) were used to create the temperature maps (see Fig. 3.30) and the data coverage was rather good for the whole European coastal zone.

Since temperature has a distinct seasonal variation (more than most other water variables), maps were created not only for average annual surface-water temperatures (using all available data), but also for the growing season (the beginning of May to the end of September) and for two individual months (January and July to represent a wide temperature range). When creating maps for individual months, data coverage was not sufficient for the whole European coastal zone, so the North Sea and parts of the Baltic Sea were chosen as an example. Note that Lindgren and Håkanson (2007) did not have access to reliable monthly data from the entire European area so the two bottom windows in Fig. 3.30 give information using data from the North Sea. From Fig. 3.30, one can note the marked geographical variation

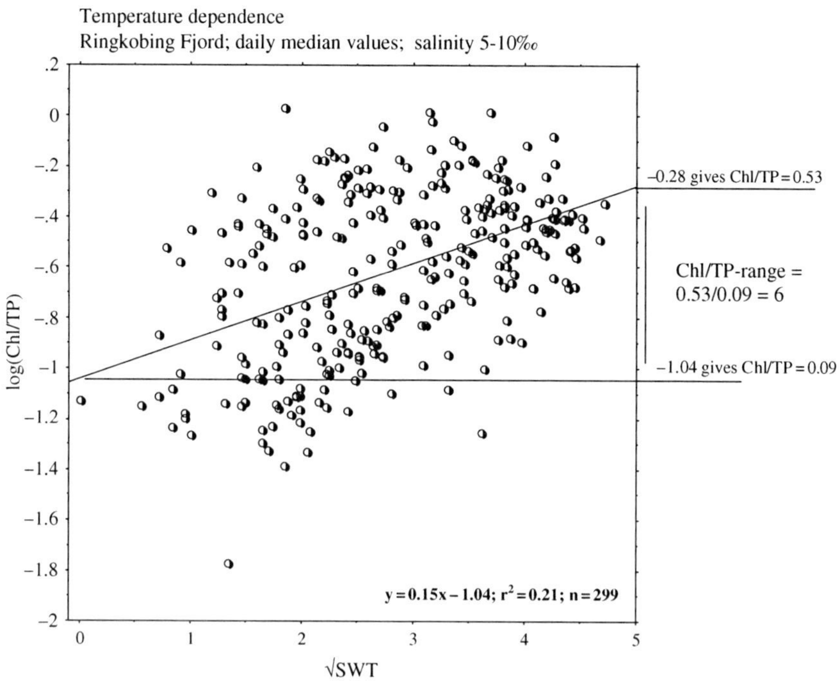

Fig. 3.31 Illustration of water temperature (surface water temperatures, SWT, in °C; the x-axis gives vSWT) influences on the ratio between chlorophyll and TP ([log(Chl/TP)] using daily median values from Ringkobing Fjord in the salinity range between 5 and 10 psu)

in surface-water temperatures. These temperature differences are fundamental for the geographical variations in primary, and hence also in secondary production (of zooplankton, zoobenthos and fish).

Figure 3.31 is included here to illustrate the role of water temperature variations for variations in the Chl/TP ratio (see also McQueen and Lean 1987). The scatter plot in Fig. 3.31 gives the ratio between chlorophyll (Chl) and TP on the y-axis (log-transformed data) and the square-root of the surface water temperatures (SWT in °C) on the x-axis.

A comprehensive data-set from Ringkobing Fjord has been used to produce this figure to normalize for variations in salinity, only data in the salinity range between 5 and 10 have been used. The actual data-set used concerns median daily data. From this figure, one can see a significant temperature dependence. If SWT varies from zero to 25, the Chl/TP-ratio will vary by a factor of 6, and for the growing season when the median temperatures would generally range between 15 and 25°C, one can expect that the Chl/TP-ratio will vary by a factor of 1.5. Note that this is a statistical relationship and that, on an annual basis there is a strong co-variation between water temperatures and the number of days with daylight.

3.6 Summary

This chapter has discussed how abiotic factors (important morphometric parameters, tides, salinity, temperature and water exchange) influence ecosystem structures and functions and also a sensitivity index for coastal eutrophication. It should be stressed that these results would fit nicely into the coastal area typologices discussed within the EC-Water Framework Directive (see Anon 2000 2003; Schernewski and Wielgat 2004), where criteria and tools, such as those discussed in this paper for coastal area sensitivity/vulnerability, are requested.

There are major differences in sensitivity or vulnerability to anthropogenic loading of nutrients (eutrophication) among different coastal areas. One aim of this chapter has been to review and discuss key criteria for coastal area sensitivity and to discuss a sensitivity index (SI). This sensitivity index is based on two morphometric parameters, which can be determined easily from simple bathymetric maps, (1) the topographical openness (or exposure) and (2) the dynamic ratio of the coastal area. The exposure of the coastal area is defined by the ratio between the section area of the coast and the enclosed coastal area. The boundaries of the coastal area should not be defined in an arbitrary manner but according to the topographical bottleneck method so that the exposure attains a minimum value. The exposure regulates the theoretical water retention time of the coastal area, which, in turn, regulates the effects of a given nutrient loading. The dynamic ratio is defined from the ratio between the square root of the coastal area and the mean depth. The dynamic ratio influences many fundamental internal transport processes (such as sedimentation, resuspension, mixing and burial), and hence also how a given coastal system responds to a given nutrient loading. Coastal management should focus remedial actions on critical coastal areas which are at hand if the nutrient loading is high and/or the sensitivity is high; even a relatively low load can cause undesired effects in highly sensitive areas. A high tidal range will generally decrease the sensitivity of coastal areas related to the nutrient loading from land-based activities. The sensitivity index is meant to be simple to use at local and regional scales to find and manage the most vulnerable coastal areas so that time and resources can be directed to such areas in a cost-efficient manner.

Chapter 4
Nutrients and Representativity of Data

The aim of this chapter is to focus on the nutrients, which could and should be targeted in contexts of coastal eutrophication when remedial actions and strategies are discussed. The factors discussed in Chap. 3 concern morphometric parameters, water exchange, salinity and water temperature. They are important in contexts of coastal area sensitivity to eutrophication, but many of those factors cannot readily be changed by man in a perspective of local to regional coastal management. This chapter will discuss mainly nitrogen and phosphorus and nutrient fractions (such as DIN and DIP, i.e., dissolved inorganic nitrogen and phosphorus), which are standard variables in coastal monitoring (see Fisher et al. 1992). Central questions concern: Should remedial actions focus on both nutrients, on nitrogen or on phosphorus? Can general recommendations be given? What are the arguments pro and con such recommendations? Those arguments also concern the variability and representativity of nutrient concentrations in coastal areas and whether or not it is possible to derive practically useful models to predict the target bioindicators discussed in Chap. 5 (Secchi depth, chlorophyll, cyanobacteria, oxygen concentration and macrophyte cover) from nutrient concentrations.

This chapter will first discuss central processes and principles related to the nutrients from these perspectives. We will discuss the variability of nutrients, the sampling formula and basic statistical tools to reveal variations in data, criteria for regressions, predictive power and spurious correlations, which is important in all contexts related to ratios, such as the Redfield ratio (see also Hassett et al. 1997), which is a central concept in matters of coastal eutrophication.

4.1 Nutrient Sources and Remedial Actions

Evidently, a focal point in remediating coastal eutrophication is to identify the nutrient sources and to rank the fluxes responsible for measured and/or predicted concentrations so that large fluxes can be separated from small and natural fluxes from anthropogenic. Table 4.1 exemplifies this using data from the Baltic Proper, the

Table 4.1 Transport of nitrogen and phosphorus to and from the Baltic Proper (tons/yr). The data from SNV (1993) concern mean values for the period between 1982 and 1989; the data from HELCOM (2000) concerns year 2000. HL is Håkanson and Lindgren (2007b)

	Total-N			Total-P		
	SNV	HELCOM	HL	SNV	HELCOM	HL
A. From countries						
Sweden	44,300	46,636		1780	1219	
Baltic states	72,600	145,697		1890	5408	
Finland	–	35,981		–	1874	
Russia	–	90,229		–	5863	
Poland	109,900	191,521		19,100	12,698	
Germany	20,000	20,602		2750	512	
Denmark	51,000	27,664		7860	1193	
Sum inflows from	297,800	558,046		33,380	28,767	
countries:		≈ 500,000			≈ 30,000	
B. From processes and water inflow from adjacent basins						
Precipitation	289,900	192,400		3420	–	
Nitrogen fixation	130,000	–		–	–	
Land uplift			480,000			160,000
Inflow from Kattegat			120,000			14,000
Inflow from Bothnian Sea			340,000			14,000
Sum from			1,261,000 – 1,359,000			191,420
processes:			≈ 1,300,000			≈ 190,000
Total inflow:			≈ 1,800,000			220,000
C. Water outflows to adjacent basins						
To the Bothnian Sea			340,000			24,000
To Kattegat			260,000			18,000
Total outflow:			≈ 600,000			40,000
D. Rest terms						
Burial in sediments			(3·180,000)* = 540,000			(220,000 – 40,000) = 180,000
Denitrification			(1800,000 – 600,000 – 540,000) = 660,000			

* the nitrogen concentration is 3 times higher than the phosphorus concentration in these sediments.

largest, southernmost basin of the Baltic Sea. The idea is to get information on fluxes that may be and not be remediated, and on uncertainties in the fluxes, so that it is possible to make predictions of how the given system would react to often very costly nutrient reductions. The fluxes given for the Baltic Proper in Table 4.1 (tributary inflow, inflow from adjacent coastal areas, internal fluxes, etc.) appear in most coastal areas but would evidently be ranked different in different coastal areas.

From the National Swedish Environmental Protections Agency (SNV 1993; see Table 4.1) and HELCOM (2000; the Helsinki Commission), one might get the impression that about 30–40 ktons of totals phosphorus (TP) on average are transported to the Baltic Proper during a year.

Evidently, there are also nutrient fluxes from defined processes, such as nitrogen fixation by certain forms of cyanobacteria and input of nitrogen and phosphorus from precipitation. It is also important to realize that the data used by SNV (1993) and HELCOM (2000) excludes several important TP and total nitrogen (TN) fluxes to the Baltic Sea, mainly nutrient input from land uplift (this is a special case for the Baltic Sea which is discussed in Sect. 9.2), and nutrient inflows from (and outflows to) adjacent basins. The latter fluxes are, however, accounted for by HELCOM (2000), see Table 4.1, but HELCOM does not include nutrients from land uplift. Håkanson and Lindgren (2007a) have produced the data given in Table 4.1. It is essential to include all major transport processes in order to understand the situation in the Baltic Proper – or in any given coastal area – if the aim is to predict how remedial measures reducing nutrient loading will likely change the nutrient concentrations in the water in the given system.

The inflow of nutrients from Kattegat to the Baltic Sea may be estimated from total annual water inflow from the mass-balance for salt. In Table 4.1, a value of 475 km^3/yr of water inflow from Kattegat was used (see Håkanson et al. 2002). The nutrient concentration in the inflowing water (which has a salinity between 15 and 22 psu) was calculated from extensive data-sets from HELCOM, which gave a typical TP-concentration in the inflowing salt water of 252 μg TN/l and 30 μg TP/l. The total inflow of phosphorus from Kattegat is then about 14 ktons per year, and the corresponding value for nitrogen is 120 ktons per year. However, the net flow of both TP and TN is from the Baltic Proper to the Kattegat. These are important data in the overall nutrient budgets for the Baltic Proper. Evidently these data are, like all data given in Table 4.1, quite uncertain, which is clear from looking at the fluxes related to the different countries given by SNV and HELCOM. This is also covered in more detail by HELCOM (2000). The nitrogen fixation is probably one of the most uncertain fluxes of all (see Elmgren 2001; Wasmund et al. 2005).

The role of the land uplift is discussed in Sect. 9.2. Land uplift relates to the subsidence of land, which rises after being depressed into the crust of the earth by the thick ice cover during the latest glaciation. The data in Table 4.1 show that the contribution of nitrogen and phosphorus from land uplift is most substantial, 160 ktons P per year and 480 ktons of N per year. One can note that these are the dominating fluxes, and it is interesting also to stress that these dominating fluxes have been disregarded in many contexts.

The nitrogen flux between the Baltic Proper and the Bothnian Sea are about the same in the two directions. The outflow of phosphorus from the Baltic Proper is, however, significantly higher than the flow to the Baltic Proper from the Bothnian Sea. This gives a net loss of about 10 ktons of TP per year from the Baltic Proper in this direction. The data on the water exchange emanate between the sub-basins from the basic mass-balance for salt (see Håkanson et al. 2007d).

HELCOM (2000) has also given data on the nutrient input from different sources to the Gulf of Finland and the Gulf of Riga (in the year 2000) as (in ktons):

Evidently, in practice, it is not possible to reduce the natural loading and all the anthropogenic loading from diffuse sources and point sources may not be removed.

	Nitrogen				Phosphorus			
	Natural	Diffuse	Point sources	Total	Natural	Diffuse	Point sources	Total
Gulf of Finland	67	45	15	127	1.5	2	1.5	5
Gulf of Riga	17	40	2	59	0.4	1.1	0.3	1.8

Generally, it is more cost-efficient to make initial reductions (if no water purification plants have been built and/or agricultural measures have been taken) than it is to remove the "last" reducible tons of the nutrients.

As a background to the data given in this overall mass-balance, one can note that the total input from Sweden to the Baltic Proper according to Table 4.1 (for the period 1982–1989) is 1780 tons P per yr, whereas the data from HELCOM for the year 2000 adds up to 1085 tons to the Baltic Proper and 3289 tons P to the Bothnian Bay and the Bothninan Sea, and a significant part of the discharges to the northern parts of the Baltic Sea will end up in the Baltic Proper. This example demonstrates that there are major uncertainties in the data caused by natural variations among years in precipitation and water and nutrient runoff, inherent analytical uncertainties in measurement data, and uncertainties regarding basic definitions of sources and contributions from different countries.

Table 4.1 also gives data on the outflows of the nutrients and on denitrification (660 ktons of TN from mainly the sediments) calculated in this overall mass-balance, which compares fairly well with figures given by Larsson et al. (2001) of between 180 and 430 ktons per year. The burial in sediments (i.e., the transport from upper sediment layers to lower "geological" sediment layer) is 180 ktons of TP and 540 ktons of TN, calculated as residual terms in this overall mass-balance.

The demands for remedial actions to improve the conditions in the Baltic Proper normally target on what can be done in individual countries. The anthropogenic fraction of TP transported by, e.g., Swedish rivers may be about 30% of the total TP-inflow from Sweden but only 0.52% of the total TP-inflow to the Baltic Proper. This means that one can not expect any major changes in the TP-concentration in the Baltic Proper from reductions in anthropogenic TP-input from Swedish rivers, Swedish urban coastal emissions, coastal industrial emissions or fish farms, since they together account for less than 1% of the total annual inflow of TP to the Baltic Proper. The importance of internal fluxes and the transport between sub-basins has also been shown by Christiansen et al. (1997) in a study of parts of the Kattegat. The calculations provided here indicate in Table 4.1 that the HELCOM goal that Sweden should reduce the emissions of TP from about 3000 tons/yr to about 1500 tons/per year will not influence the system very much, but it would be very expensive.

Table 4.2 gives estimations of the uncertainties related to all inflows of nitrogen and phosphorus to the Baltic Proper. HELCOM has recently (see http: www.helcom.fi) argued that 135,000 tons of nitrogen and 15,000 tons of phosphorus should be reduced from the annual tributary loading. It is interesting to put those data into the overall context related to the fluxes in Table 4.1, and especially related to the uncertainties in the fluxes given in Table 4.2.

Table 4.2 Uncertainties in key nutrient fluxes to the Baltic Proper

A. Nitrogen	
Into Baltic Proper	**1000·tons TN/yr**
From land uplift	350–500
From rivers	300–600
From the Bothnian Sea	300–400
From Kattegat	100–150
From wet and dry deposition on the water surface	200–300
From nitrogen fixation by cyanobacteria	100–900
Sum annual input of TN	1350–2950
Total from Sweden	40–50 (2–3%)
Realistic Swedish reductions	about 5 (<0.4%)
B. Phosphorus	
Into Baltic Proper	**1000·tons TP/yr**
From land uplift	110–160
From rivers	30–40
From the Bothnian Sea	10–15
From Kattegat	10–15
From wet and dry deposition on the water surface	1
Sum annual input of TN	160–230
Total from Sweden	1.5–2 (<1%)
Realistic Swedish reductions	about 0.75 (<0.5%)

One can then note that the uncertainty interval for nitrogen fixation is 100–900 kt of TN, the uncertainty interval for river transport is 300–600 kt TN, for land uplift 350–500 kt TN, for precipitation 200–300 kt TN and for exchange processes between Kattegat and the Bothnian Sea 400–550 kt TN. This means that given these very large uncertainties, it becomes almost impossible to predict how the system would react to such very costly nitrogen reductions as suggested by HELCOM. Sweden should reduce 20 kt TN of the 135 kt TN, and this would, according to agricultural scientists, not be feasible without drastic cutbacks in Swedish agriculture. The maximum realistic reductions would be about 5 kt TN.

Much more could be said about this, but the idea here is to highlight that similar discussions could be carried out for any coastal system, and in all such discussions, it is imperative to have reliable empirical data on all major fluxes and on the uncertainties in the data so that the suggested remedial reductions can be put into a quantitative scientific context.

4.2 Limiting Nutrient – Phosphorus or Nitrogen or Both?

The question about "limiting" nutrient is certainly central in aquatic ecology and has been treated in numerous papers and textbooks (e.g., Dillon and Rigler 1974; Smith 1979; Riley and Prepas 1985; Howarth 1988; Hecky and Kilham 1988; Evans et al.

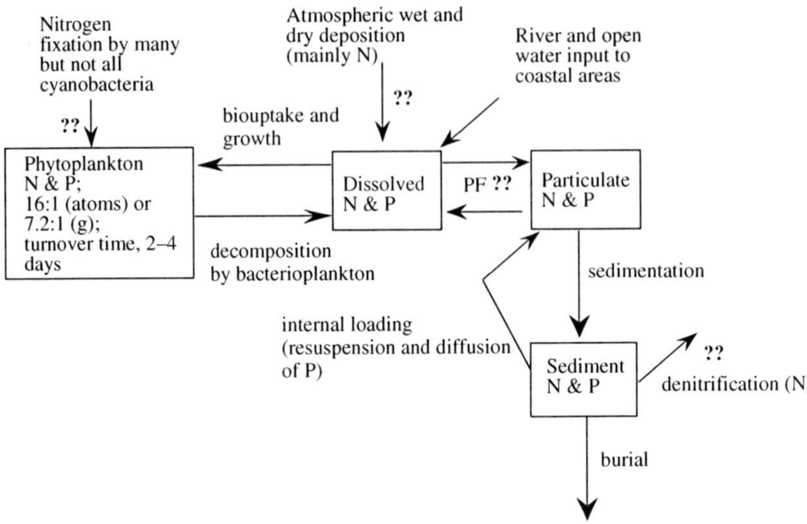

Fig. 4.1 Overview of important transport processes and mechanisms related to the concept of "limiting" nutrient

1996; Wetzel 2001; Geider and La Roche 2002; Labry et al. 2002; Newton et al. 2003; Smith et al. 2006).

Figure 4.1 illustrates key questions related to the concept of "limiting" nutrient. It shows the main processes regulating fluxes (atmospheric input, river inflow) of nutrients (nitrogen and phosphorus) to a given aquatic system, internal fluxes (sedimentation, resuspension, diffusion, denitrification and burial) including the very important relationship between the amount of the nutrient in dissolved (bioavailable) form and the amount in particulate form (the only part that can settle out due to gravity).

The average composition of algae ($C_{106}N_{16}P$) is reflected in the Redfield ratio (7.2 by mass; see Redfield et al. 1963) and 16:1 (by atoms). So, by definition, algae need both nitrogen and phosphorus and one focus of coastal eutrophication studies concerns the factors limiting the phytoplankton biomass, as expressed by chlorophyll-a concentrations in the water. Note that the actual phytoplankton biomass any given moment in a system is a function of the initial phytoplankton production minus predation on phytoplankton by herbivorous zooplankton minus the death of phytoplankton regulated by the turnover time of the phytoplankton (see Håkanson and Boulion 2002). A crucial point raised in Fig. 4.1 concerns the equilibrium between nutrients in dissolved and particulate phases, the timescales of these interactions and what is actually meant by "limiting" nutrient.

At short timescales (seconds to days), it is evident that the causal agent regulating/limiting biouptake and primary production is the available concentration of the nutrient in bioavailable forms, such as DIN and DIP, nitrate, phosphate and ammonia. Short-term nutrient limitation is often determined by measuring DIN and DIP concentrations, or by adding DIN and/or DIP to water samples or microcosms, so called bioassays. If in-situ DIN or DIP concentrations are low enough, or if bioassays show that either DIN or DIP, or both combined, produce higher algal growth than reference samples, then that gives an indication of which nutrient is limiting in the short run. However, management decisions, such as whether or not to reduce nutrient emissions, are generally based on monthly, annual or multi-annual timescales. DIN and DIP provide poor guidance for such decisions because:

- DIN and DIP are quickly regenerated and their role in regulating primary production is therefore poorly reflected by their concentrations (Dodds 2003). For example, zooplankton may excrete enough DIN to cover for more than 100% of what is consumed by phytoplankton (Mann 1982). In highly productive systems, there may even be difficulties to actually measure nutrients in dissolved forms since these forms may be picked up so rapidly by the algae (Istvanovics and Somlyody 2001). Dodds (2003) suggested that only when the levels of DIN are much higher than the levels of DIP (e.g., 100:1), it is unlikely that DIN is limiting and only if DIN/DIP < 1, it is unlikely that P is the limiting nutrient. He also concluded that DIN and DIP are poor predictors of nutrient status in aquatic systems compared to TN and TP.
- Phytoplankton and other primary producers also consume organic N and P (Huang and Hong 1999; Seitzinger and Sanders 1999; Vidal et al. 1999).
- DIN and DIP are highly variable (see Sect. 4.4.) and are very poor predictors of phytoplankton biomass and primary production (as measured by chlorophyll concentrations; see Fig. 4.30, later).
- Primary production in natural waters may be limited by different nutrients in the long run compared to shorter time perspectives. This point was first made with respect to oceanic primary production by Redfield (1958). Based on differences in nutrient ratios between phytoplankton and seawater, he hypothesized that P was the long-term regulating nutrient, while N deficits were eventually counteracted by nitrogen fixation. Schindler (1977) tested this hypothesis in several whole-lake experiments and found that primary production was governed by P inputs and unaffected by N inputs, and that previous results from bioassays were therefore irrelevant for management purposes. Redfield's hypothesis has also been successfully tested in modeling works for the global ocean (Tyrrell 1999) and the Baltic Proper (Savchuk and Wulff 1999a,b). However, Vahtera et al. (2007) have used a "vicious circle" theory to suggest that both nutrients should be abated to the Baltic Sea since they may have different long-term importance at different times of the year.

An important regression related to the discussion in this chapter is the first model, the OECD-model for average summer conditions (OECD 1982) in Table 4.3.

This empirical model is based on data from 77 lakes and the TP-concentrations range from 2.5 to 100 µg/l. An interesting aspect concerns the r^2-value ($r^2 =$ the coefficient of determination), which is 0.77. That is, 77% of the variation among these 77 lakes in mean/median summer chlorophyll values can be statistically explained by variations in TP. The slope is 0.28 and the exponent close to 1 (the exponent is

Table 4.3 Regressions illustrating the key role of total phosphorus (TP) in predicting different variables expressing primary production in lakes. Sec_{MV} = mean annual Secchi depth (in m), D_m = mean lake depth (in m), r^2 = coefficient of determination, n = number of lakes used in the regression, ww = wet weight

	Equation	Range for x	r^2	n	Units	Reference
A. Targets in this work						
Chlorophyll (summer mean)	$= 0.28 \cdot TP^{0.96}$	2.5–100	0.77	77	µg/l	OECD (1982)
Cyanobacteria	$= 43 \cdot TP^{0.98}$	8–1300	0.71	29	mg ww/m^3	Smith (1985)
B. Related variables						
Chlorophyll (summer max.)	$= 0.64 \cdot TP^{1.05}$	2.5–100	0.81	50	µg/l	OECD (1982)
Max. prim. prod. (TP>10)	$= 20 \cdot TP-71$	7–200	0.95	38	mg C/m^3·d	Peters (1986)
Max. prim. prod. (TP<10)	$= 0.85 \cdot TP^{1.4}$				mg C/m^3·d	Peters (1986)
Mean prim. prod. (TP>10)	$= 10 \cdot TP-79$	7–200	0.94	38	m C/m^3·d	Peters (1986)
Mean prim. prod. (TP<10)	$= 0.85 \cdot TP^{1.4}$				mg C/m^3·d	Peters (1986)
Phytoplankton	$= 30 \cdot TP^{1.4}$	3–80	0.88	27	mg ww/m^3	Peters (1986)
Macrophyte cover	$= 0.50 \cdot (Sec_{MV}/D_m)$			229	%	Vorobev (1977)
Nanoplankton	$= 17 \cdot TP^{1.3}$	3–80	0.93	23	mg ww/m^3	Peters (1986)
Net plankton	$= 8.6 \cdot TP^{1.7}$	3–80	0.82	23	mg ww/m^3	Peters (1986)

0.96 for the summer averages and 1.05 for the summer maximum values). In the following discussion in this chapter, data from many systems have been used, also including information on TN, different nutrient forms, temperature and salinity.

There are four highlighted spots with question marks in Fig. 4.1, indicating that it is very difficult to quantify these processes in a general manner. Three of them are (as shown in Table 4.2) denitrification, atmospheric wet and dry deposition and nitrogen fixation by certain forms of cyanobacteria. Figure 4.1 also highlights another major uncertainty related to the understanding of nitrogen fluxes in coastal systems, the particulate fraction, which is necessary for quantifying sedimentation. Atmospheric nitrogen fixation is very important in contexts of mass-balance calculations for nitrogen (see Tables 4.1 and 4.2 and Rahm et al. 2000). Without empirically well-tested algorithms to quantify nitrogen fixation, crucial questions related to the effectiveness of the remedial measures to reduce nutrient discharges to aquatic systems cannot be properly evaluated, since costly nitrogen reductions may be compensated for by effective nitrogen fixation by cyanobacteria. It also means that it is generally very difficult to understand, model and predict changes in measured TN-concentrations in the water phase, since such changes in concentrations are always mechanistically governed by mass-balances, i.e., the quantification of the most important transport processes regulating the given concentrations. Smith (1990) presented two models for predicting nitrogen fixation from TP; one model for temperate lakes ($r^2 = 0.54$) and another one for estuaries ($r^2 = 0.88$). Håkanson et al. (2007c) have presented a model for total cyanobacteria in marine systems based on TP, TN, salinity and water temperature, which gave an r^2 of 0.78 when tested against empirical data. It should be stressed that the nitrogen fixation rates have been substantially revised upwards during the last decade due to better measurement techniques and better understanding of the process (Capone 2001; Wasmund et al. 2005). This explains the wide uncertainty interval for nitrogen fixation given in Table 4.2.

The relative abundance of cyanobacteria compared to other algal groups is closely related to the TN/TP-ratio. Cyanobacteria (CB) have been found to dominate lake primary production at TN/TP $\leq$ 29/1 (by weight) and they are much less abundant at higher ratios, while nitrogen-fixing cyanobacteria tend to dominate at TN/TP $\leq$ 22/1 (Havens et al. 2003). An alternative view is based on observations that TP is a better predictor of the absolute biomass of CB than TN and TN/TP (Smith 1985, 2003; Downing et al. 2001). According to this view, TN/TP should only be used for identifying the primary growth-limiting resource, and not for predicting CB. This controversy has been further discussed by Håkanson et al. (2007c) and they showed that there seem to exist two threshold values in predicting cyanobacteria, one at the TN/TP-ratio of 15 and one at a surface-water temperature of 15°C. At lower TN/TP-ratios than 15 and at higher water temperatures than 15°C, cyanobactera may become abundant.

Figure 4.2 illustrates some of the problems related to the concept of "limiting" nutrient. Using data from the Baltic Sea, Fig. 4.2A gives a situation where the chlorophyll-a concentrations show a typical seasonal "twin peak" pattern with a pronounced peak in April. The higher the primary production, the more bioavailable nitrogen (nitrate, ammonium, etc.) and phosphorus (phosphate) are being used by the algae (the spring bloom is mainly diatoms) and eventually the nitrate concentration drops to almost zero and the primary production decreases – but the important point is that the primary production, the phytoplankton biomass and hence also the concentration of chlorophyll-a remain high during the entire growing season!

Figure 4.2B illustrates a regression between chlorophyll and TN based on empirical data from 22 Baltic coastal areas (median values for the summer period; data from Wallin et al. 1992). One can note a very high r^2 (0.89). So, 89% of the variation in median chlorophyll-a values among these 22 coastal areas could be statistically (but not necessarily causally) related to the measured variations in TN. Does this prove that nitrogen is the "limiting" nutrient in these coastal areas during this period of time? Certainly not, as will be discussed in the following text.

Figure 4.2C shows the relationship between chlorophyll-a and TN in Ringkobing Fjord, Denmark, a lagoon with a salinity in the range between 5 and 15 (data from Petersen et al. 2006). In this lagoon, there is a very poor correlation between TN and chlorophyll ($r^2 = 0.04$); and later on the correlation between chlorophyll and different forms of nutrients in this lagoon will be discussed. The point here is that this is only one coastal area, and the regression in Fig. 4.2B is based on data from "only" 22 coastal areas. In the following, the aim is to put results like these in a much wider context.

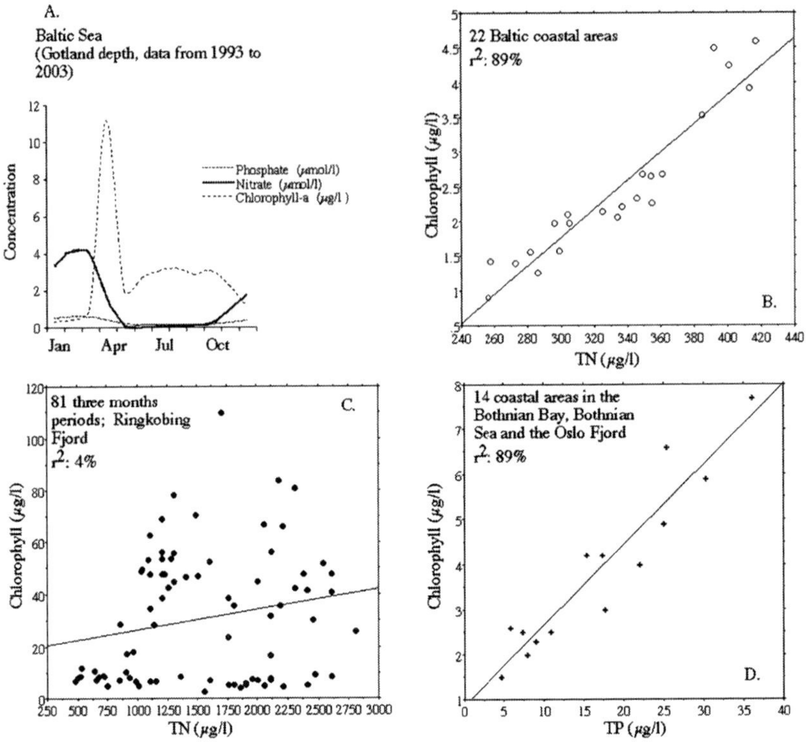

Fig. 4.2 **A**. Variations in chlorophyll-a concentrations, phosphate and nitrate in the Baltic Sea (using data from the a at the Gotland depth between 1993 and 2003; database from SMHI, Sweden). **B**. The relationship between chlorophyll and total-N (TN) using data from 22 Baltic coastal areas (data from Wallin et al. 1992). **C**. The relationship between chlorophyll and TN using monthly median values from Ringkobing Fjord for the period 1980–2004 (data from Bryhn et al. 2007). **D**. The relationship between chlorophyll and total-P (TP) using data from the Bothnian Bay, the Bothnian Sea and the Oslo Fjord (data from Magnus Karlsson, unpublished)

Figure 4.2D exemplifies that there are many coastal areas where there exist just as powerful relationships between TP and chlorophyll (Chl) as between TN and Chl in Fig. 4.2B. In the following methodological section, questions concerning the scatter around regression lines like these will be addressed, and in particular with respect to how the scatter depend on the variability and uncertainty in the empirical data.

An important argument related to the information in Figs 4.1 and 4.2 is that the concentrations of the bioavailable fractions, such as DIN and DIP in μg/l or other *concentration* units, can not as such regulate primary phytoplankton *production* in μg/l·day (or other units), since primary production is a flux including a time dimension and the nutrient concentration is a concentration without any time dimension. The central aspect has to do with the flux of DIN and DIP to any given system and the regeneration of new DIN and DIP related to bacterial degradation of organic matter containing TN and TP. The concentration of DIN and DIP may be very low and the primary phytoplankton production and biomass can be high as in Fig. 4.2A because the regeneration and/or inflow of DIN and DIP is high.

The regeneration of DIN and DIP concerns the amount of TN and TP available in the water mass, i.e., TN and TP represent the pool of the nutrient in the water, which can contribute with new DIN and DIP. In a following section, we will specifically focus on the variability in DIN and DIP. It should be stressed that phytoplankton has a typical turnover time of about 3 days and bacterioplankton has a typical turnover time of slightly less than 3 days (see Håkanson and Boulion 2002). This means that within a month there can be 10 generations of phytoplankton which would need both DIN and DIP and in the proportions given by the Redfield ratio (7.2 in grams).

An important aim of this book is to try to add a new dimension to the debate on "limiting" nutrient by using a very comprehensive data-set collected from the literature and different websites. Conditions in a wide salinity range will be discussed and the main task is to try to find general patterns in variations among areas in chlorophyll-a concentrations and in concentrations of cyanobacteria. Are such variations mainly governed by variations in phosphorus and/or nitrogen? Is it useful to account for different forms of the nutrients?

As stressed, the focus is on the ecosystem scale (i.e., on entire coastal areas and not on smaller spatial scales, such as sample sites) and on seasonal/monthly changes (rather than changes related to smaller temporal scales).

4.3 Nitrogen and Phosphorus – Hot Spots

From a management point of view, it is evidently good cost-efficient practice to spend money to improve the conditions in "hot spots", i.e., in coastal areas suffering from severe eutrophication rather than in less polluted areas. This section will stress this by presenting maps of TN and TP from northern European coastal waters (from Lindgren and Håkanson 2007).

An overview of TN data in the upper 10 m water layer from the period 1990–2005 is given in Fig. 4.3. The eastern North Sea coast shows many high values between 700 and 1000 μg/l. The Baltic Sea, the Bothnian Bay and the Bothnian Sea have lower values between 200 and 300 μg/l, while, e.g., the Gulf of Riga and the eastern part of the Gulf of Finland show higher values, ranging from 400 up to 800 μg/l. These are "hot spots" in the Baltic Sea.

Figure 4.4 gives the corresponding map for the TP-concentrations from the Baltic Sea and parts of the North Sea. The eastern North Sea coast, especially the areas around the mouths of the rivers Weser and Elbe, show the highest average values, over 90 μg/l. In the Baltic Sea, the lowest values are found in the Bothnian Bay and "hot spots" are found in the Vistula and Oder lagoons (see also Schernewski and Dolch 2004).

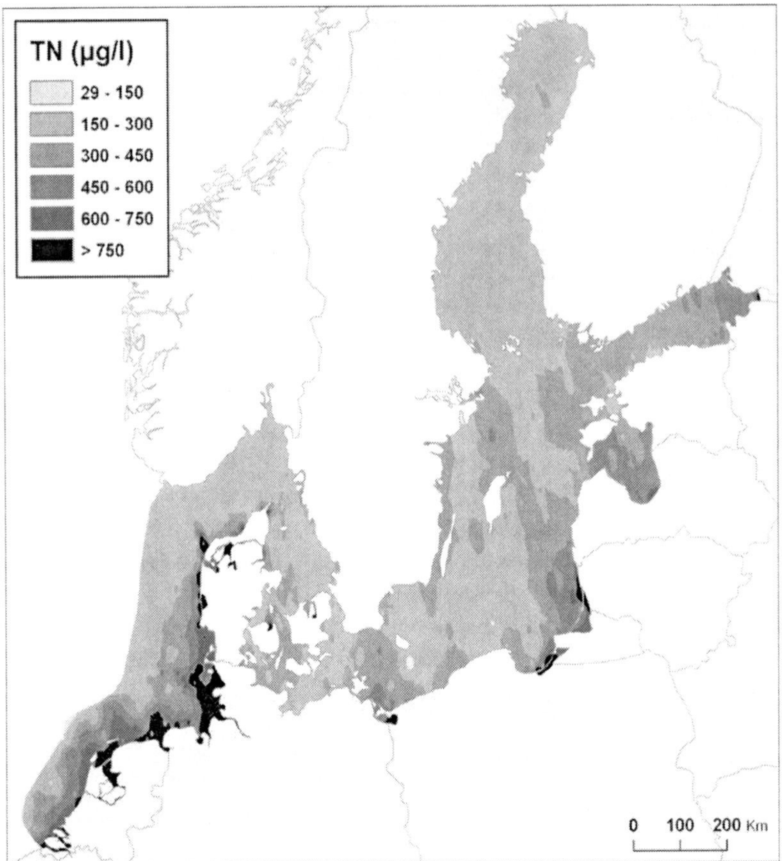

Fig. 4.3 Typical annual surface-water TN-concentrations for the Baltic Sea and parts of the North Sea in the upper 10 m water column for the period from 1990 to 2005

Should nutrient reductions focus on nitrogen or phosphorus or both in "hot spots" such as these? To answer that question, one should first try to clarify important questions related to the variability, representativity and uncertainty of empirical nutrient concentrations.

4.4 Data Variability and Uncertainty – A Review of Key Concepts

How do inherent variations and uncertainties in empirical data constrain approaches to predictions and possibilities to identify critical thresholds and points of no return? This section addresses this question in discussing and reviewing some fundamental concepts and methods for coastal ecology and management.

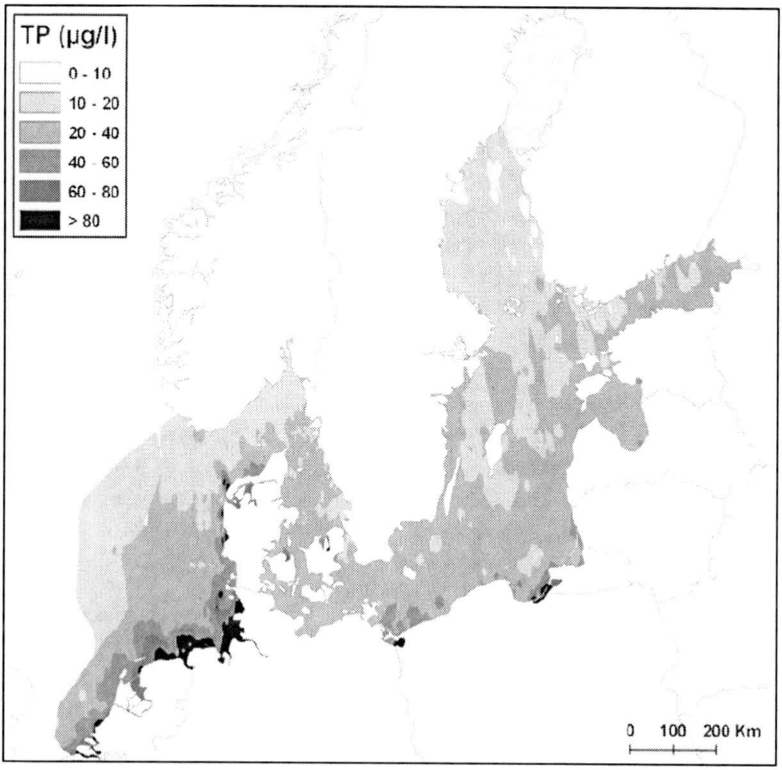

Fig. 4.4 Typical annual surface-water TP-concentrations for the Baltic Sea and parts of the North Sea in the upper 10 m water column for the period from 1990 to 2005

The main focus here is *not* on the mechanisms regulating the concentrations of the target nutrients and bioindicators but on patterns in variations in concentrations in entire lagoons, bays, estuaries or fjords (i.e., on variations at the ecosystem scale).

This section will discuss problems related to:

1. The balance between the changes in predictive power and the accumulated uncertainty as management models grow in size and include an increasing number of x-variables.
2. An approach to reduce uncertainties in empirical data. Methods to maximize the predictive power of regression models by transformations of model variables and by creating time and area compatible model variables.
3. Patterns in variations within and among coastal systems of standard water variables.
4. The concept "Optimal Model Scale" (OMS) is discussed. OMS accounts for key factors related to the predictive power at different time scales (daily to yearly prediction) and to uncertainties in predictions in relation to access to empirical data and the work (sampling effort) needed to achieve predictive power at different time scales (see Håkanson and Duarte 2007).

Thresholds, regime shifts and points of no return are fundamental concepts in ecology in general and in aquatic ecology in particular (Scheffer 1990; Scheffer et al. 2000; Carpenter 2003; Groffman et al. 2006; http://www.thresholds-eu.org/). To find methods to avoid trespassing critical ecological thresholds leading to the collapse of ecosystem structure and function is evidently an important issue for sustainable water management. Studies of ecological thresholds can be done at different temporal (days to centuries), spatial (from sampling sites to entire coastal zones) and structural scales (the level of the cell, organ, individual, species or functional group), and different methods and tools may be necessary at the different scales. This section focuses on the variability of a set of standard water variables in constraining the predictive capacity of models and their capacity to detect critical changes at the ecosystem scale.

The basic aim of this section is to discuss methods that can be used to quantify variability in water variables. These methods should also be understandable and applicable by individuals with a limited background in statistics. We see a major benefit in this, since much communication among scientists and water managers with different educational backgrounds could then, hopefully, be facilitated.

4.4.1 Basic Statistical Questions

Chlorophyll, phosphorus, nitrogen, salinity and temperature (as well as all other water variables), display a considerable temporal and spatial variability in aquatic systems, which regulate the uncertainty in the data (see Knowton et al. 1984; France and Peters 1992; Håkanson 1999; Knowlton and Jones 2006a, b). So, the content of a sampling bottle may reflect the conditions in the ecosystem very poorly, and this issue will be addressed in this section. An important question in this section is to discuss fundamental methodological aspects on how to quantify, understand and model critical events affecting the functioning and biological value of coastal ecosystems. Regime shifts represent sudden, abrupt changes in the structure and function of coastal ecosystems. An example of this is provided by the drastic regime shift in a Danish lagoon, the Ringkobing Fjord, evident in data from four measurement stations for a 21-year period (from 1984 to 2004) on chlorophyll-a concentrations (Fig. 4.5A). The conditions in this lagoon have been described by, e.g., Laursen et al. (2004) and Petersen et al. (2004). Figure 4.5B gives some key concepts related to changes in ecosystem variables, e.g., bioindicators such as chlorophyll-a concentrations in Fig. 4.5A. The threshold value in this hypothetical example is set to 20 units of the given bioindicator. The reference conditions are undesirable since the data are higher than the critical threshold value. Coastal management has four remedial options to improve the conditions and Fig. 4.5B illustrates the potential outcome of these remedial measures, which would create new situations that can be statistically and/or ecologically significant.

From this figure (which is a modified version from Di Stefano et al. 2005), one can argue that case 1 illustrates a situation where the observed mean effects are neither statistically nor ecologically significant. Cases 2 and 4 give situations when

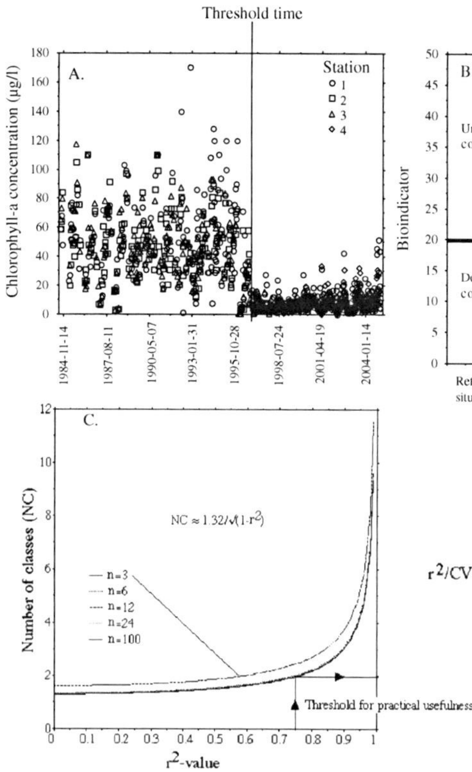

Fig. 4.5 A. An example of a regime shift in Ringkobing Fjord, Denmark, showing variations in chlorophyll concentrations at four sites between 1984 and 2004. This chapter will not address the causal reasons for this, or similar often less evident regime shifts, the aim is instead to discuss methodological aspects on how to quantify such changes. **B**. Illustrates principles aspects related to changes such as the one given in Fig. 4.5A. An undesired initial reference situation may be improved by four hypothetical remedial measures (cases 1–4). The changes may be statistically and/or ecologically significant. **C**. The relationship between the predictive power of models (as given by the r^2-value when modeled data are compared to empirical data), the uncertainty in the predictions (the lower the NC-value on the y-axis, the wider the confidence intervals for the regression line) and the number of data used in the regression (Prairie's "staircase"). A threshold value for models for practical use can be set at an r^2-value of 0.75. **D**. For any target y-variable (e.g., a bioindicator) in water management, there is an optimal size of the predictive model, as given by the ratio between the r^2-value when modeled values are compared to empirical data, and the accumulated uncertainty in the model (CV)

the effects are ecologically important but not statistically significant. Case 3 illustrates a situation when the effects are both statistically significant and ecologically important. To be practically useful, these principle arguments must also address:

1. Whether it is relevant to use a mean value or a median to represent the characteristic conditions for the given water variables. This depends on the frequency

distributions of the given variables, which will be discussed in a following section using empirical data from many coastal areas.

2. Whether the 95% confidence intervals apply for individual data or for the mean value? Methods for this using coastal data will also be discussed in the following text in the section.

Figure 4.5C illustrates an important issue related to statistical analyses of data, namely the predictive power of models and the uncertainty of model predictions. Yves Prairie (1996) has produced some useful results illustrating the practical utility of models for predictions of individual y-values, here referred to as Prairie's "staircase". If the confidence bands for the individual data are wide apart when modeled values are compared to empirical data, then the model can produce totally useless predictions for individual y-values.

The usefulness of the predictions are directly related to the number of steps, or classes, obtained in the "staircase", and this is related to the r^2-value in the regression. The number of classes (NC) is also determined by the statistics (the statistical certainty, p, and the number of data used in the regression, n). If the 95% confidence bands for the individual y-data are used, the relationship between the number of classes (NC) and the r^2-value is given by:

$$NC = 1.32/\sqrt{(1 - r^2)} \tag{4.1}$$

Figure 4.5C gives the relationship between NC and r^2 for different n-values, and one can note that if $n > 6$, (4.1) gives a good description of the relationship. The important message in Fig. 4.5C is that the number of classes increases very rapidly for r^2-values higher than about 0.75. Models yielding r^2-values lower than that are more or less useless for predictions of individual y-values. The results shown in Fig. 4.5C are applied later in this section in the definition of OMS, the optimal model scale.

Figure 4.5D illustrates another important principle related to the practical usefulness of models in coastal management. In this example, modeled values are compared to empirical data and the fit may be quantified by a regression. For any target variable in water management (e.g., a bioindicator), the ratio between the predictive power (given by the r^2-value) and the accumulated uncertainty in the model predictions (given by CV, the coefficient of variation) is governed by the model structure and the uncertainty in the individual x-variables needed to run the model (Håkanson and Peters 1995). Above the critical threshold value for model size, added model variables are likely to add less to the predictive power of the model than to the accumulated uncertainty, which means that the r^2/CV-ratio will go down. In this section, we will give examples related to coastal ecology on that matter. This example illustrates the optimal size problem, which is different from the optimal scale problem discussed later in section.

4.4.2 The Sampling Formula

If the variability within an ecosystem is large, many samples must be analyzed to obtain a given level of certainty in the mean value. There is a general formula, derived from the basic definitions of the mean value, the standard deviation and the Student's t value, which expresses how many samples are required (n) in order to establish a mean value with a specified certainty (Håkanson 1984):

$$n = (t \cdot CV/L)^2 + 1 \tag{4.2}$$

Where t = Student's t, which specifies the probability level of the estimated mean (usually 95%; strictly, this approach is only valid for variables from normal frequency distributions), and CV = coefficient of variation within a given ecosystem. L is the level of error accepted in the mean value. For example, L = 0.2 implies 20% error so that the measured mean will be expected to lie within 20% of the expected mean with the probability assumed in determining t. Since one often determines the mean value with 95% certainty (p = 0.05), the t-value is set to 1.96.

The relationship between n, CV and L is illustrated in Fig. 4.6. The figure also indicates that for many purposes, it is reasonable to regard L $\approx$ 0.2 (a 20% error in the mean value) as a threshold for practical water management. If the error is greater than that, the mean value may be too uncertain to detect ongoing changes in the system; if the L-value is smaller, the demands on the sampling program may be too high.

Tables 4.4 and 4.5 give compilations of CV-values for brackish open water sites, lakes, rivers and brackish coastal. One can note that there are some obvious patterns in these CV-values:

- Some variables generally have high CVs, e.g., DIN, DIP and the DIN/DIP-ratio, other low, e.g., salinity, TN and TP.
- There are seasonal patterns (see Table 4.5) with high CVs for DIN and DIP during the growing season.
- There are differences in CVs related to the length of the sampling period – the longer the sampling period, the higher the CV-value.
- There are also variations among aquatic systems with higher CVs in samples from rivers than from lakes.

In the following section, we will try to clarify those patterns with a focus on the nutrients. Most water variables in these coastal areas have CVs between 0.1 and 1. One can then calculate the error in a typical estimate. If n = 5 and CV = 0.33, then L is about 33%. Since few monitoring programs take more samples at a given site during a given sampling event, this calculation has profound implications for the quality of our knowledge of aquatic systems. One reason for the high CV-values in these water variables may be linked to the fact that there are large analytical uncertainties in the laboratory determinations of some of these variables (Håkanson et al. 1990). As a rule-of-thumb, one can estimate that for the nutrients (TP and TN), about 50% of the characteristic CV-value for within-system variation during a given

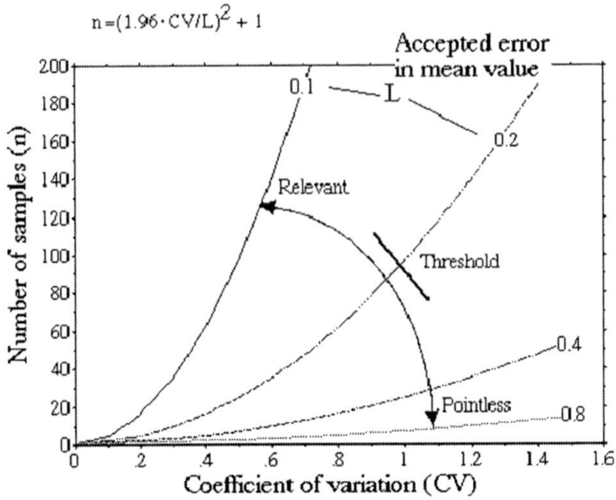

Fig. 4.6 The general sampling formula based on the definition of the mean value, the standard deviation and Student's t. This nomogram stresses that for many practical situations in water management, it is pointless to accept data which provide great uncertainties in the mean or median values, e.g., if the error in the mean value is 80% (L = 0.8). Many samples are needed for variables with great inherent uncertainties, as given by the coefficient of within-system variability (CV; defined from the ratio between the standard deviation, SD, and the mean value, MV, as CV = SD/MV). If the CV is 0.5, 97 samples are needed to get an error in the mean value smaller than 10% (L = 0.1). The figure also illustrates that for many purposes, it is reasonable to regard L ≈ 0.2 (a 20% error in the mean value) as a threshold for practical water management

month may be related to analytical uncertainties and the rest to actual variations related to biological/ecological processes (Håkanson 1999).

In summary, many factors (from methods of sampling and analysis to chemical and ecological reactions in the water system) influence the empirical values of standard water quality variables used to characterize entire coastal areas at the time scale of days to years. Since many variables vary greatly, it is often difficult in practice to make reliable, representative, area-typical empirical estimates. Data from specific sites and sampling occasions (the sampling bottle) may represent the prevailing, typical conditions in the ecosystem very poorly.

4.4.3 Patterns in Variations for Different Water Variables

Figure 4.7 exemplifies CV-values for within-system variability related to daily, 2-weekly, monthly, 3-monthly and yearly sampling periods in Ringkobing Fjord for the period from 1980 to 2004 for chlorophyll-a concentration, Secchi depths, TP-concentration, TN-concentration, salinity and water temperature. These box-and-whisker plots (showing medians, quartiles, percentiles and outliers) are meant

Table 4.4 Coefficients of within-system variation (CV) for variables from (A) from Ringkobing Fjord (data from Pedersen et al. 1995; Petersen et al. 2006), (B) CVs at a monitoring station in Chesapeake Bay, months 6–8 (from Bryhn et al. 2007), (C) CVs from lakes and rivers using data from the growing season (May–September; from Stenström-Khalili and Håkanson 2007) and (D) CVs using data from the growing season from the Baltic Sea and the Danish Straits from 1987–2006 (from Stenström-Khalili and Håkanson 2007)

A. Ringkobing Fjord	**Daily**	**Monthly**	**Yearly**	**All**
SPM	0.20	0.38	0.70	0.81
Secchi depth	0.11	0.20	0.42	0.68
Chlorophyll-a	0.18	0.30	0.56	1.00
Total-N	0.07	0.12	0.24	0.51
Total-P	0.15	0.27	0.62	0.70
Salinity	0.08	0.08	0.24	0.33
Water temperature	0.03	0.10	0.53	0.56

B. Data from Chesapeake Bay; surface-water layer

Temp	**Sal**	**DON**	**TN**	**Sec**	**DN**	**TP**	**PN**	**PP**	**SPM**	**DP**	**Chl**	**DOP**	**OrtP**	**DIN**
0.08	0.18	0.23	0.24	0.26	0.28	0.35	0.37	0.41	0.55	0.56	0.59	0.61	0.72	0.97

Deep-water layer

Temp	**Sal**	**DN**	**TN**	**DON**	**TP**	**PN**	**PP**	**DIN**	**DP**	**OrtP**	**Chl**	**SPM**	**DOP**
0.13	0.12	0.23	0.23	0.24	0.41	0.46	0.49	0.53	0.56	0.62	0.70	0.73	0.83

C. Lakes and rivers	**Period**	**TN**	**DIN**	**TP**	**DIP**	**DIN/DIP**	**TN/TP**	**n**
28 lakes	1987–2006	0.24	0.64	0.43	0.61	0.84	0.41	4963
34 river stations	1987–2006	0.36	0.70	0.48	0.64	0.87	0.49	3934

D. Baltic Sea	**TN**	**DIN**	**TP**	**DIP**	**DIN/TN**	**DIP/TP**	**DIN/DIP**	**TN/TP**	**n**
Bothnian Bay	0.08	0.25	0.27	0.62	0.14	0.11	0.65	0.41	486
Bothnian Sea	0.14	0.70	0.31	0.87	0.05	0.16	1.32	0.36	1022
Baltic Proper	0.14	0.74	0.25	0.54	0.02	0.28	1.36	0.31	2663
Kattegat + Sounds	0.24	1.13	0.39	0.74	0.06	0.26	1.28	0.44	4346
Skagerack	0.23	1.16	0.37	0.72	0.04	0.18	1.58	0.40	829

Table 4.5 Monthly CV-values for TN, DIN, TP, DIP, DIN/DIP and TN/TP in the Himmerfjärden Bay

Month	TN	DIN	TP	DIP	DIN/DIP	TN/TP
Jan	0.13	0.30	0.10	0.10	0.57	0.12
Feb	0.11	0.26	0.09	0.12	0.54	0.13
Mar	0.14	0.47	0.15	0.47	1.05	0.16
Apr	0.16	1.49	0.24	0.92	2.01	0.27
May	0.15	1.20	0.23	0.58	1.38	0.19
Jun	0.12	1.52	0.18	0.51	1.62	0.17
Jul	0.10	1.27	0.16	0.68	1.53	0.11
Aug	0.09	1.50	0.13	0.61	1.58	0.12
Sep	0.10	1.39	0.21	0.90	1.52	0.16
Oct	0.11	0.99	0.28	0.63	1.90	0.24
Nov	0.14	0.59	0.24	0.30	0.77	0.17
Dec	0.24	0.42	0.19	0.20	0.62	0.20

to illustrate general principles of variability, and hence also uncertainty, for a set of standard water variables (all these data emanate from a regular monitoring program). One can note that: (1) these CVs generally increase significantly with the length of the sampling period, (2) among the bioindicators (e.g., chlorophyll and Secchi depth), the CVs for chlorophyll are generally greater than for the Secchi depth, (3) among the nutrients, the CVs are generally greater for TP than for TN, and (4) the CVs based on annual data are, naturally and logically, very high for water temperature.

These CV-values are generally based on 3 samples per day, 6 samples per 2-week period, 8 samples per month, 24 samples per 3-month period and 88 samples per year. There are data from 21 years for chlorophyll and the box-and-whisker plots in Fig. 4.8A show that the median of these values is 0.56 and that during certain years the CV may be as high as 1 and occasionally as "low" as 0.18.

A direct comparison of CV-values for within-system variability related to monthly and yearly sampling periods in Ringkobing Fjord is shown in Fig. 4.8 for salinity (generally smallest CV), water temperature, TN, Secchi depth, TP, chlorophyll and SPM (generally highest CV).

From Table 4.4, one can note that the CVs are generally relatively high for this Danish lagoon, which is large (area = 300 km^2 and shallow, mean depth = 1.9 m) and dominated by frequent resuspension events which create not uniform and well-mixed conditions, but the contrary, great variability related to sampling depth, if the sampling is done before, during or after a resuspension event, or where in the lagoon the sample is taken. However, for this methodological discussion, we focus on the CV-values and the implications that high or low CV-values have for interpretations of data in contexts of predictive modeling and coastal monitoring rather than the causal reasons for the observed variability. To complement the data given in Table 4.4, one can note that Weston et al. (2004) have presented the following CVs for chlorophyll-a for marine open water areas (from the North Sea), yearly CV = 0.68

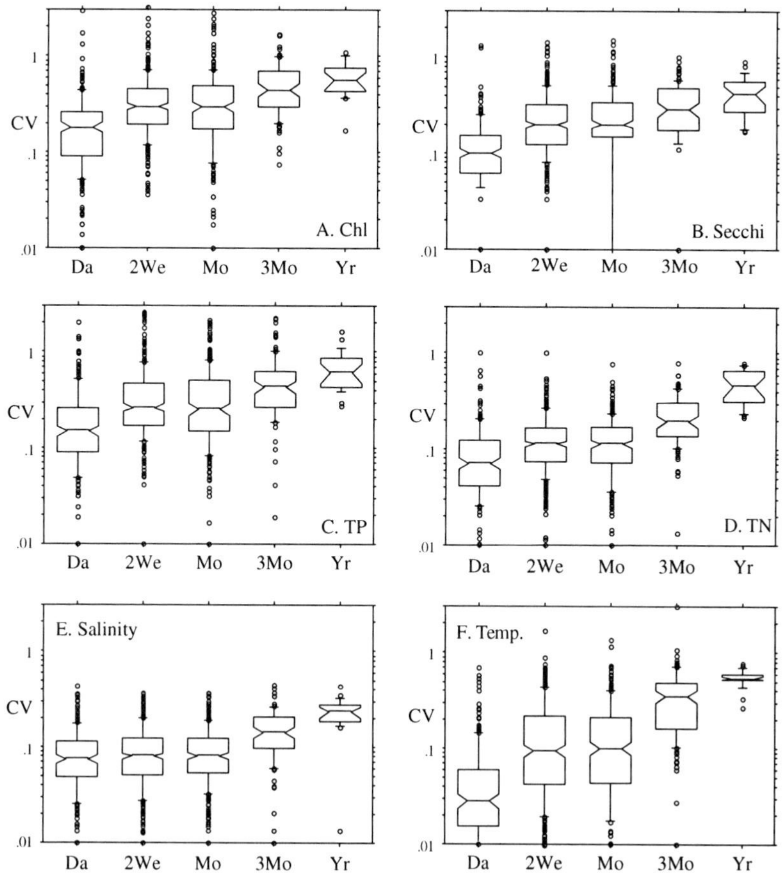

Fig. 4.7 Compilation of CV-values for within-system variability related to daily, 2-weekly, monthly, 3-monthly and yearly sampling periods in Ringkobing Fjord for (**A**) chlorophyll-a concentrations, (**B**) Secchi depths, (**C**) TP-concentrations, (**D**) TN-concentrations, (**E**) salinities and (**F**) water temperatures

and median monthly CV = 0.34. Arhonditsis et al. (2000) gave similar data from the semi-enclosed Gulf of Gera in the Mediterranean for chlorophyll (taken between June 1996 and October 1997), CV = 0.60.

The sampling program of the Himmerfjärden Bay on the Swedish side of the Baltic Proper is probably one of the most comprehensive long-term monitoring programs for coastal areas in the Baltic Sea (Elmgren and Larsson 1997; Larsson et al. 2006; Boesch et al. 2006). The data from the Himmerfjärden Bay have been used here to exemplify the variation in the monthly error (L) for the mean values for DIN/DIP and TN/TP (see Fig. 4.9).

From Fig. 4.9, one can note the seasonal patterns for the error (L, calculated from the CVs in Table 4.5 and the number of measurements each month). The CVs for

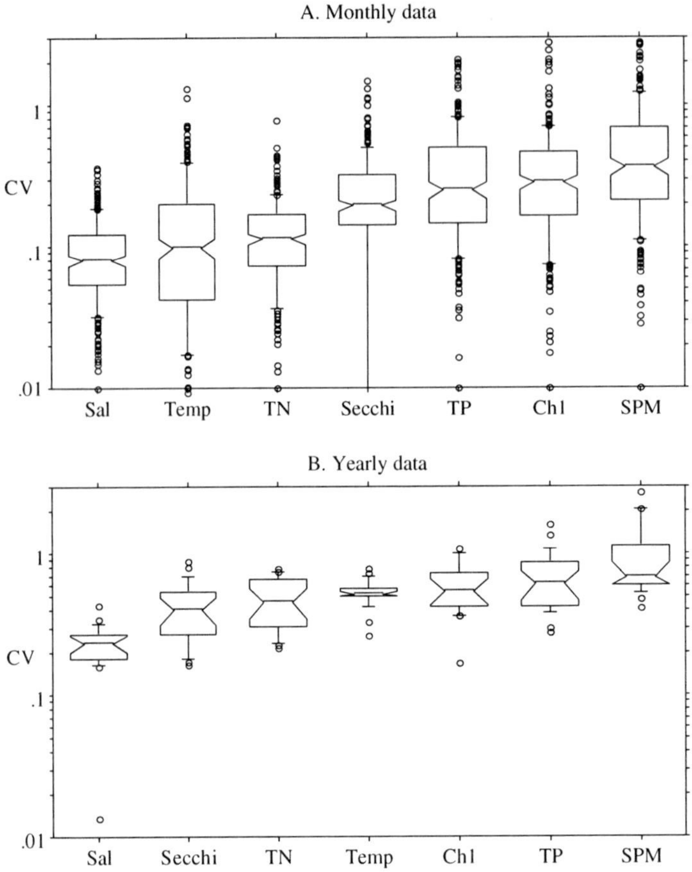

Fig. 4.8 Comparison of CV-values for within-system variability related to (**A**) monthly and (**B**) yearly sampling periods in Ringkobing Fjord for the period from 1980 to 2004 for salinity (generally smallest CV), water temperature, TN, Secchi depth, TP, chlorophyll and SPM (generally highest CV)

the DIN/DIP-ratio in this bay are very high during the growing season with peak values of about 2 in April and October. The CVs for the TN/TP-ratio are significantly lower and do not show the same temporal pattern (the average monthly CV is 0.17; see Table 4.5). This means that the error is very large for the DIN/DIP-ratio, approaching 300% in October and L is higher than 200% in several summer months (Fig. 4.9). This is logical: The high CVs for the DIN/DIP-ratio during the growing season reflect the fact that DIN and DIP represent the bioavailable fractions of the nutrients participating in fast and dynamic reactions concerning biouptake and retention of the nutrients (as stressed, typical turnover times for phytoplankton are in the order of 2.5–4 days; see Håkanson and Boulion 2002). The bacterioplankton will decompose the dead phytoplankton and this will release (regenerate) the

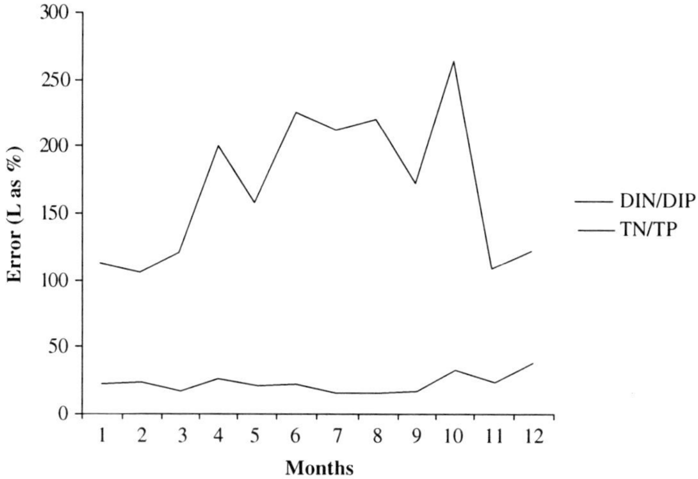

Fig. 4.9 Error (L) of the mean monthly value for DIN/DIP and TN/TP calculated from the number of samples (n) from the Himmerfjärden Bay, Sweden

bioavailable forms. Typical turnover times for bacterioplankton are 2–3.5 days (see Håkanson and Boulion 2002). Total phosphorus and total nitrogen determined from water samples include by definition TP and TN in dead and living phytoplankton and bacterioplankton. So, TN and TP in the water is the pool for DIN and DIP. As noted, the transfer of DIN to TN and DIP to TP and vice versa via biouptake and remobilisation is, thus, very quick.

Is it possible to reduce the CV so that more reliable empirical mean/median values can be obtained? Table 4.6 is included here to address that question. It shows 62 individual chlorophyll data from Ringkobing Fjord from 1984 to 1989. The column marked "MV, n = 7" shows mean values of the actual data based on 7 samples. CV for n = 7 is 0.27, which is lower than the CV for the actual data, CV = 0.48. The CV calculated when the actual data are randomly distributed in the given series is much lower, CV = 0.17. It is a standard practice to approximate the CV for a mean value according to (4.3), but then one must also, as indicated in Table 4.6, presuppose that the mean value comes from a random sample from a normal frequency distribution. The point here is that there are often seasonal or long-term trends in water variables and when this is the case (as in this example), one cannot use this approach to reduce the CV without due reservations:

$$CV_{MV} \approx CV_{ind}/\sqrt{n} \tag{4.3}$$

Where CV_{MV} is the CV for the mean and CV_{ind} is the CV for the individual data; n is the number of data used to determine the mean value.

Table 4.6 Actual data on chlorophyll concentrations (μg/l) from Ringkobing Fjord, mean values (MV) from 7 samples, the actual data redistributed randomly (RO), mean values of the redistributed data (n = 7); n = number of data, MV = mean values, SD = standard deviations, CV = coefficient of variation and CV/$\sqrt{7}$

	Actual	MV,	Random	MV,		Actual	MV,	Random	MV,
Date	data	n=7	Order (RO)	n=7	Date	data	n=7	Order (RO)	n=7
84-09-13	59		105		87-07-08	50	34	45	52
84-10-08	64		58		87-07-22	55		80	
84-11-05	84		40		87-08-05	71		48	
85-04-02	23		47		87-08-26	42		76	
85-04-16	20		51		87-09-23	71		33	
85-04-29	75		31		87-10-12	91		31	
85-05-13	34	51	32	52	87-11-18	57		49	
85-05-30	51		16		88-01-25	17	58	26	49
85-06-17	48		22		88-02-15	3		75	
85-07-17	49		71		88-03-09	16		48	
85-07-30	40		23		88-04-12	30		51	
85-08-19	53		91		88-05-19	4		53	
85-09-09	105		62		88-06-08	51		99	
85-10-08	86	62	96	54	88-06-29	41		64	
85-11-05	76		63		88-07-12	62	30	110	71
85-12-11	48		8		88-08-10	47		84	
86-04-01	23		17		88-09-01	48		48	
86-04-14	31		97		88-09-21	63		38	
86-05-05	23		44		88-10-26	97		3	
86-05-26	110		63		88-11-16	80		39	
86-06-23	48	51	110	57	88-12-13	96		45	
86-07-28	43		35		89-01-11	32	66	34	42
86-08-19	45		50		89-02-27	39		86	
86-09-15	50		54		89-03-16	47		55	
86-10-13	41		46		89-04-12	45		44	
86-11-03	71		47		89-05-09	58		20	
86-12-17	33		48		89-06-07	35		39	
87-02-04	7	41	23	43	89-06-27	23		59	
87-03-17	8		43		n:	62	8	62	8
87-04-14	23		57		MV:	48	49	52	53
87-04-23	21		84		SD:	25	13	25	9
87-05-13	61		23		CV:	0.51	0.27	0.48	0.17
87-06-02	29		66		CV/$\sqrt{7}$	0.18			
87-06-22	47		43						

4.4.4 Empirically Based Highest r^2, r_e^2

One way to determine the highest possible r^2 of a predictive model is to compare two empirical samples (see Håkanson 1999). The variables in these two samples should be as time and area compatible as possible; they should be sampled,

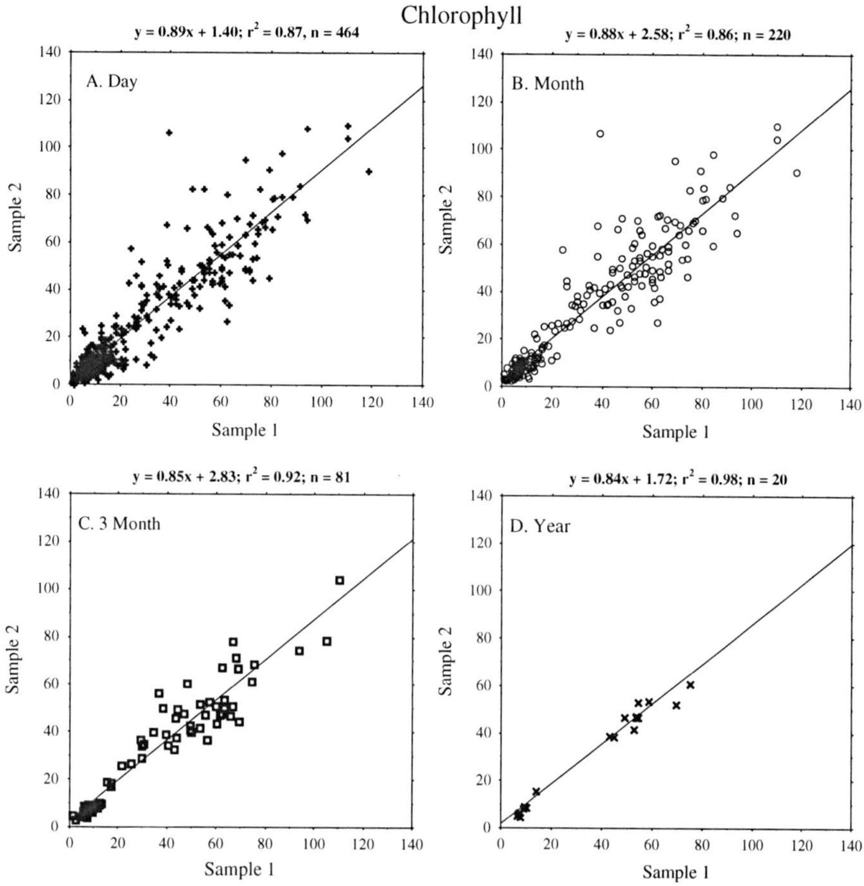

Fig. 4.10 Regressions between two empirical samples which are meant to show the same thing related to chlorophyll-a concentrations in Ringkobing Fjord (data from 1984 to 2004), (**A**) based on daily data (generally 1–3 samples for each value; the total number of sample, n = 464), (**B**) based on monthly data (generally 3–5 samples for each median value; n = 220), (**C**) based on 3-monthly data (generally 8–12 samples for each median value; n = 81) and (**D**) based on annual data (generally about 80–100 samples for each median value; n = 20 years). The figure also gives the corresponding regression lines and the r^2-values

transported, stored and analyzed in the same manner. To illustrate the basic approach to determine r_e^2 (the r^2-value obtained when two empirical dataset for the same thing are compared), we will again use chlorophyll data from Ringkobing Fjord.

Figure 4.10 gives four regressions between two empirical samples which are meant to show the same chlorophyll-a concentrations, Fig. 4.10A is based on daily data (1–3 samples for each value; the total number of samples, n = 464), the next regression is based on monthly data (3–5 samples for each median value; n = 220), the results in Fig. 4.10C are based on 3-monthly data (8–12 samples for each median value; n = 81) and the fourth regression is based on annual data (generally about 80–100 samples for each median value; n = 20 years). The figure also gives the corresponding regression lines and the r^2-values. One can note that the r^2-values increase from 0.87 to 0.98.

The point here is that one should never hope to explain all of the variation in any water variable and that the results in comparisons of this kind depend on some key statistical presuppositions, (1) the number of data used in the regression (such as n = 464 in Fig. 4.10A), (2) the number of analyses for each data-pair (such as 3–5 for the data in Fig. 4.10B), which influence the uncertainty (CV) of the data in the y- and x-directions and (3) the range of the data.

The latter aspect is especially important if predictive – and not descriptive – models are requested. Then, one must ask questions about the range of the data in the sample in relation to the range of the data in the population where the predictive model is meant to apply (Håkanson and Peters 1995).

4.4.5 Highest Reference r^2, r_r^2

An equation has been derived (Håkanson 1999) which gives the highest r^2-value that one in practice can hope to achieve as a function of (1) the number of samples (n_i) for each y_i-value in the regression, (2) the number of data points in the regression, (3) the standard deviations related to all individual data points, (4) the standard deviation of all points in the regression and (5) the range of the y-variable. The derivation of (4.4) is based on an algorithm for the 95% confidence interval (CI) for individual y for independent validations (from Håkanson and Peters 1995), giving CI as a function of range in y, n and r^2, and the sampling formula (4.2), giving the error L as a function of CV and n. If CI is set equal to L and if the n_i-values are set to 10, and if the minimum value in the range ($y_{max}-y_{min}$) is assumed to be small in comparison to y_{max}, and if y_{max} is set to 1 (which is valid for relative values), then:

$$r_r^2 = 1 - 0.66 \cdot CV^2 \tag{4.4}$$

The equation is graphically shown in Fig. 4.11. It should be noted that the CV-value in (4.4) should be the characteristic within-system variability (sometimes abbreviated as CV_y).

The practical use of r_r^2 is shown in Fig. 4.11, which illustrates that the threshold r^2-value of 0.75 (from Prairie's "staircase") corresponds to a threshold CV of 0.62. This means that if the CV is higher than 0.62, one cannot expect that a model will predict the target y-variable well. This threshold value will be used later in the discussion on optimal model scale.

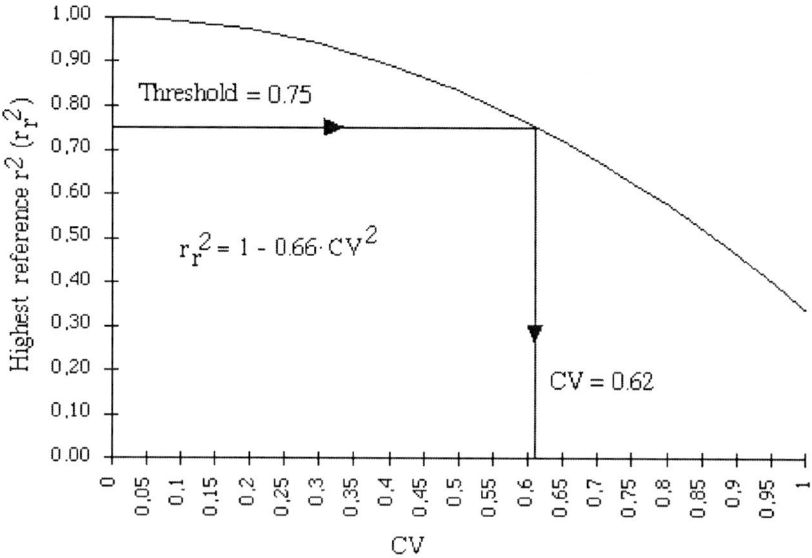

Fig. 4.11 Illustration of the relationship between the within-system variability or uncertainty for a given target variable y, as given by CV (= CV_y), and the highest r^2-value (r_r^2) that one can expect for the given y-variable in a predictive model. The threshold r^2 has been defined (from Fig. 4.5C) to 0.75. This corresponds to a threshold CV for y of 0.62. Variables with inherent CVs higher than 0.62 cannot be expected to be predicted well

4.4.6 Regressions and Confidence Intervals

Figure 4.12 addresses some important methodological issues related to regressions and uncertainties in data. All these regressions use chlorophyll as y-variable and TP as x-variable, but we would like to stress that the figure is meant to illustrate general aspects of regression analyses.

The first figure (Fig. 4.12A) gives results based on daily data (from Ringkobing Fjord). Generally, there are 1–3 data per day and a total of 623 data in this regression, which gives an r^2 of 0.55, which is well below the threshold r^2-value of 0.75 for practical usefulness. Figure 4.12A also gives the 95% confidence intervals for individual data, and indicates the way in which the number of classes or steps, NC, in Prairie's "staircase" is defined. From this figure, one can note the wide scatter, the few classes, the low r^2-value and that the slope is 1.08. Figure 4.12A2 gives two box-and-whisker plots for the CV-values for the chlorophyll concentrations for all daily data based on 3 samples, since it is not meaningful to determine the standard deviation, SD, if n is not ≥ 3. One can note that the median CV is 0.18. This figure also gives the L-value (the error for the mean values from the sampling formula, (4.2)). One can see that L is 25% if n is 3 and all individual CVs are accounted for. The figure also shows the regression based all 165 data related to n = 3, and then the r^{2-}value is 0.59 and the slope 1.31.

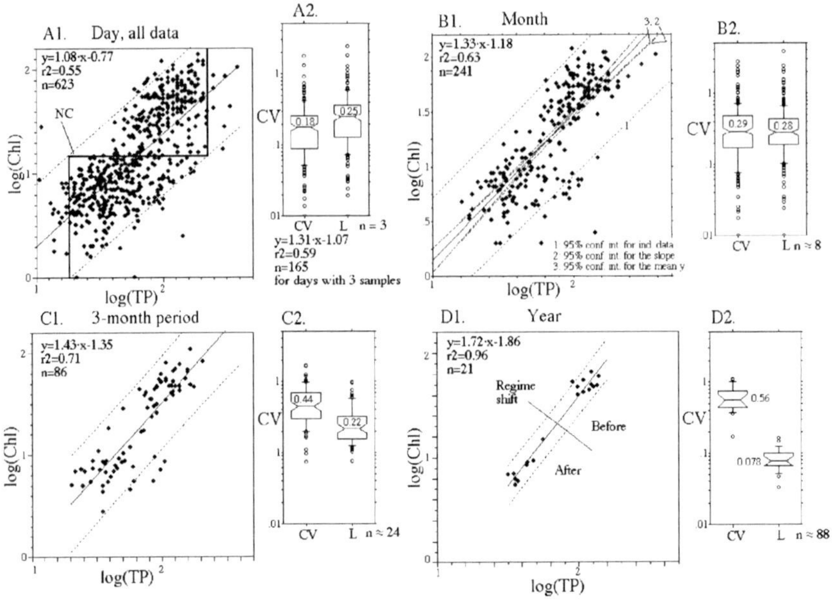

Fig. 4.12 Regressions between chlorophyll and TP using data from Ringkobing Fjord (data from 1984 to 2004). **A1.** Gives results (regression line, r^2 and n) based on daily data; generally there are 1–3 data per day and a total of 623 data in this regression, which gives an r^2 of 0.55; the slope is 1.08. **A2.** Gives a box-and-whisker plot for the CV-values for the chlorophyll concentrations for all daily data based on 3 samples. **A2** also gives a similar plot for the L-value (the error in the mean from the sampling formula). **A2** also shows the regression for all 165 data for n = 3, then r^2 is 0.59 and the slope 1.31. **B1.** Gives the same thing as A1 but for monthly data. This figure gives three standard 95% confidence intervals; the widest interval (1) relates to the variability for the individual data (n = 241 generally there are 8 analyses for each monthly median value), (2) the confidence interval for the slope (here 1.33), and (3) the confidence interval closest to the regression line, which is for the mean y-value. **B2.** Gives the box-and-whisker plots for CV (based on monthly data) and L; the median values are 0.29 and 0.28, respectively. **C1.** Gives the results using data from 3-month periods; one can note that the slope is 1.43 and the r^2-value 0.71. **C2.** Shows that the median CV for these 86 values (generally based on 24 samples) is 0.44; the corresponding L-value is 0.22. **D1.** Gives the results for the 22 annual median values (generally based on 88 samples); the slope is 1.72 and the r^2-value 0.96. **C2.** The median CV for these 22 median values is 0.56, and the corresponding L-value is 0.078

The next Fig. 4.12B1 gives the same thing as 4.12A1 but for monthly data. This figure does not provide information on NC but it gives three standard 95% confidence intervals. The widest interval (1) relates to the variability for the individual data (n = 241 generally there are 8 analyses for each monthly median value), (2) gives the confidence interval for the slope (1.33), and (3) the confidence interval closest to the regression line, is for the mean y-value. The point here is that each of these confidence intervals carries specific and useful information (the confidence interval for the individual data are used in Prairie's "staircase").

Figure 4.12B2 gives box-and-whisker plots for CV and L based on monthly data; the median values are 0.29 and 0.28, respectively. Figure 4.12C1 shows the results for data from 3-month periods; one can note that the slope is 1.43 and the r^2-value 0.71. Figure 4.12C2 indicates that the median CV for these 86 median values based on data from 3 months (generally from 24 samples) is 0.44; the corresponding L-value is 0.22. Figure 4.12D2 gives the results for the 21 annual median values (generally based on 88 samples); the slope is 1.72 and the r^2-value 0.96. The median CV is 0.56, and the corresponding L-value is 0.078.

From Fig. 4.12, one can specifically note:

- The CV-values increase from daily to annual sampling periods (median values change from 0.18 to 0.56). This is valid not just for chlorophyll, but for most water variables in most coastal areas (see Fig. 4.7 and Håkanson 2006).
- The median L-values decrease from 0.25 to 0.078 in spite of the fact that the CV-values increase; this is because the increase in CV is more than compensated for by the increase in number of data for each median (or mean value), from 3 to 88.
- The slopes increase in a logical and steady way from 1.08 to 1.72. This means that the interpretation of how changes in nutrient concentrations (here TP) influence chlorophyll, phytoplankton biomass and primary production changes drastically if different regressions are used without due consideration to the presuppositions. The main reason for this change in slopes is that the chlorophyll values in relation to the nutrient concentrations are higher during the summer because the water temperatures and the light conditions favor a higher primary production, and that there are more samples from the summer period than from the winter period in this (and probably most) monitoring.
- The r^2-values increase in steps from 0.55, 0.63, 0.71 to 0.96. The threshold r^2 is 0.75, and this figure illustrates why this threshold value is important, firstly because the 95% confidence intervals for the individual data are very far apart if $r^2 \leq 0.75$, secondly because the regime shift in the fjord related to the drastic changes in chlorophyll concentrations (shown in Fig. 4.5A) can best be identified using the annual data (Fig. 4.12D1), and this is mainly because those median value has a small inherent uncertainty, an L-value of 0.078.

The nomogram in Fig. 4.13 gives the relationship between the r^2-value, the number of data-pairs in the regression (n) and the level of statistical significance (p). If the x- and y-variables are normally distributed, then the p-value can be read directly from this graph. In this nomogram, it has been emphasized that the 95% confidence level ($p = 0.05$) is often used as a default criteria for significance, but the 90% and the 99% levels are also common.

For any regression, one could and should ask if the x-variables are related to the y-variable in a significant and meaningful manner. A simple and useful method to address this question is to conduct a random variable test (Håkanson and Peters 1995). The basic idea of random variable tests is to generate series of random data, which will be used together with the real empirical data in the regression analysis to see how the randomly generated data correlate to the target y-variable in competition with the real data. There are several statistical tests available to establish whether

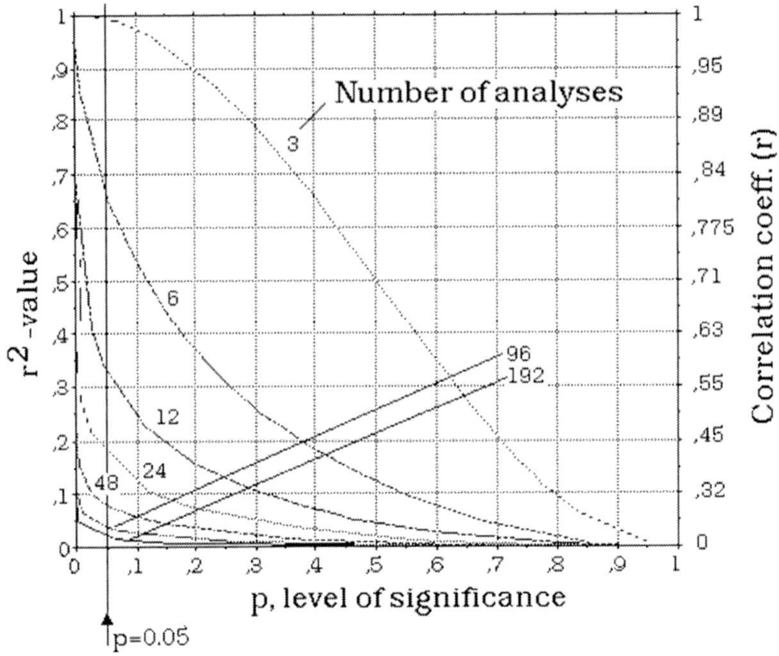

Fig. 4.13 The relationship between r^2, the number of data-pairs in a regression (n) and the statistical level of significance (p). The p-value of 0.05, corresponding to a 95% level of significance, is marked, since this an often used criterion in the ecological literature (modified from Håkanson and Peters 1995)

the correlation is significant or not (p, F, etc.); the random variable test is a simple alternative which is easy too understand and conduct. The test gives a good sense of the difference between a chance correlation and a real one.

It should be stressed that is always possible to find a regression model that would provide a 100% degree of statistical explanation ($r^2 = 1$), e.g., if n = 2! If the number of data in the regression is small, one should be very cautious about the results. For larger samples (see Fig. 4.14), the risks of misidentifying random variables or erroneous empirical data as powerful predictors, decrease. For $n > 30$ these risks are rather low, for $n > 100$ they are very low, but for $n < 10$, they are substantial. From Fig. 4.14, one can note that if n = 3 almost any correlation coefficient between –1 and 1 can be obtained and hence also r^2-values close to 1.

4.4.7 Frequency Distributions and Transformations

Regression analyses can be performed for many reasons, e.g., to compare values predicted from models (generally x) with empirical data (y), to test hypotheses about relationships, to identify thresholds and to develop statistical/empirical models.

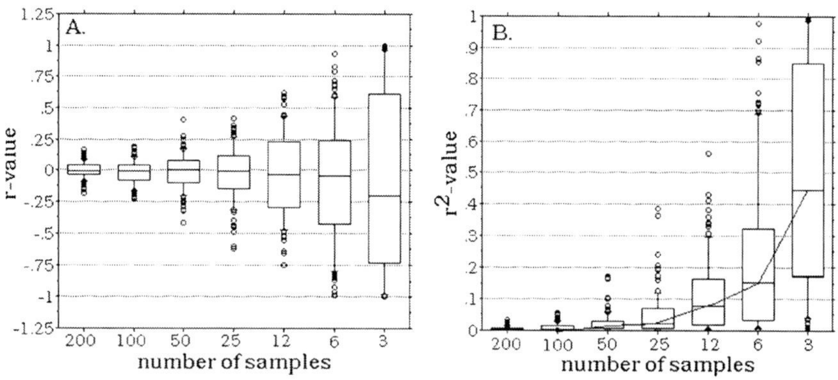

Fig. 4.14 Box-and-whisker plots illustrating correlation coefficients and coefficients of determination (r^2) in a test where hundreds of random parameters have been regressed against a target y-variable and the relationship between the number of samples (n) and the obtained r-values and the r^2-values are given. One can note that in regressions based on few data (n) the risks of getting nonsensical results are very high (from Håkanson and Peters 1995)

Many textbooks examine regression analyses (Draper and Smith 1966; Mosteller and Tukey 1977; Pfaffenberger and Patterson 1987; Newman 1993). Our aim in this section is to briefly stress some basic concepts related to regressions in contexts of water management.

In order for the r^2-value to be valid, the x- and the y-samples in a regression should be normally distributed. If this is not the case for the actual data, different types of transformations (Box and Cox 1964) can be used to obtain more compatible, and more normally distributed, x- and y-variables. Some common transformations are the logarithmic (either log10 = log or ln = logn), or different exponentials like $\sqrt{x} = x^{0.5}, x^2, x^{-1}$, or $1/(x + \text{const})$. The data could also be ranked relative to the highest and/or lowest numerical value in the series. Certain transformations (like e^x) maximize the weight of high values in regressions. Others, like log(x), minimize the weight of high values. There are linear regressions ($y = a \cdot x + b$), non-linear regressions (like $y = a \cdot \log(x) + b$) and multiple regressions (like $y = a \cdot x_1 + b \cdot \log(x_2) + c/x_3 + d$). Figure 4.16 gives frequency distributions for seven standard coastal water variables from the monitoring in the Baltic Sea, Kattegat and Skagerack. Table 4.4 gave the CVs for the variability *within* each area, e.g., for the DIN/DIP-ratio in the Kattegat (CV = 1.01), DIN in the Baltic Proper (CV = 0.75), DIP in the Bothnian Sea (CV = 0.73) and the TN/TP-ratio in the Skagerack (CV = 0.38). The variations in the mean monthly values *among* the individual stations are even higher (see the frequency distributions in Fig. 4.15): DIN/DIP-ratio (CV = 2.52), DIN (CV = 1.44), DIP (CV = 0.96) and the TN/TP-ratio (CV = 0.93). One can also note that DIN, DIP, TN/TP and DIN/DIP have positively skewed frequency distributions with mean/median (MV/M50) ratios significantly higher than 1, whereas chlorophyll, TN and TP in these systems have MV/M50-ratios close to 1. Information is also given in Fig. 4.15 about the ratio between the mean value (MV) and the median (M50), since this is a simple and useful standard

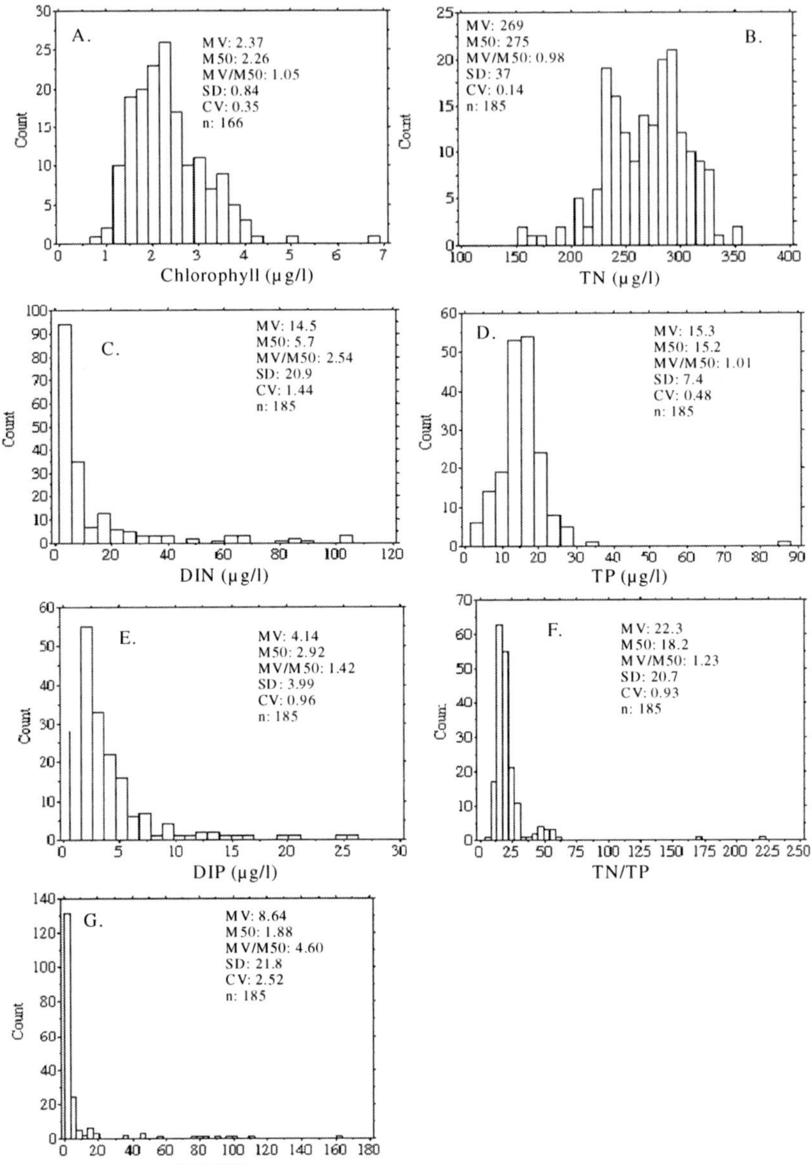

Fig. 4.15 Frequency distributions for chlorophyll-aconcentrations, TN, DIN, TP, DIP, TN/TP and DIN/DIP from the Baltic Sea, Kattegat and Skagerack based on data from the surface-water layer (< 10 m) for the growing season for the period 1987–2006 (data from HELCOM)

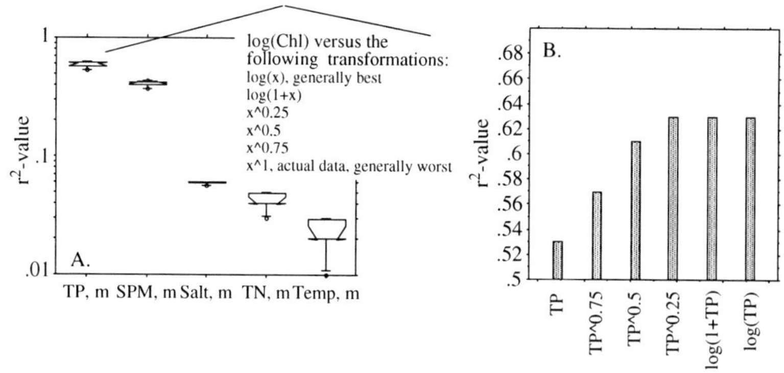

Fig. 4.16 Transformations and r^2-values using data (median monthly values) from Ringkobing Fjord (from 1980 to 2004). A. Shows box-and-whisker plots for the r^2-values based on regressions between log(Chl) and transformations (log(x), log(1-x), $x^{0.25}$, $x^{0.5}$, $x^{0.75}$and actual data, x) for TP, SPM, salinity, TN and water temperature. B. Shows the results underlying the box-and-whisker plot in A for chlorophyll versus TP

measure of the normality of the frequency distribution; the number of data used in the frequency distribution is given (n).

Figure 4.16 illustrates the role of transformations for the obtained r^2-values in regressions (using data on median monthly values from Ringkobing Fjord as examples). Figure 4.16A shows box-and-whisker plots for the r^2-values based on regressions between log(Chl) and transformations, log(x), log(1-x), $x^{0.25}$, $x^{0.5}$, $x^{0.75}$and actual data (x) for TP, SPM, salinity, TN and water temperature. Figure 4.16B exemplifies the results underlying the box-and-whisker plot in Fig. 4.16A for chlorophyll versus TP. From these results, one can note: (1) that there are major differences in the calculated r^2-values; the highest r^2-values appear between chlorophyll and TP (highest 0.63 for the log-transformations), the lowest between chlorophyll and temperature (median $r^2 \approx 0.02$), (2) that several of the transformations, and not just the log-transformation, can yield high r^2-values. However, the log-transformation can, as these examples indicate, be recommended for most water variables if normal frequency distribution are requested for regression analysis which aim to produce predictive (not descriptive) statistical models for coastal water variables until better data become available which would falsify this recommendation.

4.4.8 Multiple Regressions

Multiple regressions give y as a function of several x-variables. The aim of this section is to present some results from stepwise multiple regression using coastal data.

Table 4.7 Correlations matrix based on linear correlation coefficients for actual monthly median data from Ringkobing Fjord

	Chl, m	TP, m	Salt, m	TN, m	SPM, m	Temp, m	Sec, m
Chl, m	1	0.75	–0.28	0.21	0.61	0.04	–0.60
TP, m		1	–0.35	0.39	0.83	0.00	–0.70
Salt, m			1	–0.76	–0.31	0.50	0.43
TN, m				1	0.33	–0.69	–0.45
SPM, m					1	0.00	–0.66
Temp, m						1	0.18
Sec, m							1

A useful initial test of co-variations among variables is the correlation matrix based on linear correlation coefficients for the actual (non-transformed) data, as shown in Table 4.7 using data (monthly median values from Ringkobing Fjord).

The target y-variables in this example are chlorophyll and Secchi depth and the correlation matrix shows that the x-variable with the highest potential to statistically (but not necessarily causally) explain variations among these monthly y-values in this coastal system would be TP. The result of the stepwise multiple regression (using the most appropriate transformations according to the procedures just discussed) shown in Table 4.8.

From this table, one can note:

1. That the variations in Secchi depth can be explained (statistically) much better than variations in chlorophyll concentrations (r^2 = 0.872 after 4 steps as compared to 0.67 after 3 steps; see Table 4.8). One important reason for this is that the

Table 4.8 Ladders for (A) chlorophyll-a concentrations (in μg/l) and (B) Secchi depths (in m) using monthly median values from data Ringkobing Fjord for the period from 1980 to 2004. TP = total-P concentration (μg/l); SWT = surface-water temperatures (C); SPM = concentrations of suspended particulate matter (mg/l); TN = total-N concentrations (μg/l); Sal = salinities (psu); the results are based on stepwise multiple regression analyses according to procedures given by Håkanson and Peters (1995)

A. For chlorophyll-a concentration			
Step	r^2	Variable	Model
1	0.63	TP	log(Chl) = 1.323·log(TP) – 1.163
2	0.66	SWT	log(Chl) = 1.341·log(TP) + 0.0824·SWT – 1.441
3	0.67	SPM	log(Chl) = 1.528·log(TP) + 0.0855·SWT – 0.0032·SPM – 1.706
B. For Secchi depth			
1	0.84	TP	log(Sec) = –0.804·log(TP) + 1.454
2	0.86	TN	log(Sec) = –0.3738·log(TP) – 0.191·log(TN) + 1.441
3	0.868	SPM	log(Sec) = –0.599·log(TP) – 0.188·log(TN) – 0.128·log(SPM) + 1.829
4	0.872	Sal	log(Sec) = –0.590·log(TP) – 0.100·log(TN) + 0.133·log(SPM) + 0.0093·Sal + 1.465

inherent uncertainties in the two y-variables are quite different and much higher for chlorophyll (CV = 0.30) than for Secchi depth (CV = 0.20, see Table 4.4).

2. That the variations within this coastal area can be related to variables such as TP, TN, temperature, SPM and salinity, which are all influenced by temporal variations (in winds, precipitation, etc.). In the next section, we will address variations among coastal areas, and then it is interesting to note that there are major systematic differences in the factors causing variations within and among coastal areas. Such differences are of fundamental importance in deriving predictive models for coastal management.

There are two methodological reasons for selecting the examples in this section, the first is to stress the role of morphometric parameters in coastal science – the way in which the coastal area looks, the morphometry, will regulate the way in which the coastal area functions (see Chap. 3); the second concerns uncertainty analyses and optimal model scale.

Figure 4.17 is included here to stress one point: To analyze the relative importance of factors influencing the variability of coastal water variables, it is important to recognize the fundamental difference between the x-variables regulating the variability among and within coastal areas. Any coastal variable (y) will depend on factors causing seasonal variations within the coastal system and on factors causing variations in mean, median or characteristics values among coastal systems. Coastal morphometry will generally influence variations among coastal areas; but also many internal processes (such as sedimentation, resuspension, mixing, etc.). This means that a given load to a coast, for example of nutrients and/or toxins, may cause very different effects in coasts of different morphometry. Effect-load-sensitivity analyses are of fundamental importance in coastal management (see Chap. 2).

Table 4.9 gives an r-rank matrix (based on linear correlation coefficients) of the factors influencing the variations in median chlorophyll-a concentrations among 23 Baltic coastal areas (data from Wallin et al. 1992), including the morphometric variables defined in Chap. 3. From Table 4.9, one can note that the x-variable which shows the highest correlation towards chlorophyll is TN for these coastal areas.

Table 4.10 gives results from stepwise multiple regression analyses (using the same method as in Table 4.8) using median values from each coastal area for the growing season (months 6, 7 and 8) for Secchi depth as the target y-variable, and omitting TN from the test and instead focusing on the role of the morphometric parameters. The basic question here is: How much of the variability in the target y-variable among these 23 coastal areas can be related to the morphometric parameters discussed in Chap. 3 and Table 4.9 (excluding the most important variable, TN). From Table 4.10, one can note:

- 78% ($r^2 = 0.78$) of the variability in Secchi depth among these coastal areas may be related to coastal morphometry; first, the filter factor, Ff, then mean depth D_m, and thirdly the dynamic ratio, DR. The filter factor, Ff, quantifies how the conditions outside a defined coastal area act as energy filter for the defined inner coastal area and reduce the wave energy; the lower the Ff-value, the more islands and/or reefs, the denser the outside coastal area, the lower the influence of the

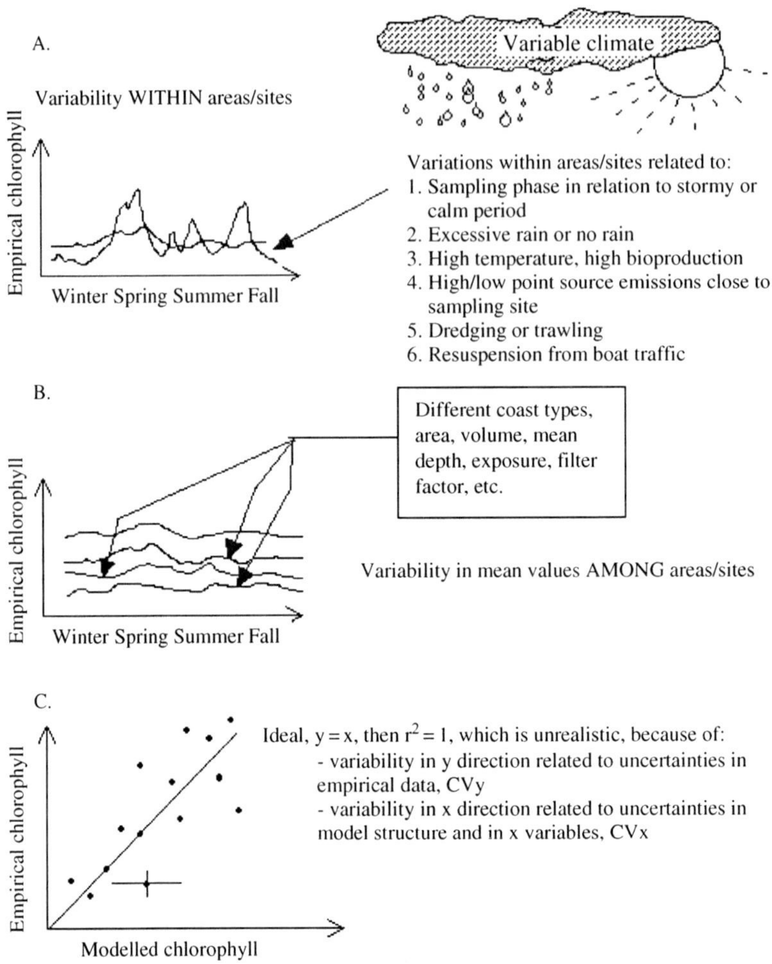

Fig. 4.17 Illustration of factors influencing within- and among-site variability in a water variable (this illustration uses chlorophyll as a proxy) and the role of coastal morphometry for variations among coastal areas and climatological variables for variations within coastal areas

winds and the waves, and more suspended particulate matter (SPM), the higher the phytoplankton production and the lower the Secchi depth in the given coastal area. So, if Ff is large, the coast is open to wind/wave influences from the sea. Then, the Secchi depth in the coastal area would be similar to the Secchi depth in the sea.

- At the second and third steps, one finds the mean depth (D_m; $r^2 = 0.67$) and the dynamic ratio (DR; $r^2 = 0.78$) The larger the mean depth and the ratio between the area and the mean depth, the more SPM will be entrapped and retained in the coastal area and the smaller the Secchi depth.
- At the fourth step, we find TP, which adds little to the regression ($r^2 = 0.80$).

Table 4.9 A ranking (r-rank) of the x-variables influencing a given y-variable (here chlorophyll-a concentrations in Baltic coastal areas; data from 23 coastal areas from Wallin et al. 1992), (see Fig. 3.3 and Håkanson 2006, for further information)

Variable	r-rank
Chlorophyll-a conc. (μg/l)	1
Total-N concentration (μg/l)	0.71
Theor. surface-water retention time, T_{SW} (days)	0.63
Sedimentation in surface-water areas, Sed_{SW} ($gdw/m^2 \cdot d$)	0.61
Theor. deep-water retention time, T_{DW} (days)	0.48
Sedimentation in deep-water areas, Sed_{DW} ($gdw/m^2 \cdot d$)	0.37
Inorganic-P (μg/l)	0.29
Inorganic-N (μg/l)	0.27
Relative depth, D_{rel}	0.19
Total-P concentration (μg/l)	0.17
Mean slope, xm	0.10
Mean depth, D_m	0.06
Maximum depth, D_{max}	0.06
Volume, Vol (m^3)	−0.03
Form factor (= volume development), Vd (dim. less)	−0.03
Area (m^2)	−0.11
Dynamic ratio, DR (dim. less)	−0.11
Total area, A_{tot} (m^2)	−0.12
Section area, At (m^2)	−0.16
Mean filter factor, MFf (km)	−0.22
Fraction of erosion and transport areas, ET (dim. less)	−0.24
Exposure, Ex (dim. less)	−0.26
Shoreline irregularity, F (dim. less)	−0.30
Filter factor, Ff (km)	−0.32
O_2-concentration in deep-water areas (mg/l)	−0.51
Secchi depth (m)	−0.75

These empirical regression models for Secchi depth demonstrate that the size and form elements of a coastal area are very important in understanding and predicting how variables like Secchi depth, and also chlorophyll, SPM, sedimentation and oxygen saturation/concentration vary among coastal areas (Håkanson 2006).

Table 4.10 Ladder (based on stepwise multiple regression analysis) for Secchi depth (Sec in m) as target variable (y) based on median values for the growing season (months 6, 7 and 8) from 23 Baltic coastal areas using data from Wallin et al. (1992). Ff = filter factor (see Chap. 3); D_m = mean depth (m); DR = dynamic ratio (= $\sqrt{Area}/D_m$; area in km^2); TP = total-P concentration (μg/l)

Step	r^2	Variable	Model
1	0.44	Ff	$\log(Sec) = 0.297 \cdot \log(Ff) + 0.319$
2	0.67	D_m	$\log(Sec) = 0.351 \cdot \log(Ff) - 0.816 \cdot \log(D_m) + 0.983$
3	0.78	DR	$\log(Sec) = 0.437 \cdot \log(Ff) - 1.474 \cdot \log(D_m) - 0.793 \cdot DR + 1.727$
4	0.80	TP	$\log(Sec) = 0.413 \cdot \log(Ff) - 1.442 \cdot \log(D_m) - 0.758 \cdot DR - 0.450 \cdot \log(TP) + 2.312$

Table 4.11 Ratios between characteristic coefficients of variation for among-site variability and (CV_a) and variations in mean values within sites (CV_w) for selected standard variables in coastal management. These data are based on studies in mainly Baltic coastal areas (from Wallin et al. 1992 and Håkanson 2006), which restrict the CV_a-values but not the principle aspects this table is meant to address

Variable	CV_a/CV_w
Total-P	1.02/0.16 = 6.25
Chlorophyll-a	1.47/0.25 = 5.88
Total-N	0.71/0.13 = 5.56
Secchi depth	0.49/0.19 = 2.56
Inorganic-P	0.57/0.28 = 2.04
O_2-concentration (deep water)	0.44/0.26 = 1.69
Inorganic-N	0.44/0.31 = 1.43
Salinity	0.07/0.05 = 1.41
SPM	0.67/0.69 = 0.97

These morphometric variables are also incorporated in dynamic models (such as the CoastMab-model in Sect. 9.1) in a mechanistic way where they influence key transport processes of nutrients, toxins and SPM.

The ratio between the CV-value for among-system variations to the CV-value for within-system variations is very informative in predictive ecology. Table 4.11 is included here to stress that point. It can be, and has been (Håkanson and Peters 1995), demonstrated that for variables showing a high CV_a/CV_W-ratio, x-variables related to morphometry, geology, catchment-area characteristics and latitude can explain statistically and causally a large part of the variability in the y-variables (as exemplified by TP, chlorophyll and TN in Table 4.11), while variables showing a low CV_a/CV_W-ratio (such as SPM in Table 4.11) are primarily related to climatological x-variables (such as temperature, light, precipitation, water discharge, etc.)

4.4.9 Stability Tests

The empirical model in Table 4.10 has been used in the following stability tests (Table 4.12) to estimate the uncertainty in the slope coefficients. The first lines in Table 4.12 give the same model as in Table 4.10, and the focus now is on the r^2-values (0.44, 0.67, 0.78 and 0.8) and the slopes (0.413 for Ff, 1.442 for D_m, –0.758 for DR and –0.45 for TP). In this stability test (see Håkanson and Peters 1995, for further information), we have sought values for the uncertainties of these slopes and this has been achieved by omitting 5 coastal areas at random ten times and recalculated the stepwise multiple regression for each round. This gives the results in Table 4.12.

At the first round, one can note that Ff, D_m and DR enter at the first 3 steps, just like in the basic reference model (for n = 23), but three new morphometric variables replace TP, namely the form factor (Vd at step 4, r^2 = 0.86, slope 0.322), the mean

Table 4.12 Stability test to estimate the uncertainties in the slopes of the statistical model given in Table 4.10; 10 rounds have been made and 5 coastal areas have been randomly omitted in each round

All	Step 1	Step 2	Step 3	Step 4	Step 5	Step 6	Intercept
n=23	Ff	Dm	DR	TP			2.312
	0.44	0.67	0.78	0.8			
	0.413	–1.442	–0.758	–0.45			
Round 1	Ff	Dm	DR	Vd	xm	At	1.5
	0.45	0.72	0.82	0.86	0.88	0.89	
	0.594	–2.204	–0.491	0.322	0.8	–2.651	
Round 2	Ff	Dm	DR	TP		2.555	
	0.48	0.62	0.71	0.74			
	0.426	–1.572	–0.86	–0.527			
Round 3	Ff	Dm	DR				1.695
	0.42	0.75	0.81				
	0.484	–1.512	–0.681				
Round 4	Ff	Dm	DR	xm	Vd		0.926
	0.5	0.71	0.81	0.86	0.88		
	0.484	–2	–0.142	1.139	0.288		
Round 5	Ff	Dm	DR	TP	Vd		2.209
	0.36	0.52	0.68	0.73	0.77		
	0.363	–1.207	–0.505	–0.684	0.197		
Round 6	Ff	Dm	DR	At	Vd		1.66
	0.39	0.67	0.78	0.8	0.82		
	0.379	–1.805	–0.823	1.443	0.216		
Round 7	Ff	Dm	xm	Vd	At		1.387
	0.42	0.78	0.85	0.89	0.92		
	0.431	–2.435	1.148	0.734	1.706		
Round 8	Ff	Dm	DR	xm	Vd		1.35
	0.39	0.57	0.73	0.76	0.81		
	0.475	–2.4	–0.447	1.132	0.264		
Round 9	Ff	Dm	DR				1.853
	0.41	0.7	0.78				
	0.431	–1.619	–0.797				
Round 10	Ff	Dm	DR	At	xm		1.806
	0.52	0.71	0.81	0.83	0.85		
	0.431	–2.255	–0.875	1.558	0.64		
	Ff	Dm	DR	TP			
MV (slope)	0.491	–2.045	–0.709	–			
SD	0.065	0.42	0.247	–			
CV	0.132	0.206	0.348	(> 0.5)			
n	10	10	9	1			

The mean values (MV), standard deviations (SD) and coefficients of variation (CV) from these 10 rounds are given and indicate the uncertainties in the slopes for the different model variables (Ff = filter factor, D_m = mean depth, DR = dynamic ratio, TP = total-P concentration; Vd = volume development = form factor, xm = mean slope of the coastal area, At = section area). The first, upper, part of the table gives the results when all 23 coastal areas are used in the stepwise multiple regression analysis.

slope (xm at step 5, $r^2 = 0.88$, slope 0.8) and the section area (At, step 6, $r^2 = 0.89$, slope –2.651).

An important conclusion from this test is that had data been available only from the 15 coastal areas used in round 1, the regression would look differently, and the interpretation of the factors influencing the target y-variable would also be different.

From Table 4.12, we can note that the CV-values for the slopes are 0.132 for x_1 (= Ff), 0.21 for x_2 (= D_m), 0.35 for x_3 (= DR) and > 0.5 for x_4 (TP). This means that the uncertainty in the slope in this case, and also more generally, increase for each x-variable added in the regression model. One should also note that the stability test is a method to estimate the uncertainties of all slopes in multiple regressions, but that the results depend on how many samples are omitted and the strategy to omit the samples.

4.4.10 The Optimal Size of Practically Useful Predictive Models

In this section, we will illustrate how uncertainties in model variables (x) and model constants, such as slopes in regression models, or rates (= rate constants) in dynamic models, influence the accumulated uncertainty in the target y-variable. Two main approaches to uncertainty analysis exist, analytical methods (Cox and Baybutt 1981; Beck and Van Straten 1983; Worley 1987) and statistical methods, like Monte Carlo techniques (Tiwari and Hobbie 1976; Rose et al. 1989; IAEA 2000). In this section, we will only discuss Monte Carlo simulations. Uncertainty tests using Monte Carlo techniques may be done in several ways, using uniform CV-values, or more realistically, using characteristic CV-values (e.g., from Table 4.4). For predictive empirical or dynamical models based on several uncertain model variables (rates, slopes, etc.), the uncertainty in the prediction of the target variable (y) depends on such uncertainties. The cumulative uncertainty from many uncertain x-variables may be calculated by Monte Carlo simulations, and that is the focus of this section.

Monte Carlo simulation is a technique to forecast the entire range of likely observations in a given situation; it can also give confidence limits to describe the likelihood of a given event. Uncertainty analysis (which is a term for this procedure) is the same as conducting sensitivity analysis for all given model variables at the same time. The procedure is illustrated in Fig. 4.18. The figure shows different frequency distributions with different CV-values showing uncertainties in x-variables which may be used in statistical and dynamic models to predict a given y-variable and how these uncertainties in the x-variables cause an uncertainty in the target y-variables.

Table 4.13 illustrates a calculation of optimal model scale using the regression model for Secchi depth in Table 4.10, the uncertainties in the slopes from Table 4.12 and characteristic coefficients of variation for the model variables (CV_x) (filter factor, Ff, mean depth, D_m, dynamic ratio, DR, and TP-concentration). The CV-values are from Håkanson (2006). The uncertainties in the target y-variable (CV_y) have been calculated using Monte Carlo techniques (MC). Table 4.13A gives results

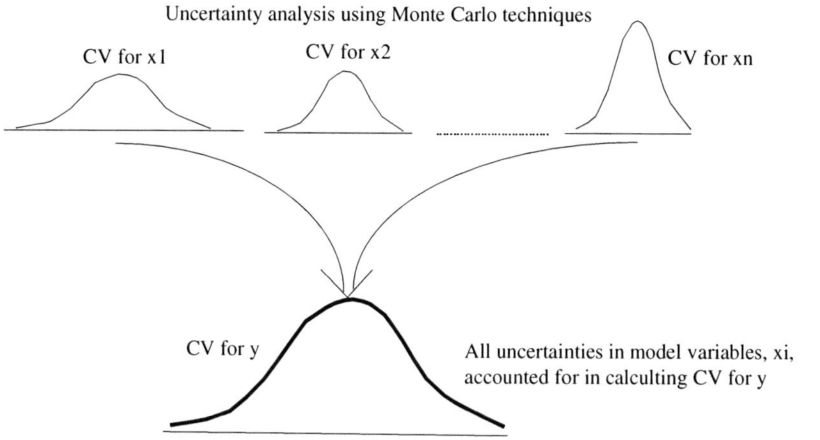

Fig. 4.18 Illustration of the principles of uncertainty analysis using Monte Carlo simulations

when the uncertainties in slopes and the model variables are accounted for, and Table 4.13B when only the uncertainties in model variables are accounted for.

The optimal size must also include a consideration to the predictive power, here given by the r^2-value, when empirical data on Secchi depth are compared to modeled values. The threshold value for practical use is set at an r^2-value of 0.75 (see

Table 4.13 Calculation of optimal model size using the regression model for Secchi depth in Table 4.10, the uncertainties in the slopes from Table 4.12 and characteristic coefficients of variation for the model variables (CV_x) (filter factor, Ff, mean depth, D_m, dynamic ratio, DR, and total-P concentration, TP)

	Step 1 Ff	Step 2 D_m	Step 3 DR	Step 4 TP	Note
CV for x, CV_x	0.10	0.05	0.075	0.16	
CV for slope, CV_s	0.13	0.21	0.35	0.50	
r^2	0.44	0.67	**0.78**	**0.80**	>0.75, threshold value
NC	1.76	2.30	**2.81**	**2.95**	NC>2.64, threshold value; see Fig 4.5C
A. Including uncertainties in variables and slopes					
CV for y, CV_y	0.071	0.39	0.74	0.96	from MC calculations
r^2/CV_y	6.20	1.72	1.05	0.83	
NC/CV_y	24.8	5.9	3.8	3.1	
$r^2\cdot(1\text{-}CV)$	0.409	0.409	0.203	0.032	
$r^2\cdot(1\text{-}CV^2)$	0.44	0.57	0.35	0.06	
B. Including only uncertainties in model variables					
CV for y, CV_y	0.033	0.061	0.099	0.14	from MC calculations
r^2/CV_y	13.3	11.0	7.9	5.7	
NC/CV_y	53.5	37.7	28.4	21.1	NC from Fig 4.5C
$r^2\cdot(1\text{-}CV)$	0.43	0.63	0.70	0.69	
$r^2\cdot(1\text{-}CV^2)$	0.44	0.67	0.77	0.78	
			optimal size		

Table 4.14 Illustration and definition of OMS, the optimal model scale, accounting for the variability/uncertainty in target y-variables and the accessibility of data for predictive models for y. These examples use data for chlorophyll from Ringkobing Fjord as an illustration. CV = the coefficient of variation; n = the number of samples (from the sampling formula) for L = 0.2, the "threshold" error for the mean value (see Fig 4.6); 2 is Student's t ($2 \approx 1.96$); 0.62 is the CV for $r_r^2 = 0.75$, the threshold r^2 for practical utility

Time	Time units (T)	CV (Chl)	$n=(CV \cdot 2/0.2)^2$	T·n	OMS = 100·(1/T·n)·(0.62-CV)
Daily	365	0.18	3.2	1168	0.037
Weekly	52	0.29	8.4	437	0.075
Monthly	12	0.30	9.0	108	0.30
Yearly	1	0.56	31.4	31	0.19

Fig. 4.5C). This means that using these criteria, the optimal size is obtained after 3, not 4, steps (see Table 4.13).

There is at least one further dimension to the problem of optimal size of practically useful predictive models in water management. A central problem in communications among scientists, and a key reason for much misunderstanding, has to do with scale. To model in great detail at the scale dealing with hourly or daily changes for target variables in water management is a very difficult task indeed for one system, and do this in a general, predictive manner for many systems is even more difficult. This is illustrated in Fig. 4.20. The definition of the optimal model scale (OMS) is given in Table 4.14, where the concepts are further elaborated at four time scales, daily, weekly, monthly and yearly units.

$$OMS = 100 \cdot (1/T \cdot n) \cdot (0.62 - CV) \tag{4.5}$$

The optimal model scale involves a trade-off between predictive power and the necessary work in terms of data collection and money and time spend to develop and use the given model. Using the threshold CV of 0.62 related to the threshold r^2 for practical use of 0.75, the factor (0.62-CV) defines the predictive success; if the CV-value for the target y-variable is higher than 0.62, the model is more or less useless for predictions in individual systems – the lower the CV-value, the smaller the expected uncertainty in the model predictions.

The accessibility of the data, the cost for predictive success, is given by the factor (1/T·n), where T is the time scale of the modeling (T = 365 for daily predictions, since this is the number of days in a year, etc.), n is the number of data needed to achieve the threshold for accepted error (L = 0.2 in the sampling formula if also t = 1.96 is set to 2). This means that T·n is an objective measure of the demands for data in the modeling; the higher T·n, the more costly and the less useful for practical water management, the model is likely to be.

The curves in Fig. 4.19 give the optimal model scale (OMS) and are calculated from the given formula (4.5) for salinity, temperature, TN, Secchi depth, TP, chlorophyll and SPM using data from Ringkobing Fjord (the CV-values are given in Table 4.4). From Table 4.14, one can note:

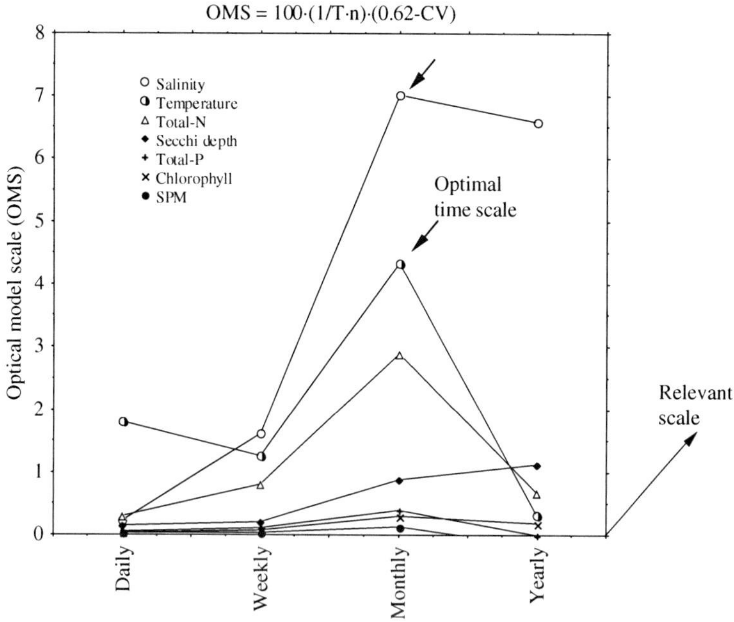

Fig. 4.19 Illustration of factors regulating the optimal scale of practically useful predictive models in water management. The curves give the optimal model scale (OMS) as calculated from the given formula, which is based on two main factors, an expression for the predictive power of the model (0.62-CV; from the threshold r^2-value of 0.75 for practical use for predictions of individual y-target y-values, see Fig. 4.5) and, on the other hand, the accessibility of data needed to run and use a given model

- OMS is a general approach to define the optimal scale of any model for water management for any coastal area, and it is interesting to see that the daily scale is sub-optimal for the given variables in Ringkobing Fjord, and probably in most similar cases, because of the heavy data requirements. The monitoring in Ringkobing Fjord is probably one of the most comprehensive in the world, and not even in this fjord would there be enough data for a meaningful modeling at the finer daily time scale.
- There is also a z-direction added in Fig. 4.19 called "relevant scale", and the idea is simply to stress that for many biological variables, it may not be very relevant to model on an annual basis because the key issues concern seasonal variations, and/or extreme events during the year. So, even if OMS would indicate that the annual scale would be best, as it does for Secchi depth in Ringkobing Fjord, the scientists and managers responsible for the sustainability of this coastal area could argue that they need to know the Secchi depth during the tourist season in the summer time. So, the monthly model would be the optimal scale if one adds that perspective.

4.4.11 *Variations and Spurious Correlations Related to DIN, DIP, TN, TN, DIN/DIP and TN/TP*

As stressed, substances in the water column can be divided into two main parts, the dissolved phase and the particulate phase, relating to their fates and transport routes (pelagic versus benthic). The distribution (= partition = partitioning) coefficient of substances depends on the association to suspended particulate matter (SPM; see Håkanson 2006). Particulate bound substances are, by definition, subject to gravitational sedimentation. Hence, they are to a high degree retained within a given system (e.g., a coastal area) and affect benthic habitats. The dissolved fraction, on the other hand, is more related to the pelagic pathways. Operationally, the particulate fraction is generally defined as the non-filterable remains on a filter. For such determinations, one would often use a pore size of 0.45 μm or pore sizes in the range from 0.2 to 0.9 μm (e.g., Seritti et al. 1980; Bloom and Effler 1990; Mason et al. 1995).

Filtration is often a justifiable method from many ecological and mass-balance modeling perspectives. Substances bound to colloids (i.e., bound to particles smaller than approximately 0.45 μm) are, hence, often operationally included in the dissolved fraction, although, they are *not* truly dissolved in a chemical sense. Chemical fractions, such as phosphate, nitrite, nitrate, ammonium, DIN, DIP, DOC, etc. would often not correspond to the dissolved fraction as determined from filtration (DF = 1 – PF). This means that if SPM and the particulate fraction (PF) are operationally determined from filtration, the dissolved fraction derived from the chemical fractions (e.g., DIN = nitrate + nitrite + ammonium) is not the same as DF.

A general rule is that nutrient limitation is decided by the Redfield ratio, R, estimated from:

$$R = TN/TP \text{ or } R = DIN/DIP \tag{4.6}$$

TN = concentration of total nitrogen (mg/l)
TP = concentration of total phosphorus (mg/l)
DIN = concentration of dissolved inorganic nitrogen (mg/l)
DIP = concentration of dissolved inorganic phosphorus (mg/l).

One well-known and general approach to describe the affinity of all types of substances to carrier particles is by means of the partition coefficient, K_d (l/g dw). K_d is generally defined as the ratio of filter-retained to filter-passing concentrations of a certain substance calculated as:

$$Kd = (C_{part}/SPM)/C_{diss} \tag{4.7}$$

Where SPM is the suspended particulate matter concentration (g dw/l), C_{diss} is the dissolved (filter-passing) concentration (g dw/l) and C_{part} is the particulate concentration (g dw/l). Physically, K_d describes the particle affinity and represents the equilibrium of numerous processes such as sorption onto particulate matter, precipitation and dissolution (Salomons and Förstner 1984). Examples of substances for which K_d have been determined are trace metals (Balls 1989; Benoit 1995; Turner 1996), organic micropollutants (Turner et al. 1999; Zhou et al. 1999), phosphorus

(Håkanson 1999) and radionuclides (Santschi and Honeyman 1991; Carroll and Harms 1999). There are, to the best of our knowledge, no operationally and well-tested algorithms available to predict Kd or the particulate fraction (PF) for nitrogen in coastal areas (see Fig. 4.1).

The variabilities or uncertainties of all variables evidently influence the statistical reliablility of the data (e.g., the mean value and the standard deviation) and the predictive power of models. The coefficient of variation (CV) varies among different variables. For example, CV for PF varies significantly less than CV for Kd and PF is therefore more suitable in predictive models. On average CV for K_d is 3.0 times larger than CV for PF (see Johansson et al. 2001).

Spurious correlations is a fundamental problem in situations where the y-variable is a function of x, such as ratios u/x versus x or u, or products, such as u·s versus x or u. The theory of spurious correlations was developed by Pearson (1897) and Reed (1921); see also Kenney (1982); Jackson et al. (1990); Krambeck (1995); Berges (1997); Johansson et al. (2001) or Håkanson (2006). In the aquatic sciences, there are many papers dealing with DIN/DIP or TN/TP-ratios and mechanistic interpretations along gradients of TN, TP, DIN, DIP, chlorophyll or trophic level (see, e.g., Downing 1997; Turner et al. 2003; Jeppesen et al. 2005; Smith 2006). Such mechanistic interpretations are often spurious and this will be specifically discussed in this section.

An important aspect related to spurious correlations concerns to the fact that both nutrients (N and P) appear in plankton cells. This means that one generally finds a marked co-variation between phosphorus and nitrogen concentrations in aquatic systems (see Fig. 4.20 and Wallin et al. 1992).

Since both nutrients generally co-vary, one would also often find regressions between TN and chlorophyll and TP and chlorophyll yielding high r^2-values (see Fig. 5.12, later). There are numerous regressions between bioindicators and nutrients in the literature since they address fundamental questions concerning the factor influencing target bioindicators (such as chlorophyll) and measures to reduce.

From this background, it should be stressed that the main objective in this section has been to conduct random parameter tests to scrutinize and try to clarify the problem of spurious correlations related to ratios in contexts of mainly marine eutrophication, i.e., TN/TP and DIN/DIP-ratios versus primary production.

4.4.11.1 Random Parameter Tests

Figure 4.21 gives a scatterplot of 10,344 data randomly generated for the first round of tests. Both variables (x and u) have lognormal frequency distributions (see Fig. 4.22), since many water variables, not just in Fig. 4.15, but also more generally (see Håkanson and Peters 1995; Håkanson 1999) are lognormally distributed. Note the high skewness (MV/M50 = 6.92) and the high CV (7.01) for the u/x-ratio in Fig. 4.22. No specific filters have been applied except that neither x nor u have been permitted to become 0. In this section, two cases will be tested, u/x versus x and u/x versus u.

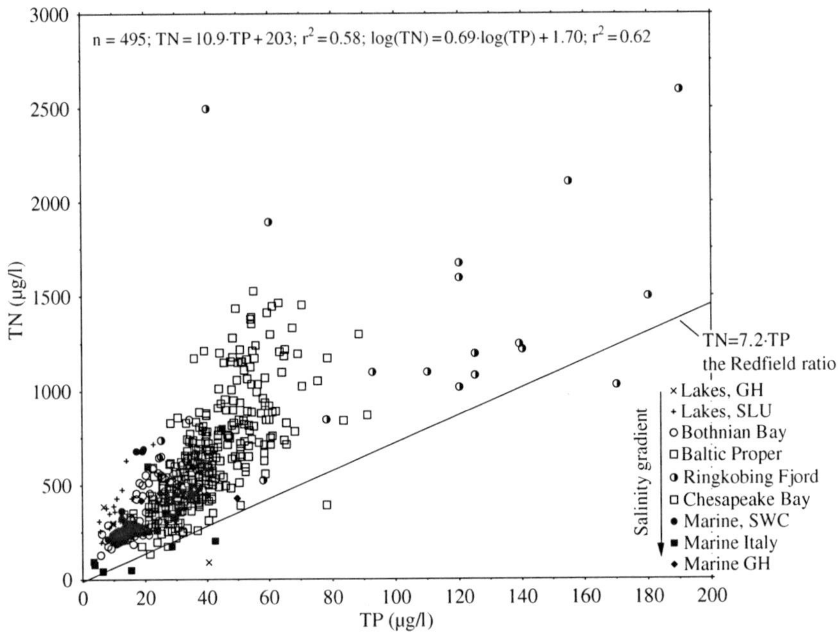

Fig. 4.20 Scatter plot between concentrations of total-P (TP) and total-N (TN) for the growing season from 9 sub-groups constituting a salinity gradient. The figure also gives regressions for the actual data and log-transformed data for the 495 data points

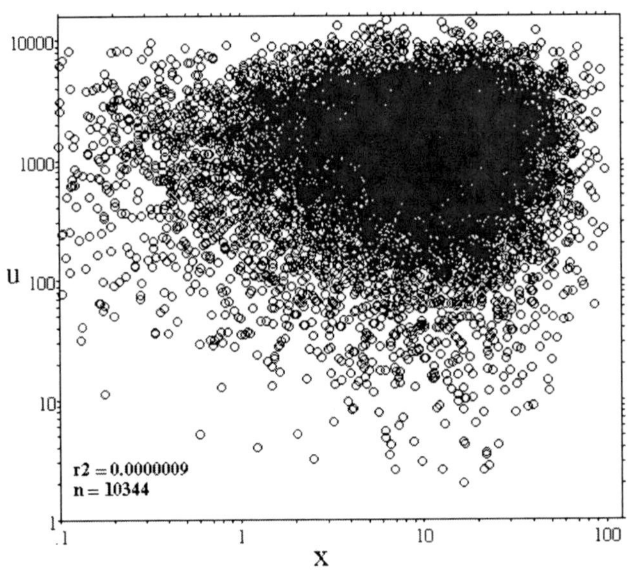

Fig. 4.21 The relationship between randomly generated u and x-values (in the first round). n = 10,344

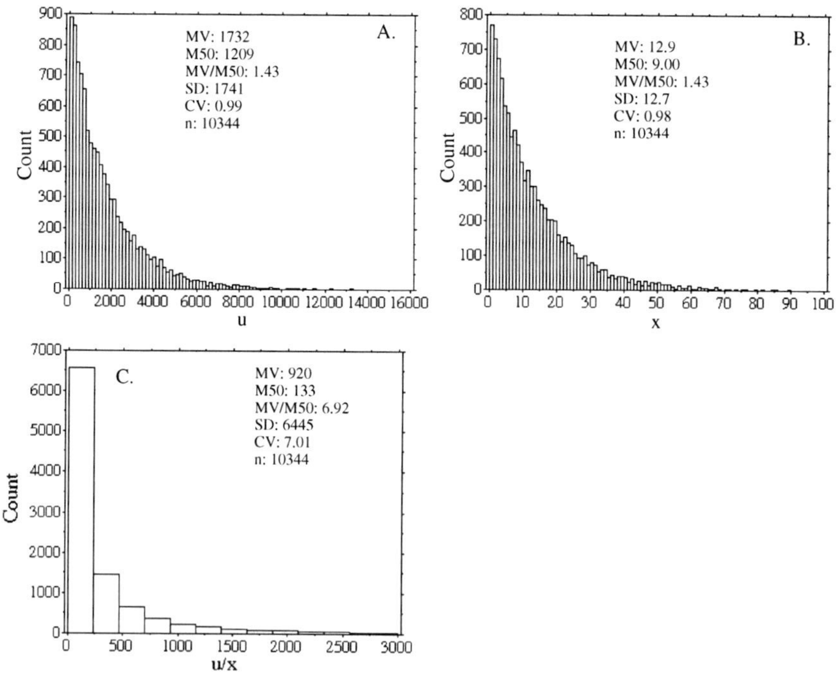

Fig. 4.22 Frequency distributions for u, x and u/x (in the first round)

The data for case one (u/x versus x) are shown in Fig. 4.23. The regression line yielding the best fit is given by:

$$\log(1+u/x) = -0.96 \cdot \log(x) + 2.97 \quad (r^2 = 0.51) \tag{4.8}$$

Very high ratios (u/x) will appear at low x-values. Equation (4.8) and the high r^2-value is, evidently, spurious.

Figure 4.24 gives the results for the second type, u/x versus u. The best-fit regression line is given by:

$$\log(u/x) = 0.98 \cdot \log(u) - 0.79 \quad (r^2 = 0.48) \tag{4.9}$$

Also (4.9) is also clearly spurious, but the relationship between u/x and u is highly significant ($p < 0.0001$). Equations (4.8) and (4.9) are given in Fig. 4.25 and compared to the TN/TP-ratio calculated from the empirical regression between TN and TP given by Fig. 5.12 and (4.10) (from 58 coastal systems; $r^2 = 0.88$):

$$\log(TN) = 0.70 \cdot \log(TP) + 1.61 \tag{4.10}$$

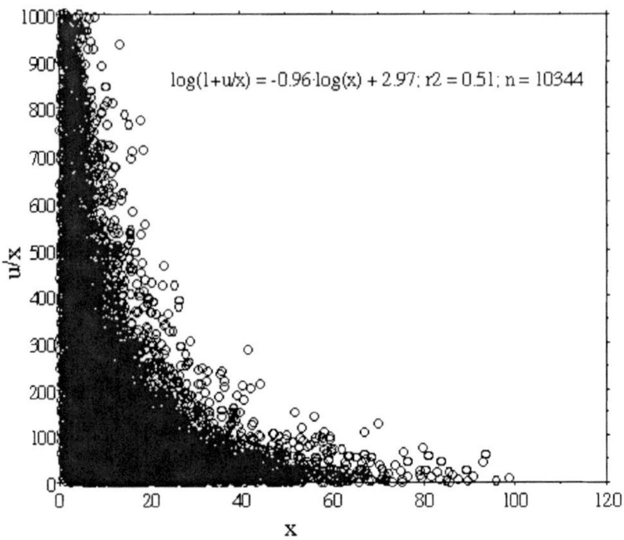

Fig. 4.23 The relationship between the ratio u/x and x (first round results). The figure also gives the best-fit regression line and the r^2-value

Figure 4.25 gives the three curves along a trophic level gradient. TP-data in the range from 1 to 300 µg/l have been recalculated into TN-values by means of (4.10). The TN/TP-ratios are shown in Fig. 4.25 (the triangles). One can note that in hypertrophic systems, i.e., systems dominated by phytoplankton, (4.10) asymptotically

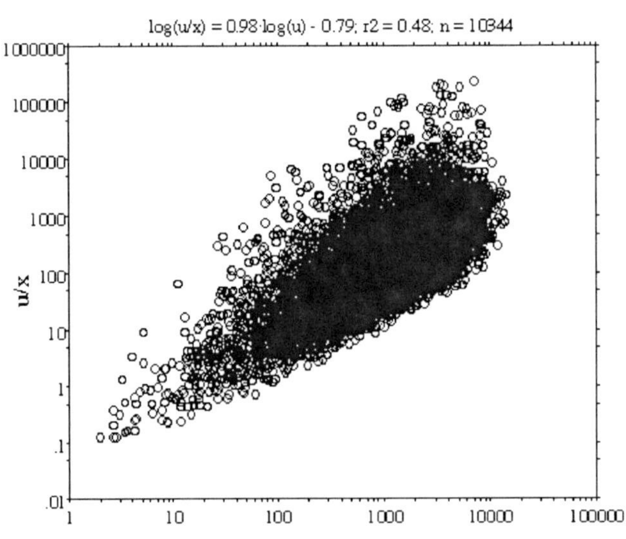

Fig. 4.24 The relationship between the ratio u/x versus u (first round results). The figure also gives the best-fit regression and the r^2-value

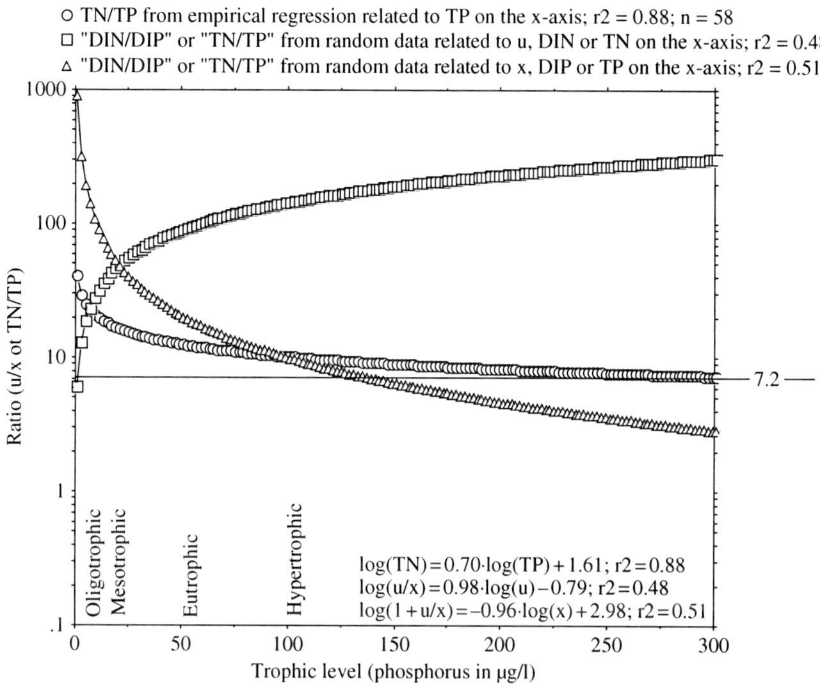

Fig. 4.25 A comparison between the ratio u/x (or randomly generated data called DIN/DIP and TN/TP) versus u (or randomly generated data called DIN or TN) and x (or randomly generated DIP and TP) using the two best-fit regressions from the first round experiment and TN/TP calculated from the empirical regression between TN and TP based on data from 493 aquatic systems covering very wide ranges in trophic level and salinity

reaches the Redfield ratio, 7.2. This demonstrates the relevance of the Redfield ratio and that (4.10) reflects the relationship between TN and TP very well (this is also shown by the high r^2-value, 0.88, for (4.10)).

From Fig. 4.25, one can also note the typical characteristics of (4.8) and (4.9). In spurious relationships of the type “u/x versus x”, high ratios appear for low x-values; and in spurious relationships of the type “u/x versus u”, the opposite is valid, i.e., low ratios for low u-values. There should be nothing confusing about this. The confusion appears when real empirical data replace u and x. One example of this (for case one) has been presented by Jeppesen et al. (2005), who discuss the scientific content of this particular relationship (TN/TP versus TP). Another example from Turner et al. (2003) concerns the second type DIN/DIP versus DIN in rivers.

The next round will illustrate results when the randomly generated data from round one have been changed by multiplication with constants.

The results behind (4.8) depend on the choice of the frequency distribution, i.e., on the definition of the mean, standard deviation, skewness and kurtosis (peakedness) of the frequency distribution. For the second round, we have altered the mean values but kept the initial characteristics of the frequency distribution in Fig. 4.22.

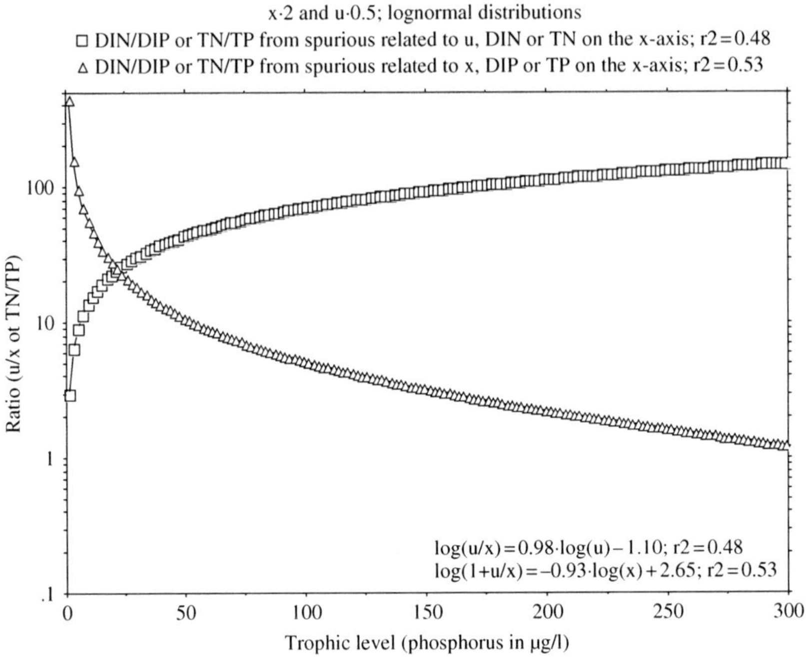

Fig. 4.26 A comparison between the ratio u/x versus u and x using the two best-fit regressions from the second round experiment when the randomly generated x-values from the first round have been multiplied by 2 and the random u-values by 0.5

All randomly generated x-data from the first round have been multiplied by 2 and all randomly generated u-data from the first round by 0.5. The two new regression lines are shown in Fig. 4.26. These results may be directly compared to the results in Fig. 4.25.

One can note the same general pattern: low u/x-ratios for low u-values and high u/x-ratios for low x-values. The range between the two regression lines (yielding r^2-values of 0.48 and 0.53, respectively) for low u or x-values is in the range from 1 to 100, which corresponds to the results presented by Dodds (2003), who concluded from studies based on empirical data in many aquatic systems that only when DIN/DIP was higher than about 100 or lower than about 1 would it be meaningful to use these ratios to discuss nutrient limitation.

The next test uses not lognormal frequency distributions but normal ones (see Fig. 4.27). Note then that the frequency distribution for the ratio (u/x) is positively skewed in spite of the fact that both u and x have normal frequency distributions and that the CV for the ratio is significantly higher (0.54) than the CVs for x (0.25) and u (0.33). The two types of relationships (u/x versus u and x) are shown in Fig. 4.28 using the randomly generated data in this case. One can note the same general pattern: low u/x-ratios for low u-values and high u/x-ratios for low x-values.

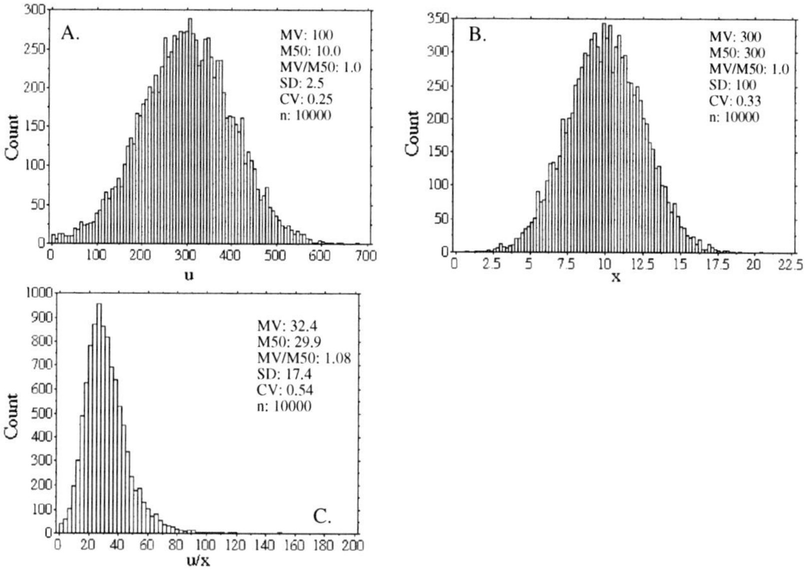

Fig. 4.27 Frequency distributions for u, x and u/x (in the third round)

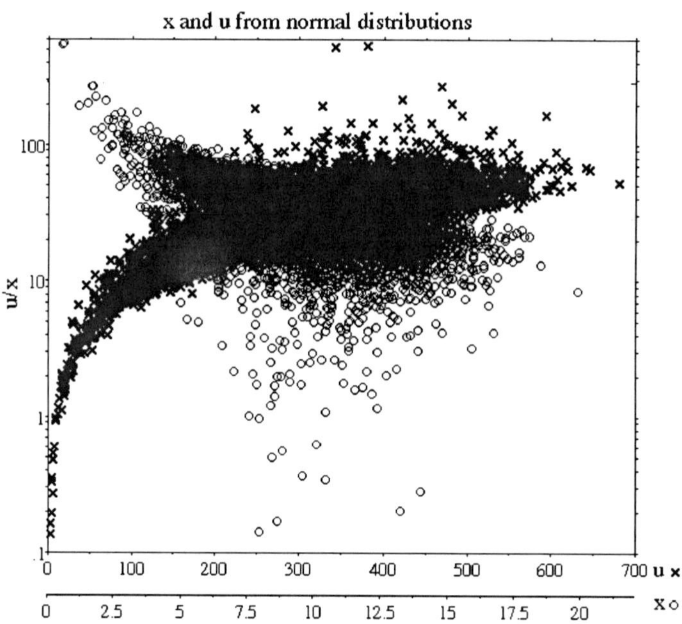

Fig. 4.28 A comparison between the ratio u/x versus u and x using 10,000 randomly generated data from normal frequency distributions (mean values for x and u = 10 and 300 with standard deviations 0.25 and 33, respectively)

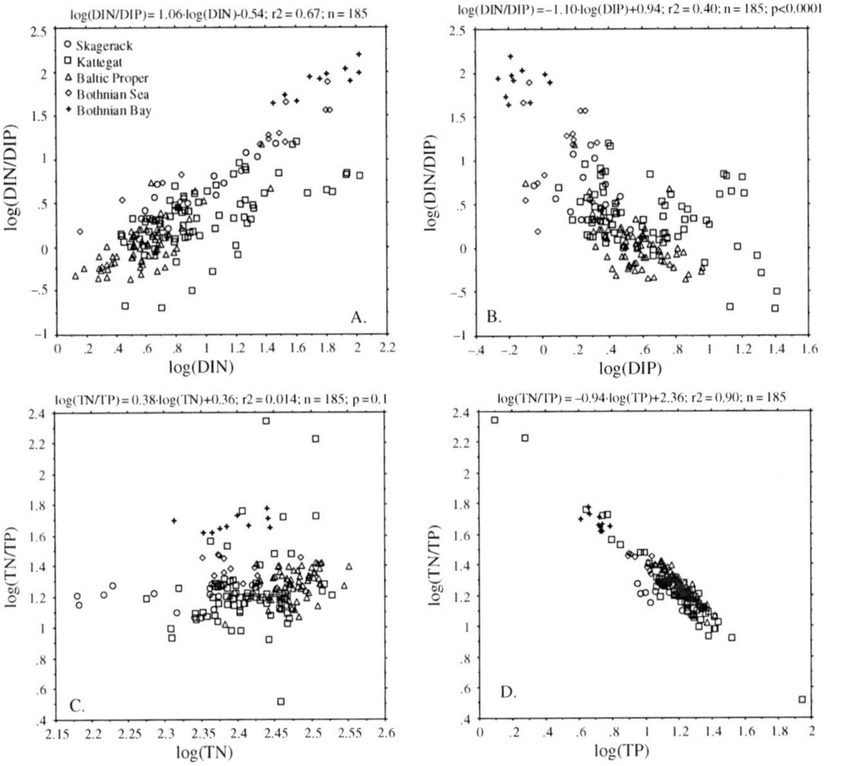

Fig. 4.29 Empirical data from the Baltic Sea, Kattegat and Skagerack on DIN/DIP and TN/TP (logarithmic values) versus empirical data (log) on DIN, DIP, TN and TP, respectively. The figure also gives the equations for the regressions and the corresponding r^2-values. All of these relationships are spurious

We will now use the data collected from sites in the Bothnian Bay, the Bothnian Sea, the Baltic Proper, Kattegat and Skagerack. For these systems, we have determined mean monthly values from 185 stations and every mean value is supported by many individual samples.

The results are shown in Fig. 4.29: (A) gives DIN/DIP versus DIN, (B) DIN/DIP versus DIP, (C) TN/TP versus TN and (D) TN/TP versus TP (all logarithmic to obtain as normal frequency distributions as possible). All these relationships are spurious. The r^2-values vary from 0.90 (TN/TP versus TP), 0.67 (DIN/DIP versus DIN), 0.40 (DIN/DIP versus DIP) to 0.014 (TN/TP versus TN), but also the last one is statistically significant at the 90% level ($p = 0.1$). Very little mechanistic understanding about "limiting nutrient" and/or the role of TN/TP or DIN/DIP-ratios can be deduced from these results – and similar results from other systems – because they are spurious and a mathematical consequence of comparing a variable on the x-axis against itself or a form including the given x-variable.

4.5 Should Nitrogen or Phosphorus be Removed to Combat Coastal Eutrophication?

Phosphorus is often recognized as the most crucial limiting nutrient for lake primary production in most but not all lakes (see Table 4.3; Schindler 1977, 1978, 2006; Bierman 1980; Chapra 1980; Boynton et al. 1982; Wetzel 2001; Persson and Jansson 1988; Boers et al. 1993). Nitrogen is regarded as a key nutrient in some marine areas (Redfield 1958; Ryther and Dunstan 1971; Nixon and Pilson 1983; Howarth and Cole 1985; Howarth 1988; Hecky and Kilham 1988; Ambio 1990; Nixon 1990; Livingston 2001).

4.5.1 Key Questions

In responding to the question of this section, there are a few general arguments or questions that apply to all coastal areas that should be addressed.

1. Is it possible to predict the outcome of a given (often costly) reduction by quantifying how the given reduction would compare to other nutrient fluxes to, within and from the given coastal area? To answer that question, one would need at least a budget of the kind exemplified in Table 4.1 for nitrogen and phosphorus for the Baltic Proper, or even better a validated mass-balance model handling all major fluxes to, within a and from the given coastal area so that the reductions and be put in a holistic process-based perspective and the environmental benefits related to the cost be properly evaluated (see the CoastMab-model in Sect. 9.1). This issue will also be discussed in a three scenarios in Chap. 6.
2. What is the basic aim of the remedial action? To increase the Secchi depth (from what to what?), the phytoplankton biomass expressed by chlorophyll-a concentrations (from what to what?), to reduce the risks of harmful algal blooms expressed by concentrations of cyanobacteria (from what to what?), to minimize the oxygen concentration in the deep-water layer during the growing season (from what to what?)? To produce quantitative goals require quantitative models. Such management models should be validated and shown to predict well. In Chap. 6, we will give case-studies to stress this quantitative aspect. In one case-study, we will specifically focus on an approach to estimate reference values or targets for remedial actions related to "good" water quality.
3. There are pro and cons related to all actions, also the action of doing nothing. What are the pros and cons related to reductions in nutrient loading? This will be discussed in another scenario in Chap. 6. There are a few classical situations. Lowering primary production or biomass by lowering the nutrient loading will often lead to increased Secchi depths (positive effect), lower concentrations of cyanobacteria (positive), but also to higher concentrations of toxins in biota (biological dilution; see Håkanson 1999) and a lower total fish production related to a lower total primary production. This will be discussed in a third scenario in Chap. 6.

4. A central question in coastal monitoring and modeling is to collect reliable data so detect possible ongoing changes, to detect critical thresholds and to suggest remedial methods to avoid such critical thresholds. Such aspects relate to the uncertainty and variability in the empirical data from the monitoring. That issue has been addressed in Sect. 4.4.
5. An important question concerning reductions in nitrogen and phosphorus is highlighted in Fig. 4.20, which is based on data from, 495 aquatic systems (from Håkanson and Eklund 2007a) covering a very wide range in salinities and TN and TP-concentrations (i.e., in trophic status). This figure gives a scatter plot where TN-concentrations are put on the y-axis and TP-concentrations on the x-axis. One can note the highly significant co-variation between the two nutrients ($r^2 = 0.58$ for actual data and 0.62 for log-transformed data). There is a large scatter among the systems, but generally the TN-concentrations are high if the TP-concentrations are high. This is mainly because these two nutrients have common sources, i.e., they are transported to the aquatic systems from land, urban areas and industries, and by the phosphorus-driven atmospheric nitrogen fixation by cyanobacteria. When there is a major difference from the general relationship shown by the regression line in Fig. 4.20, there should be specific causal reasons for this, if one first accounts for the scatter related to the inherent uncertainties in the data. This has also implications for the remedial measures since most reductions related to agricultural practices and waste-water treatment reduce both nutrient. There are few methods related to the major fluxes (see Table 4.1) that could be made substance-specific.

Only 9 out of a total of 495 cases in Fig. 4.20 have TN/TP-ratios lower than 7.2. Note that in Fig. 4.20, median values of TN and TP for the growing season have been used. The results in Fig. 4.20 concern the conditions in the surface-water layer for the entire growing season and they do *not* imply that nitrogen cannot limit phytoplankton production at specific sites and for shorter periods of time (see Savchuk and Wulff 1999a,b).

4.5.2 Arguments, Data and Results Pro and Con N and P

As stressed, plankton cells need both nitrogen and phosphorus to grow, but does that imply that remedial measures should focus on both nutrients? First, there are some impiortant reasons why remedial actions should not be directed to nitrogen.

To the best of our knowledge, there are no general, validated algorithms, which could be used within the framework of existing general well-tested mass-balance models, that can quantify nitrogen fixation either from the atmosphere or from sources within a given aquatic systems in a practically useful manner. One reason for this is the lack of well-tested, practically useful approaches to predict the concentration of nitrogen-fixing cyanobacteria. Table 4.1 exemplified that atmospheric nitrogen fixation may be very important in contexts of mass-balance calculations for nitrogen. Lacking empirically well-tested algorithms to quantify atmospheric fallout and nitrogen fixation, crucial questions related to the effectiveness of remedial measures to reduce nutrient discharges to aquatic systems cannot be properly evaluated. It also means that it is generally very difficult to understand, model and predict changes in measured TN-concentrations in the water phase, since such changes in TN-concentrations are always mechanistically governed by mass-balances, i.e., the quantification of the most important transport processes regulating the given concentrations. The problem to understand and predict TN-concentrations in marine systems is accentuated by the fact that there are no (to the best of our knowledge) practically useful models to quantify the particulate fraction for nitrogen in saltwater systems (but such approaches are available for phosphorus in lakes and brackish systems, see Sect. 9.1). In mass-balance modeling, it is imperative to have a reliable algorithm for the particulate fraction of nitrogen, since the particulate fraction (PF) is the only fraction that by definition can settle out due to gravity. From previous modeling work (see, e.g., Floderus 1989), one can conclude that it is also very difficult to quantify denitrification. Denitrification depends on sediment redox conditions, i.e., on sedimentation of degradable organic matter and the oxygen concentration in the deep-water zone, but also on the frequency of resuspension events, on the presence of mucus-binding bacteria, on the conditions for zoobenthos and bioturbation. Given this complexity, it is easy to understand why empirically well-tested algorithms to quantify denitrification on a monthly basis do not exist to the best of our knowledge. The atmospheric wet and dry deposition of nitrogen may (as indicated in Table 4.1) also be very large (in the same order as the tributary inflow) and patchy (Wulff et al. 2001a), which means that for, e.g., large coastal areas and relatively smaller systems far away from measurement stations, the uncertainty in the value for the atmospheric deposition is also generally very large. Since nitrogen concentrations in coastal systems cannot generally be predicted from nitrogen emissions, the scientific case for nitrogen abatement is very week.

Figure 4.30 gives two scatterplots between chlorophyll and DIN (A) and between chlorophyll and DIP (B). These data represent 166 monthly mean values from the surface-water layer (< 10 m), from the growing season (May to September), from the period 1989 to 2005 from the Bothnian Bay, the Bothnian Sea, the Baltic Proper, Kattegat and Skagerack.

The r^2-value between log(Chl) and log(DIN) is 0.01, which is not statistically significant ($p = 0.24$). Almost the same results can be seen for the relationship between chlorophyll and TP ($r^2 = 0.0085$, $p = 0.19$).

Table 4.15 gives results for many regressions between chlorophyll (as y-variable) and potential x-variables based on data from Ringkobing Fjord (from Bryhn et al. 2007). From these results, one may conclude:

- The results depend very much on the season of the year; the best results are generally obtained for data from the summer period.
- Better correlations were obtained for log-median values than for log-mean values (data not displayed) because most frequency distributions for most water variables are not normal but log-normal.

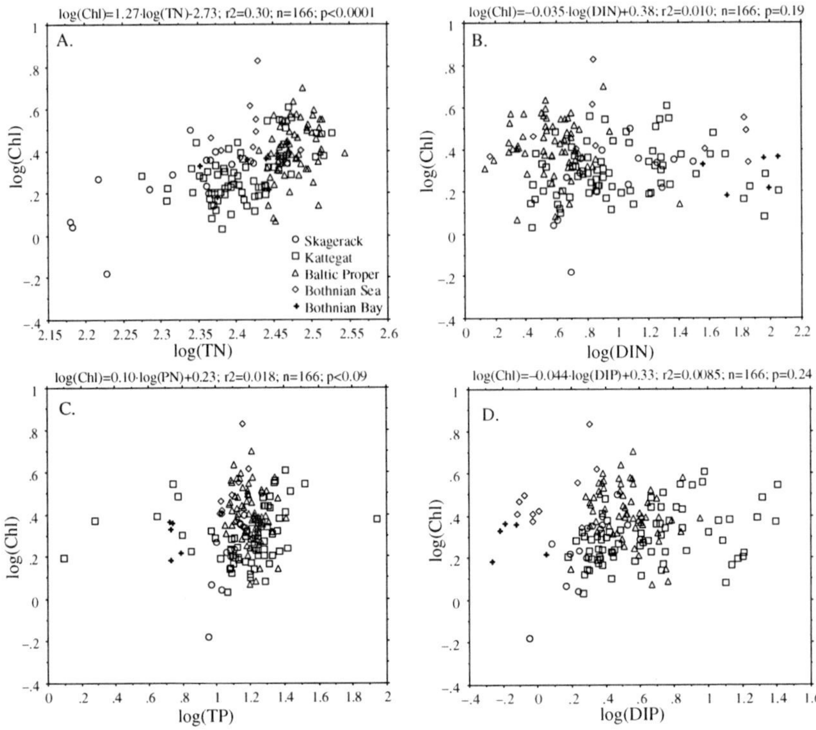

Fig. 4.30 Empirical data from the Baltic Sea, Kattegat and Skagerack on mean monthly chlorophyll-a concentrations (logarithmic data) versus empirical data (log) on DIN and DIP, respectively. The figure also gives the equations for the regressions and the corresponding r^2-values

- There are major differences among the x-variables in how they correlate to chlorophyll in this lagoon. TP is by far the best predictor for chlorophyll.
- Nitrogen or ratios based on nitrogen or different forms of nitrogen generally co-vary with the two target bioindicators less well than TP in this lagoon.

Table 4.16 gives similar results relating chlorophyll and Secchi depth to various nutrient forms (DIN, DN, DON, DOP, OrtP, PN, PP, TN, and TP) in Chesapeake Bay. In this estuary, TP is not always superior to TN as a predictor as it was in Ringkobing Fjord. Those nutrients or nutrient fractions which were the best predictors were generally total or particulate nitrogen or phosphorus, followed by dissolved organic nutrients and dissolved inorganic nutrients. From the results in Table 4.17, it seems that various fractions of both nitrogen and phosphorus are needed to explain areal variations in Chl and Secchi depth in Chesapeake Bay. Given the high inherent variability in chlorophyll in Chesapeake Bay (a CV of 0.77; see Bryhn et al. 2007), one can conclude that Chl cannot be predicted with much higher certainty than in Table 4.16 (r^2= 0.79).

To put these regressions between nutrients and bioindicators of eutrophication into a wider comparative context, it is necessary to use data from many systems.

Table 4.15 Relationships between different nutrients, different forms of nutrients and nutrient ratios versus chlorophyll-a in Ringkobing Fjord. Six different averaging methods have been used on median log values: (1) annual values of all data, (2) all data adjusted to give equal weight to each of the four seasons, (3) spring values, (4) summer values, (5) autumn values and (6) winter values. All correlations are based on log-values. Data are from the period 1986–2004, although some series that only include TP, NH_X and Chl also cover 1985 (except for winter values)

	All data	Season adjusted	Spring	Summer	Autumn	Winter
r^2 (TP vs Chl)	0.96 p	0.94 p	0.86 p	0.92 p	0.93 p	0.56 p
r^2 (OrtP vs Chl)	0.23 n	0.32 n	0.03 n	0.31 p	0.57 m	0.30 m
r^2 (TN vs Chl)	0.33 p	0.24 p	0.06 n	0.76 p	0.58 p	0.13 n
r^2 (NO_X vs Chl)	0.50 m	0.06 n	0.00 n	0.45 m	0.27 m	0.00 n
r^2 (NH_X vs Chl)	0.42 m	0.59 m	0.55 m	0.23 m	0.52 m	0.35 m
r^2 (DIN vs Chl)	0.56 m	0.09 n	0.00 n	0.21 n	0.33 m	0.02 n
r^2 (TN:TP vs Chl)	0.81 m	0.77 m	0.73 m	0.73 m	0.63 m	0.46 m
r^2 (DIN:OrtP vs Chl)	0.45 p	0.00 n	0.01 p	0.55 p	0.15 p	0.33 p

Significance at the $p < 0.05$ level; p = positive, m = negative, n = not significant, r^2 = coefficient of determination, TP = total phosphorus, TN = total nitrogen, OrtP = orthophosphate, Chl = chlorophyll, NO_X = nitrate + nitrite, NH_X = ammonium + ammonia, DIN = dissolved inorganic nitrogen = NO_X + NH_X. Due to changes in methods data for Chl, for OrtP and for TN, NO_x, DIN are more reliable from October 1984, from January 1986 and from May 1986, respectively.

In such a comparative study, Guildford and Hecky (2000) found a much stronger correlation (r^2= 0.60 compared to 0.08) between TP and Chl than between TN and Chl at several ocean sites. Rather similarly, Håkanson et al. (2007c) found that TP and salinity in combination correlated slightly more strongly with Chl (r^2= 0.71) than TN and Sal (r^2= 0.68) in a wide range of aquatic systems. Conversely, Smith (2006) found TN to be a better predictor of Chl than TP (r^2= 0.84 compared to 0.60) and that TN and TP are strongly mutually correlated (r^2= 0.55).

Table 4.16 r^2-values for chlorophyll and Secchi depth versus nutrients from surface water variables in Chesapeake Bay (data from months 6–8)

	DIN	DN	DON	DOP	DP	OrtP	PN	PP	TN	TP	Chl
Chl	0.01 n	0.06 p	0.27 p	0.26 p	0.03 p	0.00 n	0.69 p	0.52 p	0.32 p	0.22 p	
Sec	0.07 m	0.31 m	0.34 m	0.23 m	0.12 m	0.14 m	0.36 m	0.51 m	0.50 m	0.38 m	0.38 m

Significance at the $p<0.05$ level; p = positive, m = negative, n = not significant; D for dissolved, N for nitrogen, P for phosphorus, O for organic, Ort for orthophosphate, PP for particulate phosphorus, Chl for chlorophyll.

Table 4.17 Stepwise multiple regression models based on surface water data from Chesapeake Bay (months 6–8)

y-variable	n (stations)	Step	r^2	x-variable	Model
log (Chl)	191	1	0.69	x_1=log(PN)	$y=1.548+1.022 \cdot x_1$
		2	0.79	x_2=log(PP)	$y=1.530+0.7614 \cdot x_1+0.4445 \cdot x_2$
log (Sec)	186	1	0.57	x_1=log(SPM)	$y=1.596-0.6341 \cdot x_1$
		2	0.75	x_2=log(TN)	$y=3.410-0.4692 \cdot x_1+0.6927 \cdot x_2$
		3	0.76	x_3=log(DOP)	$y=3.113-0.4765 \cdot x_1-0.5169 \cdot x_2-0.1929 \cdot x_3$

These regressions clearly demonstrate that concentrations of dissolved inorganic nutrients are of very limited use in predictive coastal management. Even though batch experiments in laboratories often show that dissolved inorganic nutrients is what phytoplankton need, concentrations of dissolved inorganic nutrients poorly reflect their availability since they are very rapidly regenerated (Dodds 2003). Instead, Chl can be predicted much better from TN, TP, PN or PP. Other disadvantages with DIN and DIP were discussed in Sect. 4.2.

Using median monthly TN/TP-values in the Baltic Proper, Fig. 4.31 shows that this ratio has been higher than 7.2 all months since 1994. Compilations of extensive empirical data by Håkanson et al. (2007c) have demonstrated that there is an increasing risk of harmful algal blooms (of) when the TN/TP-ratio decreases below 15. This means that 15 may be regarded as a threshold value in these contexts and the TN/TP-ratio in the Baltic Proper is often between 7.2 and 15.

There are also clearly increasing risks of harmful algal blooms if the water temperatures increase above 15°C (Edler 1979; Wasmund 1997; Håkanson et al. 2007c). Given the situation in the Baltic Proper, as revealed by these empirical data, a lowering of the TN/TP-ratio would likely imply greater risks for blooming of

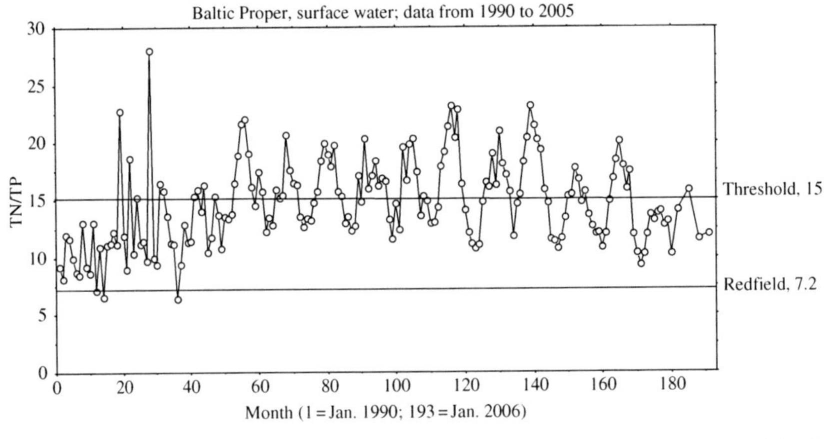

Fig. 4.31 Variations in median monthly TN/TP-ratios in the surface-water zone of the Baltic Proper from 1990 to 2005 in relation to the Redfield ratio of 7.2 and the threshold ratio of 15

cyanobacteria. So, from this perspective, to reduce nitrogen inflow to the Baltic Proper is a counterproductive and expensive approach to improve the conditions in the Baltic Proper. The focus should be set on reductions of the major anthropogenic fluxes of phosphorus.

Using the model for cyanobacteria given by Håkanson et al. (2007c) for the conditions in the Baltic Proper, one can also illustrate (see Fig. 4.32) what would likely happen if the four x-variables in this model (i.e., the monthly median values of temperature, salinity, TN and TP) are reduced by 25%. A *reduction* in TN would significantly *increase* the predicted biomass of total cyanobacteria considerably, while reductions in salinity, temperature and TP would lower the predicted values; the clearest response would be from reductions in temperature. According to these results, the high values of cyanobacteria in 2005 may be attributed to relatively high TP-concentrations and high temperatures in the summer and fall that particular year (see also Hansson 2006).

Figure 4.33 gives another angle to the problem of using the TN/TP or the DIN/DIP-ratios in contexts related to "limiting" nutrient and if remedial actions should focus on nitrogen or phosphorus. The data in Fig. 4.33 emanate from the Himmerfjärden Bay on the Swedish east coast in the Baltic Proper. From the TN/TP-data, one could be tempted to say that nutrient reductions should focus on phosphorus because the TN/TP-ratio is higher than the Redfield ratio of 7.2; using the DIN/DIP-ratio, the argument could be that nutrient reductions should focus on

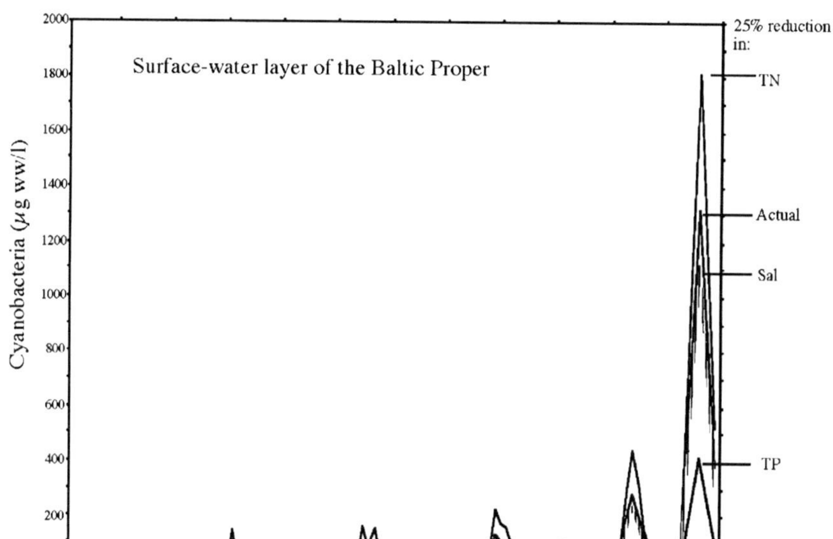

Fig. 4.32 Simulations of how 25% reductions in total-N, total-P, water temperature and salinity would likely influence the concentrations of cyanobacteria in the surface-water layer of the Baltic Proper using data from 1996 to 2005

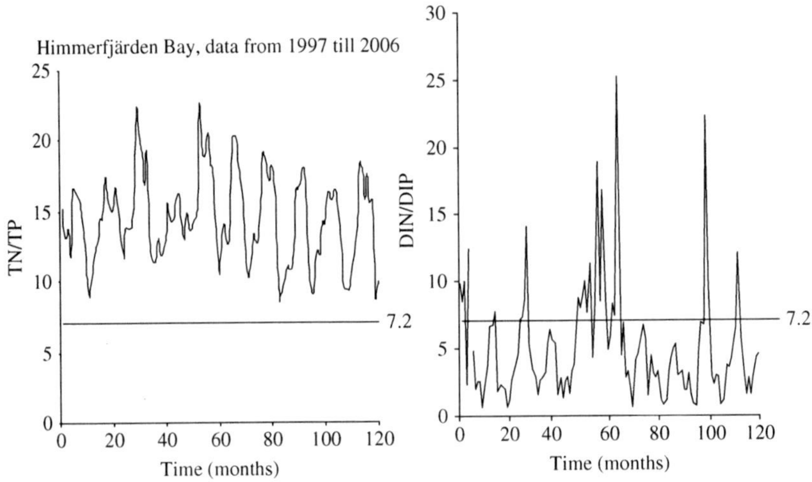

Fig. 4.33 Monthly mean values on the TN/TP-ratio and the DIN/DIP-ratio from the surface-water layer in the Himmerfjärden Bay (Swedish Baltic Proper) using data from 1997 to 2006

nitrogen because the DIN/DIP-ratio is often lower than 7.2. Our point is that neither of these arguments are valid because the primary *production* (biomass per time unit) is not regulated by concentrations or ratios based on concentrations but by the availability (emissions plus regeneration) of DIN and DIP.

Figure 4.34 illustrates the development in Ringkobing Fjord, Denmark between January 1989 and January 2005. According to the theory related to the Redfield ratio, one would have expected that there would have been higher concentrations

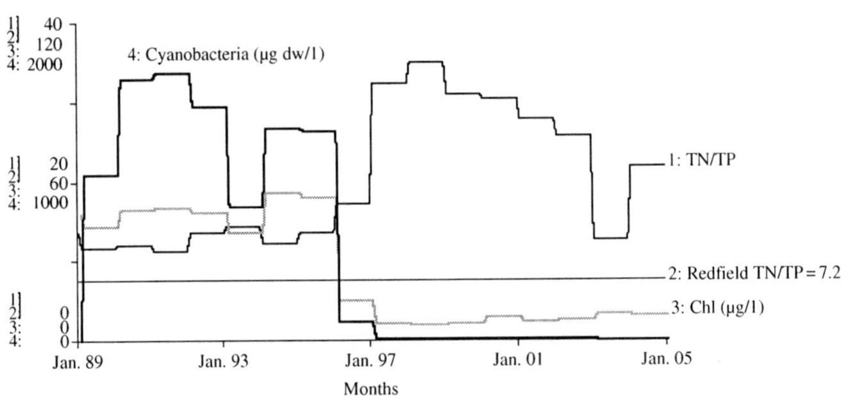

Fig. 4.34 Variation in the TN/TP-ratio (curve 1), empirical median annual data in chlorophyll-a concentrations (curve 3) and empirical annual data on the concentration of cyanobacteria in Ringkobing Fjord (curve 4) in the years from January 1989 to January 2005. The figure also gives the Redfield ratio (TN/TP = 7.2). Ratios are dimensionless and units of the other variables are stated in the figure

of cyanobacteria during periods when the Redfield ratio is lower than 7.2, since this would have favored algae which can take up and utilize atmospheric nitrogen (Sellner 1997; Gruber and Sarmiento 1997). Figure 4.34 shows that there is no clear relationship between the TN/TP-ratio (curve 1 gives measured annual median values since there are no monthly or seasonal data available to us on cyanobacteria from the growing season in this lagoon) and empirical chlorophyll concentrations (curve 3, annual values) or concentrations of cyanobacteria (curve 4, annual values). The TN/TP-ratio is never lower than 7.2 (curve 2 gives the Redfield ratio). In this lagoon, there were no cyanobacteria when the TN/TP-ratio was higher than 15.

4.6 Example of Mass-Balance and Foodweb Modeling in a Target Ecosystem at the Local Scale

The idea here is to give a case-study using data from a real coastal area to exemplify that data on morphometry (coastal area, mean depth, maximum depth, section area, dynamic ratio and form factor) may be used in a mass-balance model for phosphorus (CoastMab; see Sect. 9.1) and in the foodweb model for functional groups of organisms (CoastWeb; see Håkanson and Lindgren 2007b). The idea of this eutrophication scenario is to study how a hypothetical stepwise (3-year steps) increase of the TP-concentration in the sea outside a coastal area would likely influence a given local coastal ecosystem. In this case, data have been used from the Haverö coastal area in the Finnish Archipelago Sea (latitude 61 N); note that the data below (from Wallin et al. 1992) are sufficient to run the model and that all of these data may be accessed from the system discussed in this work.

Water area (km^2)	D_{max} (m)	D_m (m)	At (km^2)	Chl (μg/l)	Salinity (psu)	C_{TPsea} (μg/l)	Sec_{sea} (m)
2.3	22.5	8.6	0.0172	2.1	6.5	24	3.4

D_{max} denotes the maximum depth; D_m the mean depth; At the section area; Chl the concentration of chlorophyll-a; C_{TP} the TP-concentration in the sea outside the coastal area; and Sec is the Secchi depth in the sea outside the coastal area.

Results of the simulations are presented in Fig. 4.35. The default TP-concentration in the sea outside this coastal area is 24 μg/l, and tests have been carried out to see of how values of 0.75·24, 24, 1.5·24 and 2·24 would change modeled values of (A) TP within the coastal area for, (B) chlorophyll, (C) Secchi depth, (D) oxygen saturation in the deep-water zone, (E) the normal and actual biomasses of zoobenthos, (F) herbivorous zooplankton, (G) prey fish and (H) predatory fish. Modeled values of TP, chlorophyll, Secchi depth and oxygen saturation are also compared with empirical data and the uncertainty bands for the empirical data (MV = mean value, SD = standard deviation). One can note:

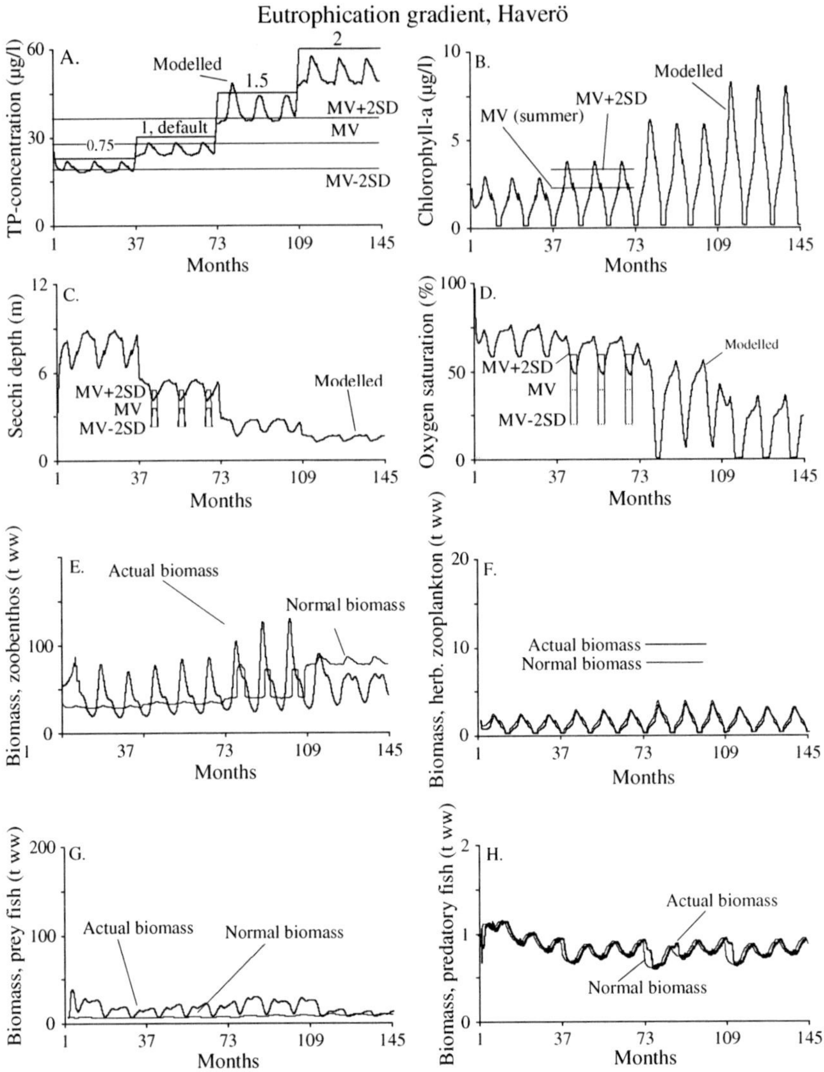

Fig. 4.35 Case-study on coastal eutrophication using data from the Haverö coastal area. There are changes in 3-year steps in the TP-concentration in the sea adjacent to the coastal area. The default TP-concentration in the sea is 24 $\mu g\ l^{-1}$ and this value has been set to 0.75·24, 24, 1.5·24 and 2·24 (i.e., 18, 24, 36 and 48 $\mu g\ l^{-1}$) and the consequences calculated for (**A**) the TP-concentration in the given coastal area, (**B**) chlorophyll, (**C**) Secchi depth, (**D**) oxygen saturation in the deep-water zone (all compared to empirical mean values and inherent uncertainties in the mean values; the chlorophyll mean value is for the summer period) and actual and normal biomasses of (**E**) zoobenthos (**F**) herbivorous zooplankton, (**G**) prey fish and (**H**) predatory fish

- There is generally very good correspondence between modeled and empirical data. Note that the empirical chlorophyll value is the mean value for the growing season.
- The increased eutrophication of the sea outside the coastal area will drastically increase TP also in the given coastal area (which is logical because the retention time of the surface water in Haverö is only 3–5 days, calculated from the empirical model in Table 3.2). This leads to higher chlorophyll values, reduced water clarity (Secchi depth) and significantly lower oxygen saturation in the deep-water zone, which will influence zoobenthos more than, e.g., zooplankton.
- Since there is much more zoobenthos in the system than zooplankton (about 50 tons ww as compared to about 3–5 tons ww), zoobenthos is an important source of food for coastal prey fish and reductions or changes in zoobenthos biomass will have clear effects on the prey fish, and changes in prey fish biomass will in turn influence the predatory fish, which eat prey fish.
- The zoobenthos within the areas of continuous fine sediment accumulation will die if the oxygen saturation is lower than 20%, but the oxygenation of the areas dominated by fine sediment erosion and transport above the theoretical wave base will maintain a low biomass of zoobenthos in the more shallow parts of the coastal area.

So, the increased eutrophication will imply several changes to the water quality and foodweb characteristics of this coastal area. Many of these changes could be expected without a model, but the point here is that they have been quantitatively predicted using a general comprehensive foodweb model that includes a dynamic mass-balance model for phosphorus. This foodweb model (CoastWeb) accounts for many abiotic and biotic interactions and feedbacks and it is meant to give the "normal" response of the system to the given change in the TP-concentration in the sea. The model accounts for different types of compensatory effects that would be difficult to quantify without a model. Note that, the CoastWeb-model simulates functional groups and hence does not include responses related to individual species (see Håkanson and Lindgren 2007b, for more details).

4.7 Summary and Conclusions

If the scientific task is to gain better understanding about how aquatic systems work, there are few more rewarding avenues than comparative studies and this chapter has used data from many aquatic systems. We would like to stress that the selected examples are meant to illustrate general principles.

The focus in this chapter has been on simple statistical techniques to evaluate data and uncertainties in data, which may be used to detect and quantitatively evaluate possible thresholds and to run and test predictive models for sustainable coastal management. We have addressed problems related to the balance between the changes in predictive power and the accumulated uncertainty as model grow in size, inherent uncertainties in empirical data, regressions and approaches to maximize predictive power of regression models, variations in standard water variables within and among coastal systems. We have also discussed the concept "Optimal Model Scale" (OMS), and an algorithm to calculate OMS, which accounts for the most important factors related to predictive power at different time scales (daily to yearly predictions).

TN, TP, TN/TP and DIN/DIP are key variables in practically all contexts of aquatic sciences and management dealing with eutrophication, primary production, phytoplankton biomass, "limiting nutrient" and predictions of bioindicators. The focus in this chapter has been set at the ecosystem scale, i.e., on the conditions in entire coastal areas and on seasonal/monthly changes (rather than variations at smaller temporal scales). So, the focus is on the conditions in entire coastal areas, which may be reflected in data from several sample sites. The ecosystem scale is also the scale of main interests in water management and in discussions on remedial measures and strategies.

This work has presented CV-values and patterns in CV-values. There are very high CVs for DIN during the summer period, very high CVs for the DIN/DIP-ratio compared to the TN/TP-ratio and the very high CVs for the DIN/DIP-ratio in the marine areas compared to lakes. In the literature, there are many papers discussing how the relationship between DIN/DIP and TN/TP vary among systems of different trophic level neglecting the hazards of spurious correlations.

High CVs imply that one must take many samples to obtain representative mean values for the given system and it also means that one must take samples from many systems in regressions where the aim is to find more generally how a given x-variable with a high CV-value may influence a target y-variable, which could also have a high inherent CV-value.

Given the inherently high CV-values for many key variables in eutrophication studies, it must be stressed that more samples than generally taken in most regular monitoring programs are needed if scientific unassailable conclusions are to be made concerning interrelationships among the variables and to produce scientifically meaningful information to detect critical ecosystem changes.

The random parameter tests have highlighted the problem of spurious correlations related to ratios in contexts of eutrophication, i.e., TN/TP- and DIN/DIP-ratios versus primary production. It has been shown that the patterns in spurious relationships can be identical to patterns in empirical data that have been interpretated in mechanistic terms. The ratios TN/TP and DIN/DIP have higher CVs than TN, TP, DIN and DIP individually. This explains why DIN and DIP generally are poor predictors in contexts of nutrient limitation and in modeling of primary production. Based on empirical data from many systems, Dodds (2003) argued that when the levels of DIN are much higher than the levels of DIP (e.g., 100/1), it is unlikely that DIN is limiting and only if DIN/DIP < 1, it is unlikely that DIP is the limiting nutrient. These random parameters tests support those arguments.

Nutrient reductions for coastal areas should generally focus on phosphorus rather than on nitrogen. Reductions should be done in a cost-efficient manner, which means that one should focus primarily on the major anthropogenic emission sources. The main reasons for combating coastal eutrophication with phosphorus abatement are:

- TP-concentrations in the water can be predicted from TP-emissions by means of general dynamic models, but it is neither possible to predict TN-concentrations, nor the effects of reduced TN-emissions, in a general manner.
- Lower TN-concentrations increase the risk of cyanobacterial blooms.
- Arguments for nitrogen abatement are often backed by experiments with DIN and DIP – variables which for mechanistic and statistical reasons provide very scant information that may be relevant for management purposes.
- Primary production in most coastal zones seems to be limited by phosphorus in the long-term (monthly, annual and multi-annual) time perspective.

These results will hopefully appear clear, sound and reliable to many oceanographers, marine geologists, coastal engineers and biologists who have worked with different types of aquatic ecosystems. Some marine biologists may, however, have a hard time accepting eutrophication models based on phosphorus as the key nutrient. We would still advise those colleagues to continue reading the book and regarding TP as a proxy variable of "nutrients".

Chapter 5
Operational Bioindicators for Coastal Management

Operational bioindicators for water management should be comprehensible without expert knowledge. In fact, one reason to develop such measures is so that politicians and the general public can understand the present condition and possible future changes in the environment. All bioindicators should not just be understandable by coastal managers and scientists, it should also be possible to predict them using validated models and to calculate changes in the bioindicators and relate such changes to changes in fundamental abiotic driving variables, such as salinity, temperature, nutrient loading/concentrations and water exchange.

5.1 Secchi Depth and Suspended Particulate Matter (SPM)

To most people, clear waters with few suspended particles and with large Secchi depths seem more attractive than turbid waters. Many factors are known to influence the Secchi depth (the "sight" depth where a white disc lowered through the water is lost from eye sight; Secchi 1866; Vollenweider 1958, 1960; Carlson 1977, 1980; Højerslev 1978; Preisendorfer 1986). The Secchi depth is a direct reflection of the amount of SPM scattering light in the water and hence closely related to concepts like turbidity and water clarity and it is also used as a key component in an algorithm expressing a water quality index for coastal waters (see Vollenweider et al. 1998). SPM and Secchi depth depend on, (1) autochthonous production (the amount of plankton, detritus, etc. produced in the coastal water; more plankton, etc. mean a lower Secchi depth); (2) allochthonous materials (such as humic, fulvic and minerogenic substances supplied to the coastal area from outside sources, such as tributaries); and (3) the amount of resuspended material (materials resuspended from the bed via wind/wave activity, slope processes, etc.). These factors are not independent: High sedimentation leads to high amounts of resuspendable materials; high resuspension leads to high internal loading of nutrients and increased production; a high amount of colored substances in estuaries means a smaller photic zone and a lower production; a high input of allochthonous substances and a high production would mean a high sedimentation, etc. Wallin et al. (1992) showed that the Secchi depth should be much greater than that observed if only plankton cells were responsible for the light extinction. This observation means that particles other than plankton cells may be the most important factors for determining the Secchi depth in many coastal areas. It was also concluded that the empirical relationship between Secchi depth and chlorophyll-a largely depends on the chlorophyll-a concentration co-varying with the total amount of suspended particles. This correlation has been discussed also in other studies (Kiefer and Austin 1974; Tilzer 1988). The maximum depth of the photic zone is generally set to about two times the Secchi depth, and the effective (or mean) depth of the photic zone, a key regulator of macrophyte expansion, is often set equal to the Secchi depth (see Håkanson 2006).

Several studies have quantified and ranked variables of significance to predict how Secchi depths vary among water systems (see Wallin et al. 1992; Nürnberg and Shaw 1998). Table 5.1 gives empirical models for the Secchi depth (mean values for the growing season) using data from 23 Baltic coastal areas. In this brackish estuary, where the range in the data is rather narrow, one can note the significant relationships between Secchi depth and total-N (TN) and coastal morphometry.

The ladder in Table 5.1 predicts Secchi depth in three steps:

- First from mean TN-concentrations for the growing season ($r^2 = 0.83$; r^2 = the coefficient of determination); the higher the TN-concentration, the higher the primary production, the more turbid the water and the lower the Secchi depth.
- Then from the form factor (Vd; $Vd = 3 \cdot D_m/D_{max}$; D_m = the mean coastal depth in m; D_{max} = the maximum coastal depth in m; $r^2 = 0.89$); Vd relates to resuspension; if the coastal area is very shallow, i.e., if Vd attains a small value, the coastal area will be dominated by resuspension areas with relatively coarse sediments and little fine sediments and little resuspension of SPM. So, the Secchi depth will be high. The main reason for this seems to be that for open coastal areas the fine suspended particles will be transported out of the area and not be trapped in the same manner as in coastal lagoons.
- The last step gives the section area (At in km^2) as a model variable; then r^2 reached 0.91. If At is large, the water exchange between the coastal area and the sea will be very dynamic and one should expect that the Secchi depth in the coastal area would be close to the Secchi depth in the sea.

Table 5.1 Results of the stepwise multiple regression for Secchi depth (Sec; mean value for the growing season) using data from a Baltic coastal database (see Wallin et al. 1992). n = 23 coastal areas; y = Secchi depth (m). TN = total-N concentration (µg/l); At = section area (m^2); r^2 = the coefficient of determination (r = the correlation coefficient). Vd = the form factor (volume development defined by $Vd = 3 \cdot D_m/D_{max}$; D_m = the mean depth, D_{max} = the maximum depth)

Step	r^2	x-variable	Model
1	0.83	TN	$y = -0.033 \cdot x_1 + 14.5$
2	0.89	Vd	$y = -0.033 \cdot x_1 - 1.65 \cdot x_2 + 16.3$
3	0.91	At	$y = -0.033 \cdot x_1 - 1.51 \cdot x_2 + 18.3 \cdot x_3 + 15.9$

These empirical regressions for Secchi depth demonstrate that there are some morphometric parameters (Vd, At and D_m) that are important in understanding and predicting how variables like Secchi depth, and hence also SPM and oxygen saturation/concentration vary among coastal areas (see next section). Numerous factors can potentially influence Secchi depth. It is easy to speculate and qualitatively discuss such relationships. With statistical analyses based on empirical data, it is possible to quantitatively rank such influences and derive predictive models based on just a few, but the most important factors influencing variability in mean Secchi depth among coastal areas.

A "natural" or reference Secchi depth may be estimated from the models given in Table 5.1. Such a prediction would account for differences in coastal morphometry and TN. If the actual mean long-term Secchi depth differs from such a predicted reference value, such divergences may be discussed in a quantitative manner. Note that the empirical model in Table 5.1 only apply for coastal areas of the same type and within the ranges of the variables used in the statistical analyses. In other words, empirical models like these are not generally valid. But within their range of application, empirical models often give better predictions than dynamic models (Ahlgren et al. 1988; Peters 1991). Furthermore, it is likely that the same basic principles concerning factors influencing water clarity and Secchi depth apply for most coastal areas, although the parameter constants may have to be altered to predict Secchi depth for coasts of other types.

Figure 3.25 gave a nomogram illustrating the relationship between Secchi depth, SPM and salinity (from Håkanson 2006). This nomogram has been constructed using (5.1).

$$\begin{aligned}\mathrm{SPM_{SW}} = 10\wedge(-0.3 - 2\cdot(\log(\mathrm{Sec}) - (10\wedge(0.15\cdot\log(1+\mathrm{Sal_{SW}})+0.3)-1))/\\((10\wedge(0.15\cdot\log(1+\mathrm{Sal_{SW}})+0.3)-1)+0.5))\end{aligned} \tag{5.1}$$

SPM regulates the two major transport routes, the dissolved transport in the water (the pelagic route) and the particulate sedimentation (or benthic) route, of all types of materials and contaminants (Håkanson 2006). SPM is directly related to many variables of general use in water management as indicators of water clarity, e.g., the depth of the photic zone (see Jørgensen and Johnsen 1989; Wetzel 2001; Kalff 2002; Panagiotopoulos and Sempere 2005). Suspended particles will settle out on the bottom and the organic fraction will be subject to bacterial decomposition (= mineralization). This will influence the oxygen concentration in the sediments and hence also the survival of zoobenthos, an important food for fish. SPM influences primary production of phytoplankton, benthic algae, macro algae and macrophytes, the production and biomass of bacterioplankton, and hence also the secondary production, e.g., of zooplankton, zoobenthos and fish. The effects of SPM on recycling processes of organic matter, major nutrients and pollutants determine the ecological significance of SPM in any given aquatic environment. Understanding the mechanisms that control the distribution of SPM in aquatic systems is an issue of both theoretical and applied concern, as physical, chemical and biological processes ultimately shape aquatic ecosystems (Håkanson 2006).

Many models for SPM are basically meant to be sub-models in, e.g., ecotoxicology and radioecology to quantify fluxes of toxic substances (see, e.g., Monte 1997). Many contaminants (such as radionuclides, heavy metals and nutrients) may be removed from the water column due to their sedimentation with SPM and burial in the bottom sediments (see IAEA 2000). So, SPM is an important variable in water management and aquatic ecology, but the main focus may not be on SPM but on bioindicators influenced by SPM, such as the Secchi depth (see (5.1)) and the oxygen saturation in the deep-water zone (see Sect. 5.2).

Another version of the nomogram between Secchi depth, SPM and salinity is shown in Fig. 5.1. The curves for the salinities 0, 5 and 20 psu are shown in Fig. 5.1

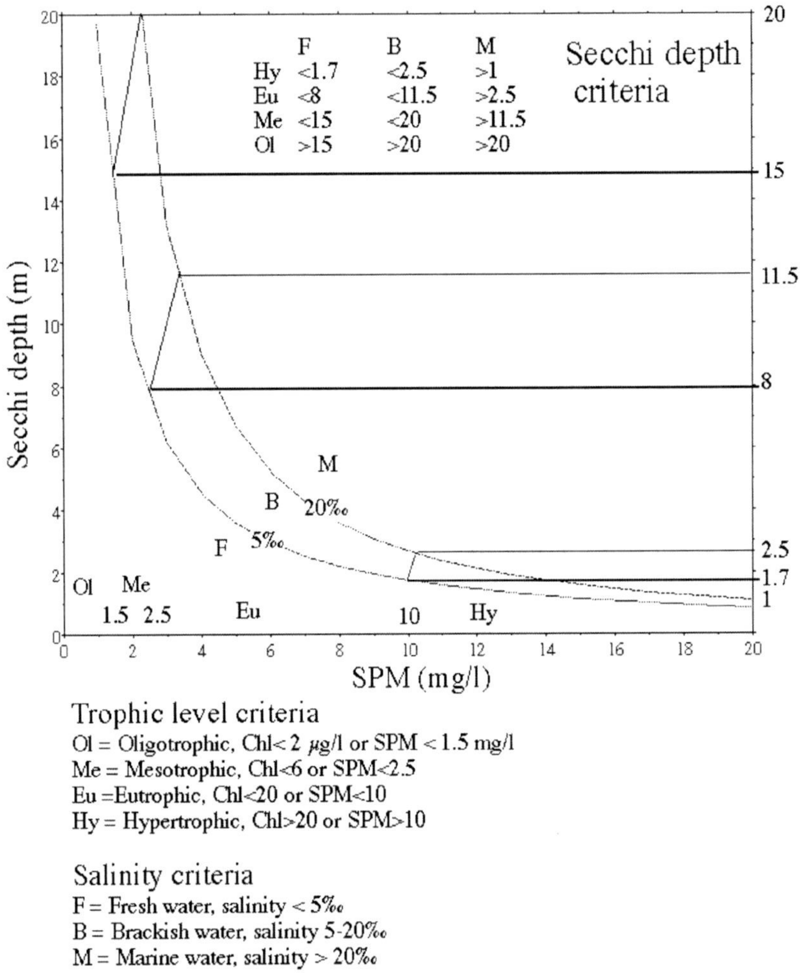

Fig. 5.1 Illustration of the relationship between Secchi depth, SPM in surface water and salinity in surface water and trophic categories for fresh water systems (salinity < 5 psu), brackish systems (salinity 5–20 psu) and marine coastal systems (salinity > 20 psu)

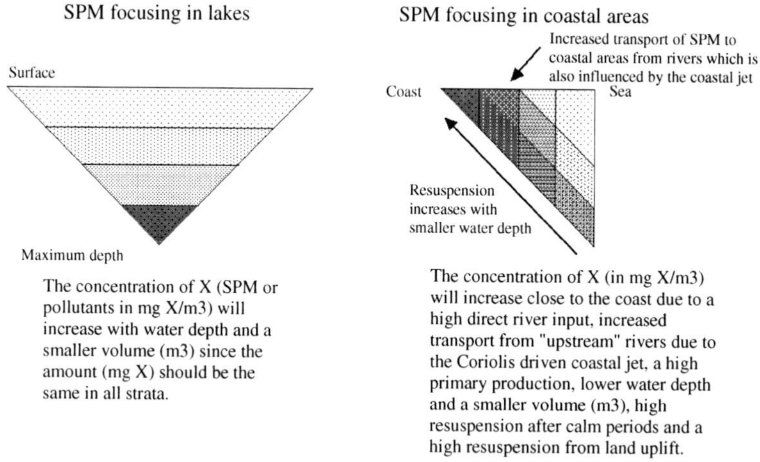

Fig. 5.2 SPM focusing in lakes/lagoons and coastal areas (from Håkanson 2006)

and the corresponding Secchi depths at the limits for the different trophic categories (oligotrophy, mesotrophy, eutrophy and hypertrophy). Note that if the Secchi depth is calculated to be higher than 200 m in very saline and low-productive systems, there is a boundary rule which never permits the value to be higher than 200 m.

Figure 5.2 illustrates SPM-variations in aquatic systems in a more general context. It is well known that sedimentation increases from about zero at the water depth separating areas of fine sediment transport and accumulation, to maximum values at the deepest part of the basin, and the concentration of SPM increases due to this and by the fact that the volume (defined for a certain water depth) decreases with increasing water depths, a phenomenon often referred to as "sediment focusing" in closed lagoons and lakes (see Fig. 5.2, left).

The concept "coastal focusing" may be used in analogy with sediment focusing (see Fig. 5.2, right). SPM (and related substances) generally appear with relatively high values in coastal areas where the river input of SPM may be significant and accentuated by coastal currents and small water depths. Resuspension will also add to the SPM-fluxes. The pattern shown in Fig. 5.2, right, is typical for SPM, but this pattern may be modified by many factors. The amount of SPM always depends on three main causes, river inflow, production in the coastal system and resuspension. It is easy to imagine that storms leading to high waves causing resuspension of previously deposited materials have a major influence on coastal ecosystems. The wave base separating bottom areas where transport processes dominate the bottom dynamic conditions from areas of continuous sedimentation of fine materials, depends on the fetch (see Fig. 3.15), and the duration and velocity of the winds. Resuspension implies that the carbon, nitrogen and phosphorus contained by the previously deposited sediments, as well as metals and mineral particles, will again enter the water system.

Table 5.2 SPM-variations among sites (mean values from 17 sampling situations for 5 sites) in the open Baltic Sea (brackish) during mixed, homothermal conditions. D = water depth of the given site (m); Temp = water temperature (°C); Days,5 = Number of days with winds less than 5 m/s. y = log(SPM). From Håkanson and Eckhell (2005)

Step	r^2	x-variable	Model
1	0.59	log(Temp)	$y=1.17 \cdot x_1-0.77$
2	0.78	log(Depth)	$y=0.89 \cdot x_1-0.636 \cdot x_2-0.59$
3	0.80	(Days,5)^1.3	$y=-0.764 \cdot x_1-0.524 \cdot x_2-0.0221 \cdot x_3+0.569$

Table 5.2 gives empirical models for SPM in open brackish areas. This table ranks the factors influencing variations in mean SPM-values among sites when the water mass is homothermal (i.e., for mixed conditions). Data for SPM from 17 sites/events (even from storms) were used (Håkanson and Eckhell 2005). The mean SPM varied from 0.67 to 4.86 mg/l.

One can note that:

- After three steps, the r^2-value is 0.80.
- The mean water temperature at the site was the most important factor explaining the SPM-variations among sites ($r^2 = 0.59$) – the higher the temperature the higher the SPM-value. This is logical, although the causal reason may be obscured since temperature correlates to light and influences both bioproduction and stratification/mixing.
- The second variable is the water depth (D) at the site. Accounting for D increases r^2 from 0.59 to 0.78.
- The third factor is days with wind speeds lower than 5 m/s (r^2 increases from 0.78 to 0.80).

Shallow and wind/wave exposed sites/areas are generally dominated by processes of fine particles erosion and transport. The relationships between winds, waves and SPM-concentrations have been investigated in several studies (see Fig. 3.23 and Hellström 1991; Kristensen et al. 1992). A high wind speed might increase SPM and decrease the Secchi depth. However, not only the prevailing wind situation is of importance, but also the frequency of resuspensions (Floderus and Håkanson 1989). If there are many resuspensions per month, it is likely that there is less material on the bottom to be resuspended. Also the wind direction is of interest. If the fetch is large, the wave base (i.e., the water depth down to which the wave orbitals can resuspend fine particles) can be deep. Studies by Andersson (2000) demonstrate that resuspension correlates with winds higher than 7 m/s in four different archipelago areas in the Baltic Proper, whereas studies by Eckhéll et al. (2000) indicate that wind speeds higher than 14 m/s correlate best with resuspension. Burban et al. (1989, 1990) have demonstrated that changes in water turbulence, SPM and salinity are also key regulatory factors for the aggregation and flocculation of suspended particles and hence also for the fall velocity. A typical fall velocity for SPM in lakes is 10–100 m/yr depending on the material and this can increase significantly along a salinity gradient (Kranck 1973, 1979; Lick et al. 1992). In the summer, when the water

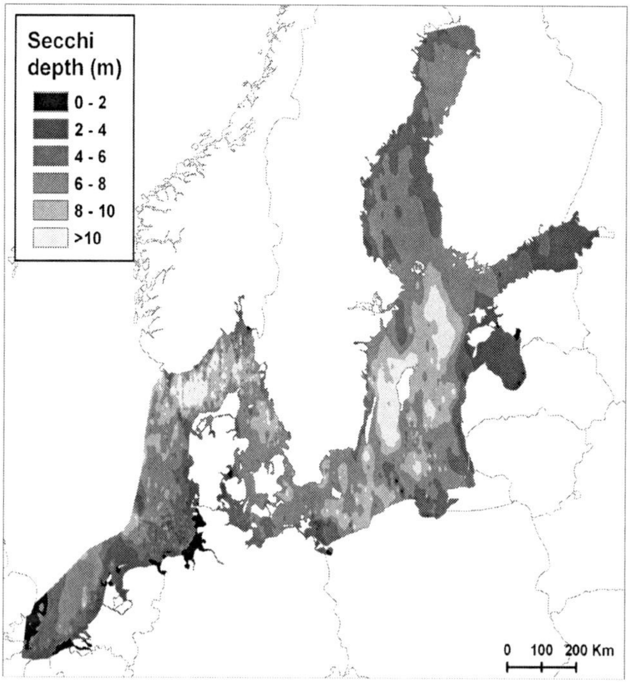

Fig. 5.3 Average annual Secchi depths in the Baltic Sea and parts of the North Sea in the upper 10 m water column for the period from 1990 to 2005

temperature is high, the biological production is also high and this affects SPM and the Secchi depth.

The general map of the average Secchi depth for the Baltic Sea and the south-eastern parts of the North Sea is shown in Fig. 5.3. Some areas with lower Secchi depths can be observed, e.g., in the Gulf of Riga and along the North Sea coasts of Holland, Belgium and Germany.

5.2 Oxygen Saturation in the Deep-Water Zone (O_2Sat in %)

There are positive as well as negative effects of aquatic eutrophication. Positive effects are, e.g., increased catches of fish in the Baltic Sea between 1920 and 1980 (Hansson and Rudstam 1990). The catch per unit effort has decreased since the late 1970s (Baden et al. 1990) and this is probably mainly caused by overfishing (see Håkanson and Gyllenhammar 2005).

Negative effects are, e.g., decreased caches for certain species of fish and altered fish communities (see, e.g., Kautsky and Kautsky 1989; Elmgren 1989; Hansson and Rudstam 1990; Kautsky 1991), changes in the structure and composition of other functional groups of organisms, like large perennial algae (*Fucus vesiculosus*). It should be noted that the important cod population in the Baltic Sea has decreased significantly since the early 1980´s. It certainly depends on many factors in a complicated system of interactions, but one of those is linked to the oxygen conditions (Voipio 1981). Each female cod lays many eggs. In salty water, the roe remains suspended in the surface, oxygen-rich, water. In the brackish surface waters of the Baltic Sea, the salinity and hence also the density is not high enough to allow the roe to float. It sinks to the saline deep-water layer before this descent is stopped by the greater density of the water. If this deep water is oxygen-rich, there will be a good spawning year for cod, but if the oxygen content is low, the eggs will die and the spawning will fail.

Numerous reports during the early 1950s demonstrate that the higher animal life in the bottoms in large areas was being eliminated as a result of oxygen deficiency. In those parts laminated sediments were encountered (Jonsson 1992). Laminated deposits dominate in anaerobic sediments, i.e., in oxygen-depleted sediments, where mixing due to bioturbation is negligible due to the extinction of the bottom fauna. The formation of anaerobic sediments is a natural process occurring predominantly in deep, stagnant, stratified and highly productive aquatic systems. But this natural phenomenon has been happening more often and in larger areas due to the anthropogenic eutrophication of the Baltic system. Fine, darkish, thin layers are generally laid down during winter, and thicker browner layers during summer. By counting the number of annual layers in sediment cores from the Baltic Sea, it has been shown how the oxygen-free bottom areas have grown during the last 100 years (Jonsson 1992). The sediments also provide evidence of the increasingly difficult conditions in which the bottom animals live. In the sediments, it is often possible to record the transition period between homogeneous and oxygenated sediments and those which are varved/laminated and oxygen-free. The bottom areas are periodically covered by extensive carpets of sulphurous bacteria. Sediment removed from such places generally has an unpleasant odour of hydrogen sulphide, a gas which is toxic to higher life.

Oxic sediments generally have an abundant macrofauna which cause a mixing of the sediments. The result is a more or less homogeneous mixture with a color that varies in different shades of brown (from grey/brown to black/brown). When oxygen-rich water returns to the bottom areas, the zoobenthos can start to recolonize the sediments. The laminated deposits may then be more or less homogenized, and so on in cycle after cycle. Jonsson (1992) gave a compilation of several changes in the Baltic Sea related to eutrophication, increased catch of fish, decreased oxygen conditions in the bottom water, increased areas with no bioturbation ("dead bottoms") and increased sedimentation of organic materials and nutrients.

The surface water is generally well oxygenated with O_2-concentrations between 7 and 9 ml/l; the values from January are often higher than for September depending on the high production of organic material during the summer months and because more oxygen can remain dissolved in colder water. When the algae and plankton die, bacterioplankton will decompose the organic materials and in this decomposition/mineralization of dead organic matter oxygen is utilized (Jonsson 1992). This results in lower oxygen values. From about 60–70 m water depth in the Baltic Proper, one can generally note a very steep O_2-gradient; and

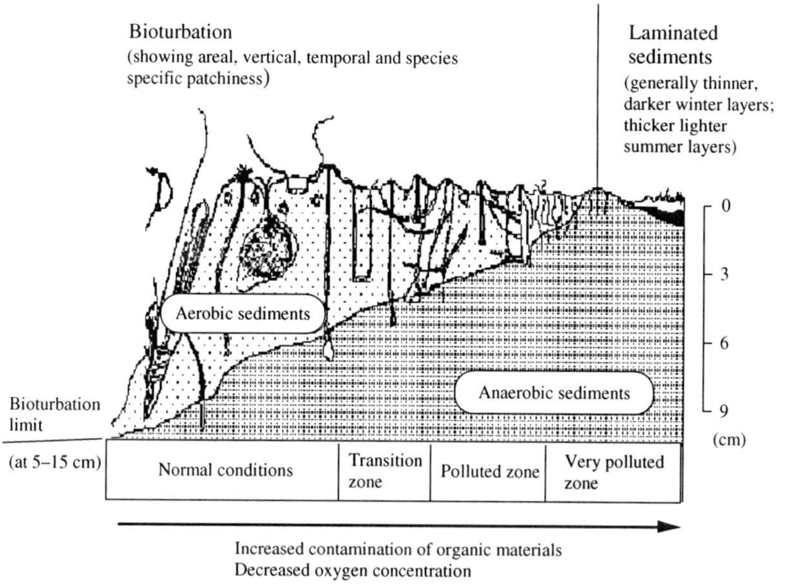

Fig. 5.4 Under aerobic (= oxic) conditions, zoobenthos may create a biological mixing of sediments down to about 15 cm sediment depth (the bioturbation limit). If the deposition of organic materials increases and hence also the oxygen consumption from bacterial degradation of organic materials, the decreasing oxygen concentration may reach a threshold value of 2 mg/l, when zoobenthos die, bioturbation ceases and laminated sediments appear (figure modified from Pearson and Rosenberg 1976)

at a water depth of about 140 m, there is often no oxygen at all left; this is the so-called redoxcline. Beneath the redoxcline, hydrogen sulphide (H_2S; a compound that smells like rotten eggs) is produced in microbiological and chemical processes.

The rationale for using a chemical measure such as the oxygen concentration or the oxygen saturation as an operational effect variable in coastal management is related to the linkages between the conditions for the bottom fauna, the oxygen concentrations in the sediment/water zone and the load of organic matter to the sediments (Fig. 5.4; from Pearson and Rosenberg 1976). When the O_2-saturation is lower than about 20% ($\approx$ 2 mg O_2/l), key functional benthic groups are extinct (Cornett and Rigler 1979). This is a threshold value. This is also stressed in the European Water Framework Directive (Anon 2003).

An effect-load-sensitivity (ELS) model for the mean summer oxygen saturation in the deep-water zone (the water beneath the wave base) is shown in Fig. 5.5. It should be noted that Fig. 5.5 has been derived using data from coastal areas in the Baltic Sea, which is a brackish water system.

The following working hypotheses were tested and verified when the model in Fig. 5.5 was constructed (see Håkanson 1999):

1. No single factor could explain the variability in mean O_2Sat that exists among these coastal areas. Several factors must be included and each factor is likely to add only a limited predictive power.

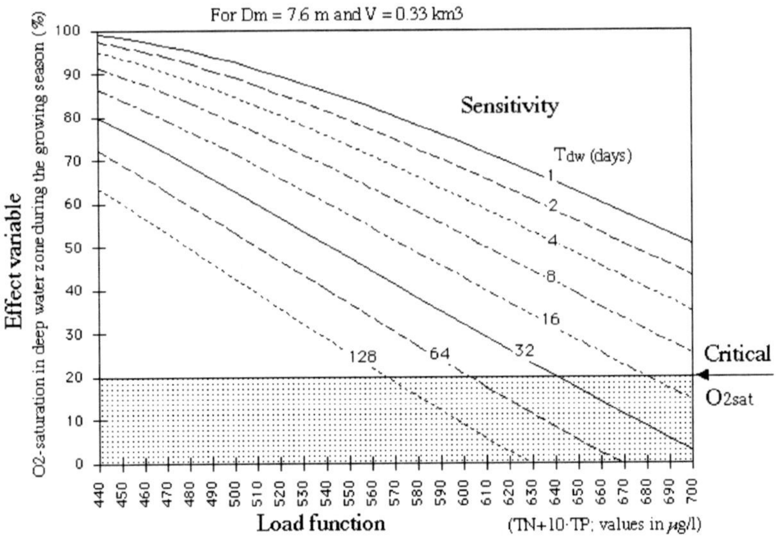

Step	r2-value
Step 1: Nutrient response function (TN+10·TP)	0.49
Step 2: Theor. deep water retention time (TDW)	0.76
Step 3: Mean depth (Dm)	0.89

Data from 23 Baltic coastal areas

Model:
$O2sat=100 \cdot SIN(14.83-4.46 \cdot \log(TN+10 \cdot TP)-0.40 \cdot \log(1+Tdw)-1.05 \cdot \log(Dm)))$

Fig. 5.5 Effect-load-sensitivity (ELS) model using the oxygen saturation in the deep water (mean values for the growing season) in Baltic coastal areas as the operational bioindicator (modified from Håkanson 1999)

2. O_2Sat depends on both load factors (related to nitrogen and phosphorus concentrations in water) and sensitivity factors linked to the size and form (the morphometry) of the coast.

It should be noted that the mean O_2Sat-value is not a constant, but a variable, and that a model based only on morphometric parameters cannot be used for site-specific predictions of time-dependent y-variables. The empirical model in Fig. 5.5 is based on nutrient concentrations and morphometry, and the aim is to predict summer averages of O_2Sat for entire and well-defined coastal areas defined according to the topographical bottleneck approach. One can note from Fig. 5.5 that O_2Sat may be predicted quite well ($r^2 = 0.89$) from a load function ($TN + 10 \cdot TP$) plus morphometric (sensitivity) parameters, which are easily accesses from bathymetric maps: The higher the load to the system of both N and P, the lower O_2Sat; the longer the theoretical deep-water retention time (T_{DW}), the deeper the coastal area, the lower

O_2Sat. However, the model in Fig. 5.5 should not be used for, for instance, coastal areas dominated by tides.

When the zoobenthos die at low oxygen concentrations, the biological mixing (= bioturbation) of the sediments will halt. Generally, various types of zoobenthos live in sediments down to about 5–15 cm sediment depth (see Fig. 5.4). When bioturbation is negligible, for example, during anaerobic conditions, the diffusion of phosphorus from the sediment can be very high (see Sect. 9.1).

5.3 Chlorophyll-a

Chlorophyll is a standard measure of phytoplankton biomass and has a direct influence on the entire foodweb in a given coastal area. Thus, chlorophyll is a key bioindicator in aquatic sciences. As stressed, the phytoplankton biomass also influences the Secchi depth. This means direct effects on the depth of the photic zone and hence also on the primary production of phytoplankton, benthic algae and macrophytes. The phytoplankton biomass also affects the oxygen conditions in the system since the bacterial decomposition of dead phytoplankton is an oxygen consuming process. Deep-water oxygen conditions in turn influence the habitat for benthic organisms and important biogeochemical processes in the sediment-water interface.

Calculating phytoplankton biomass from chlorophyll-a is a focal issue in aquatic sciences. Generally, chlorophyll-a concentrations are predicted from water temperature or light conditions and nutrient concentrations (e.g., Dillon and Rigler 1974; Smith 1979; Riley and Prepas 1985; Evans et al. 1996).

Empirical models to predict chlorophyll-a concentrations (Chl in µg/l) are exemplified in Fig. 5.6. One model is based on empirical median TN-concentrations from the summer period (see Fig. 5.6A; $r^2 = 0.92$), the other on empirical median TP-concentrations (see Fig. 5.6B; $r^2 = 0.89$). From Fig. 5.6C, one can also note that over a wide range (these data emanate from Baltic coastal areas and from the Ringkobing Fjord, Denmark; see Bryhn et al. 2007), there is a strong correlation between TP- and TN-concentrations ($r^2 = 0.88$).

Typical chlorophyll-a concentrations for the Baltic Sea and parts of the North Sea are shown Fig. 5.7. Values lower than 2 µg/l (oligotrophic conditions) are found in the northern parts of the Bothnian Bay and the outer parts of the North Sea, while values higher than 20 µg/l (hypertrophic conditions) and more are often found in, e.g., the Vistula and Oder lagoons. The hot spots shown on the map outside the British coast may be a result of data from situations when algal blooms are overrepresented. This map shows that at water depths smaller than 10 m, the Baltic Sea has typical chlorophyll concentrations between 2 and 6 µg/l during the growing season (May–September), which correspond to the mesotrophic class (see Table 3.1).

Figure 5.8 exemplifies chlorophyll data in the entire Baltic Proper (the southern and most eutrophic part of the Baltic Sea) at different depth intervals and one can note that the highest values, as expected, are to be found in the upper layer. As a reference to the depth of the photic zone, Fig. 5.8 also gives information that

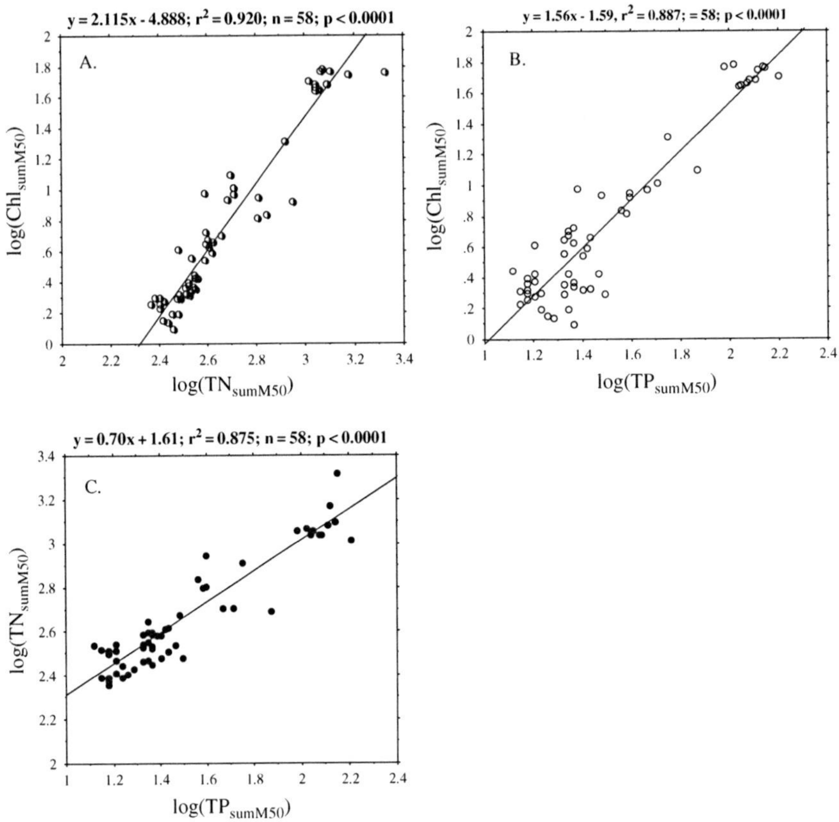

Fig. 5.6 The relationship between chlorophyll-a concentrations, total nitrogen (TN) and total phosphorus (TP) concentrations in brackish coastal areas based on logarithmic data and median values (M50) for the growing season. The figure also gives the regression lines and basic statistics (r^2-value = the coefficient of determination, r = the correlation coefficient, n = number of data in the regression and p = the statistical certainty). Data from Wallin et al. (1992); Meeuwig et al. (2000); Nordvarg (2001) and Bryhn et al. (2007)

the average Secchi depth in the Baltic Proper is 7 m (the standard deviation is 3.3 based on 14,306 data from the period 1990 to 2005 using data from the HELCOM database).

Naturally, there might be trends in data from individual areas. Using data from the very comprehensive HELCOM database, Fig. 5.9 gives a trend analysis of chlorophyll-a concentrations for the period from 1974 to 2005 for the conditions in the Baltic Proper. Note that the data in Fig. 5.9 emanate from the entire surface-water layer down to the theoretical wave base of 44 m. From this trend analysis, one can observe:

1. That the eutrophication, as expressed by the chlorophyll data, has not really changed during the last 30 years – if anything, the situation is improving rather

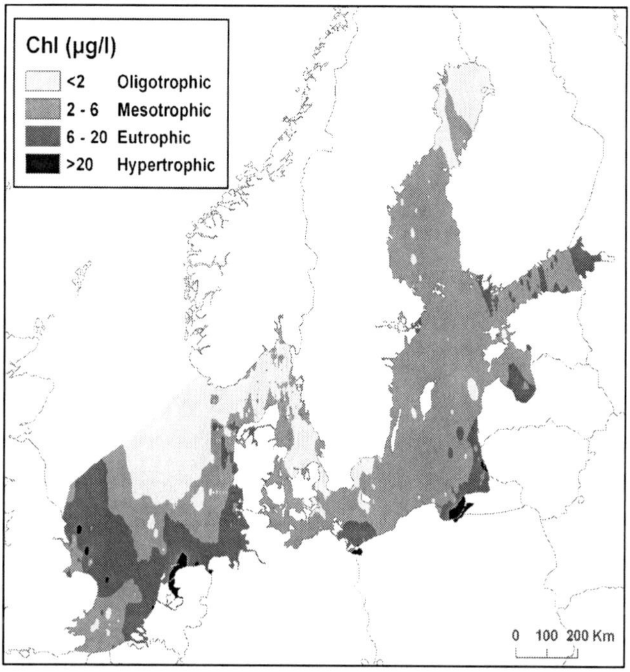

Fig. 5.7 Typical chlorophyll concentrations in the Baltic Sea and parts of the North Sea during the growing season (May–September) in the upper 10 m water column for the period from 1990 to 2005

than the opposite. However, the greatest increase in anthropogenic nutrient loading may have occurred earlier than that, starting before World War II, although there are few reliable time-series of data on riverine loading from this period (see Sect. 6.3).

2. On the whole, the Baltic Proper is not eutrophic; the mean chlorophyll-value of about 2 µg/l in the surface-water layer (down to the theoretical wave base at 44 m) is close to the boundary between oligothrophy and mesotrophy (using the classification system in Table 3.1).

It has been claimed that there has been a marked increase in TP-amount/concentrations after 1995 in the Baltic Sea when there was no corresponding increase in TP-inflow. This has been called "the flip" (Swedish Environmental Advisory Council 2005; Wulff 2006; Vahtera et al. 2007) and has been attributed to a regime shift which causes accelerating internal loading when there are no corresponding increases in external TP-loading. Regime shifts and thresholds are "hot" topics in aquatic science and management (Scheffer 1990; Scheffer et al. 2000; Carpenter 2003; Groffman et al. 2006; Thresholds 2006). The data on chlorophyll in Fig. 5.9 do not support the "flip" theory.

The variations in the Chl/TP-ratio versus the salinity shown in Fig. 3.26 can be modeled in a simple way, see Fig. 5.10. Figure 5.10a gives the model (using

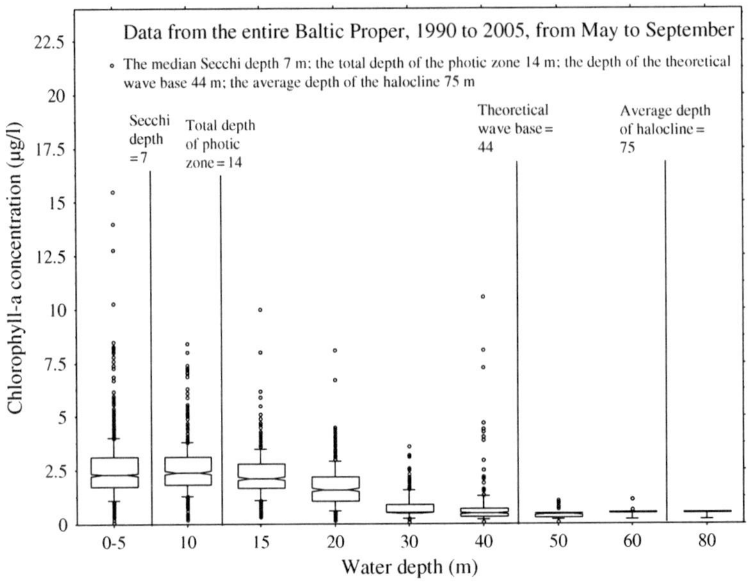

Fig. 5.8 Chlorophyll-a concentrations at different water depths in the Baltic Proper. The median Secchi depth this period (1990–2005) was 7 m, the theoretical wave base 44 m, and the average depth of the halocline 75 m

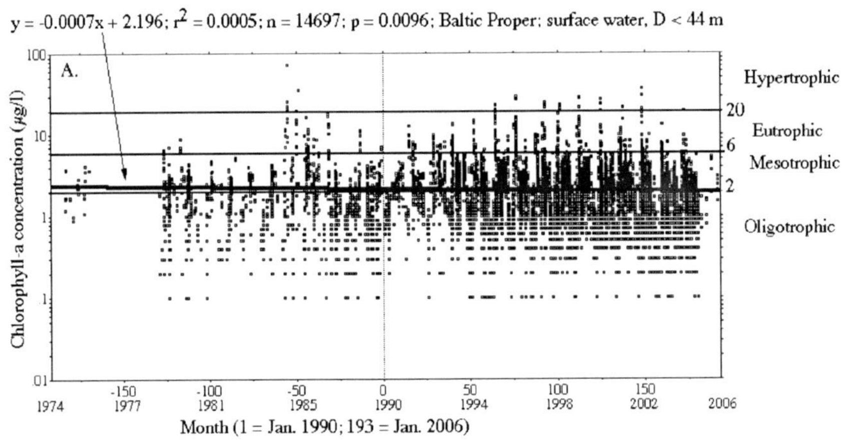

Fig. 5.9 Trend analysis based on 14,697 chlorophyll data from the surface-water layer (< 44 m) in the Baltic Proper for the period 1974–2006 and values indicating the trophic status related to different chlorophyll concentrations. The figure also gives the regression line, the coefficient of determination (r^2), the number of data (n) and the statistical certainty (p) (from Håkanson and Lindgren 2007a)

Model for influences of salinity on primary production

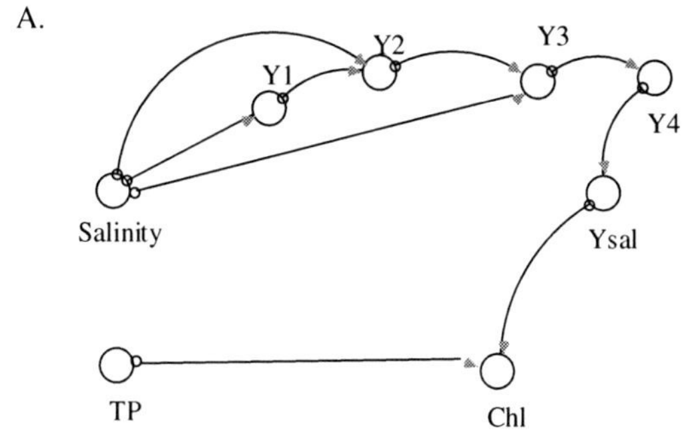

Chl = Ysal·TP[chlorophyll-a concentration in μg/l]
TP = [TP-concentration in water in μg/l]
Ysal = Y4 [dimensionless moderator for the influence of salinity on chlorophyll]
Y1 = if Salinity < 2.5 psu then (0.20-0.1·(Salinity/2.5-1)) else (0.20+0.02·(Salinity/2.5-1))
Y2 = if Salinity < 12.5 then Y1 else (0.28-0.1·(Salinity/12.5-1))
Y3 = if Salinity > 40 then (0.06-0.1·(Salinity/40-1)) else Y2
Y4 = if Y3 < 0.012 then 0.012 else Y3

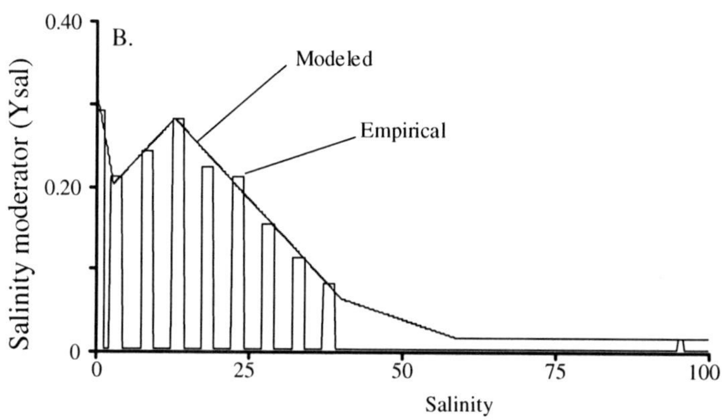

Fig. 5.10 A. Illustration of the model for how salinity influences the Chl/TP-ratio (the Y_{sal}-moderator) and the equations. **B**. Illustration of how the model describes the empirical median values for the salinity moderator

"if-then-else" statements and a technique with dimensionless moderators; see Håkanson and Peters 1995). The correspondence between modeled data and empirical data for the Chl/TP-ratio for the growing season is shown in Fig. 5.10B. The model is simple to use. It predicts chlorophyll from TP and salinity as Chl = Y_{sal}·TP, where Y_{sal} is the dimensionless moderator quantifying how variations in salinity would generally influence variations in chlorophyll. By accounting for variations in salinity, one can increase the predictive power and applicability range of regression models between chlorophyll and nutrients.

5.4 Cyanobacteria

Photolithoautotrophic bacteria (sometimes called bluegreen algae; here referred to as cyanobacteria), play two key roles in eutrophication contexts: they can form extensive nuisance blooms that may be toxic (Smith 2003), and many cyanobacterial species can fix large amounts of dissolved gaseous nitrogen of atmospheric origin (Rahm et al. 2000; Tõnno 2004). In the Baltic Sea, they constitute the dominating form of harmful algal blooms, although other harmful algae may be more important in other coastal areas.

Figure 5.11 gives the (log–log) regression between chlorophyll-a concentration and cyanobacteria (CB = median values for the growing season (using data from lakes since similar data have not been available to us from marine systems)). One can note a highly significant and mechanistically understandable strong positive covariation between these two measures of algal biomass in water. Evidently, this relationship may look different had it been based on daily, weekly or monthly values. Håkanson et al. (2007a) also tested whether it was possible to use a lake dataset to predict the biomass of nitrogen-fixing cyanobacteria (CB_{fix}) but that turned out to be more difficult than predicting total CB, in part because there were much fewer empirical data available. There was a division line at TN/TP = 24, below which CB_{fix} increased sharply. The least uncertain way to predict CB_{fix} was to calculate the median CB_{fix}/CB-ratio from the five lakes in Smith (1985), which gave $CB_{fix} = 0.33 \cdot CB$. A regression of empirical data against modeled data gave an r^2-value of 0.36 (n = 29), although this value increased to 0.84 when two data points from one of the lakes were excluded. Possibly, these data points may emanate from summers with unusual weather conditions.

Figure 5.12A gives a regression between CB (transformed as $CB^{0.25}$) and TN (log-transformed data) for 86 systems with CB > 0.

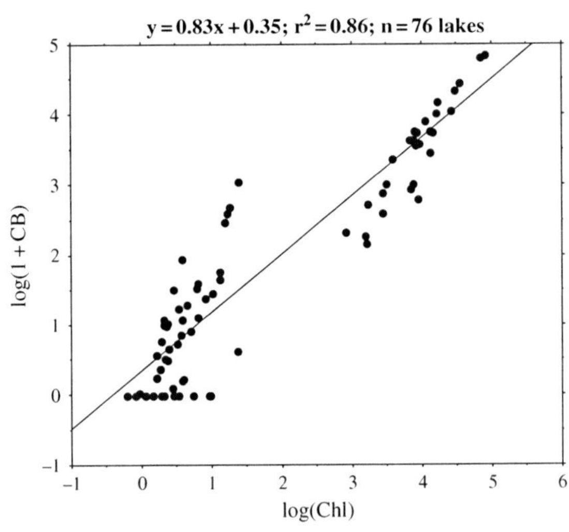

Fig. 5.11 Regression between log(1+CB) and log(Chl) based on data from 76 lakes. The figure also gives the coefficients of determination (r^2) and the regression lines. Cyanobacteria (CB) in μg ww/l, chlorophyll-a concentrations (Chl) in μg/l

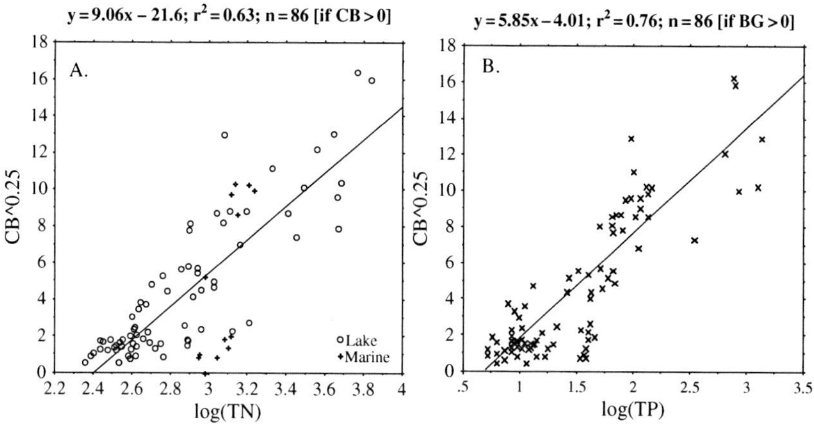

Fig. 5.12 **A**. Regression between CB (transformed into $CB^{0.25}$; CB in µg ww/l) and log(TN) (TN in µg/l) using median values for the growing season from 86 systems with CB-values higher than zero. **B**. A similar regression between $CB^{0.25}$ and log(TP) (TP in µg/l). The figure also gives the number of systems used in the regressions (n), the coefficients of determination (r^2) and the regression lines

One can note a highly significant co-variation ($r^2 = 0.63$) but that the scatter around the regression line is high. The scatter is even higher for the actual (non-transformed) data.

It should be stressed that the following model to predict concentrations of cyanobacteria from TP does not concern cyanobacteria produced in the benthic zone and it does not differentiate between cyanobacteria fixing atmospheric nitrogen and non-fixing species. To expand this modeling in such directions, one would need a more comprehensive set of reliable data (that meet the criteria given by the sampling formula).

The model has been derived in the following steps (see Håkanson et al. 2007c).

A basic regression between CB and TP is given in Fig. 5.12B.

$$CB^{0.25} = 5.85 \cdot \log(TP) - 4.01 \quad (r^2 = 0.76; n = 86) \tag{5.2}$$

It should be stressed that this regression includes data from many more systems than the equation from Smith (1985). It also gives a higher r^2-value (0.76 as compared to 0.71) so it is more general and provides better predictive power. Given the inherent uncertainties in the empirical CB-data, one should not expect to derive predictive models for cyanobacteria that give much higher r^2-values.

Figure 5.13A shows a scatter plot between log(CB) and the TN/TP-ratio.

One can note that the scatter is high but also that some interesting general conclusions can be made:

- High CB-values only appear in systems with relatively low TN/TP-ratios.
- The regression between log(CB) and TN/TP attains a maximum value ($r^2 = 0.73$; $\log(CB) = 0.142 \cdot TN/TP + 5.47$; n = 61) if one only use data from systems with

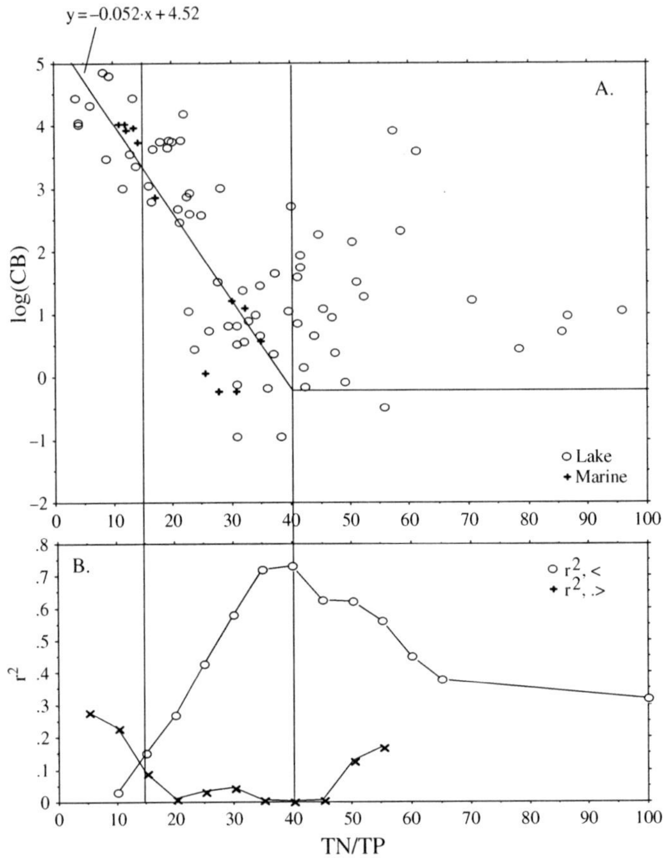

Fig. 5.13 CB (in μg ww/l) and the TN/TP-ratio. **A**. Scatter plot between log(CB) and the TN/TP-ratio. The given regression line is based on data from systems where TN/TP < 15. **B**. Compilation of r^2-values between log(CB) and TN/TP using data from systems when successively smaller (<) and higher (>) TN/TP-ratios have been omitted. The maximum r^2 is obtained for systems with smaller TN/TP-ratios than 40; very low r^2-values are obtained if the regressions are done for systems with TN/TP-ratios higher than 15

TN/TP-ratios smaller than 40. The upper curve (circles) in Fig. 5.13B gives the r^2-values when only systems with TN/TP-ratios smaller than 10, 15, 20, ….. 100 were used in the regressions. The lower curve in Fig. 5.13B gives similar results when only system with higher TN/TP-ratios were used and then one can note that there is no statistically significant ($p < 0.01$) relationship between log(CB) and TN/TP if TN/TP is higher than 15.

- So, two critical TN/TP-ratios, 15 and 40, have been identified. This information is used in the following where the basic regression between CB and TP given by (5.2) is complemented with (multiplied by) a dimensionless moderator (see Håkanson 1999, for discussions on modeling using dimensionless moderators) Y_{TNTP} defined as:

$$\text{If TN/TP} < 15 \text{ then } Y_{TNTP} = (1 - 3\cdot((TN/TP)/15 - 1)) \text{ else } Y_{TNTP} = 1 \quad (5.3)$$

This means that for systems with TN/TP (based on median values for the growing season) higher than 15, one can use the basic regression (5.2) without any correction for the TN/TP-ratio (i.e., (5.2) multiplied by 1), but for systems with lower TN/TP-ratios, (5.3) will provide a correction factor. If, e.g., TN/TP = 7.2, then Y_{TNTP} = 2.56, and the CB-value a factor of 2.56 higher than the value suggested by (5.2). The use of the TN/TP ratio as a determinant of the biomass of cyanobacteria has been questioned by some authors (see Smith 2003 for a review). Equation 5.2 for Y_{TNTP} is supported by the data in Fig. 5.13 and future data and testing may clarify the general validity of this approach.

In the literature, temperatures between 15 and 17°C have been reported as the minimum for cyanobacteria blooms in freshwater systems and in the Baltic Sea (Reynolds and Walsby 1975; Edler 1979; Wasmund 1997). Laboratory experiments on cyanobacteria also support this conclusion (Konopka and Brock 1978; Lehtimäki et al. 1994, 1997) since many species of cyanobacteria have a slow growth rate below 15°C. To complicate matters, there are also reports that cyanobacteria have a requirement of temperatures of about 20–21°C to form blooms and become dominant in a system. Those reports are from a freshwater lake in Canada (McQueen and Lean 1987), from the North Pacific Ocean (Marumo and Asaoka 1974) and from an estuary in Australia (Lukatelich and McComb 1986).

The optimal growth temperatures are species-specific for cyanobacteria, but around 25°C for many species from temperate areas according to laboratory experiments (Konopka and Brock 1978; Robarts and Zohardy 1987; Lehtimäki et al. 1997), but these experiments often use species from temperate areas. With higher temperatures, the growth rate usually starts to decrease (Konopka and Brock 1978). In field data from the Baltic, this decrease in growth rate is not shown because there are few occasions with temperatures higher than 20°C Sea (Wasmund 1997). Tilman and Kiesling (1984) studied the cyanobacteria response to temperature in Lake Superior and concluded that they were disfavoured at temperatures under 17°C and dominating at temperature over 24°C. Data from the Pacific Ocean area (Marumo and Asaoka 1974; Lukatelich and McComb 1986) indicate that the cyanobacteria in this area may have another response to temperature. Evidently, it would be interesting to study if cyanobacteria react to temperature changes differently in tropical areas and saline environments as compared to fresh and brackish waters in the temperate zone.

Figure 5.14 gives data on the relationship between CB (here log(1 + CB)) and surface water temperatures (SWT in °C) from 74 systems. One can note that all systems with higher CB-values than 100 (median values for the growing season) have temperatures higher than 15°C.

As already stressed (Fig. 5.11), there exists a close co-variation between chlorophyll-a and CB-concentrations in lakes. Since no more data relating CB to temperature have been accessible to us than those shown in Fig. 5.14, the database from Ringkobing Fjord has been used to investigate the role of surface water temperatures (SWT) versus chlorophyll, or rather versus the ratio between

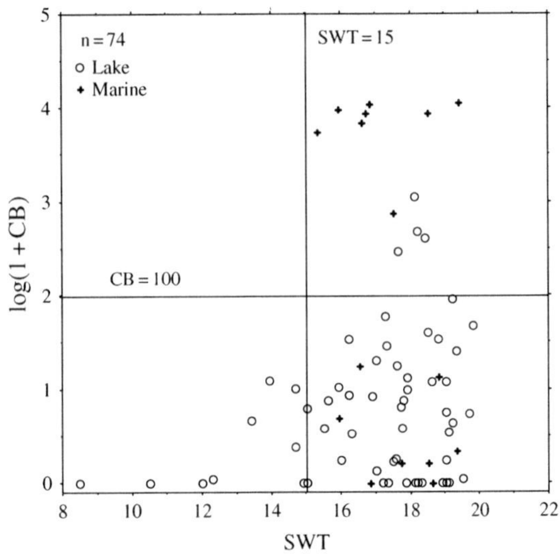

Fig. 5.14 Scatter plot between cyanobacteria (transformed into log(1+CB); CB in μg ww/l) versus surface water temperatures (SWT in °C) based on data from 74 systems

chlorophyll (Chl) and TP (see Fig. 5.15). Note that the aim has been to try to minimize the possible influences of salinity by taking data only in the salinity range between 5 and 10. The regression in Fig. 5.15 is based on daily median values (n = 299) and there is a highly significant relationship ($p < 0.0001$) between Chl/TP and SWT (but the scatter around the regression is also evident and the r^2-value is 0.21).

From the results in Fig. 5.15, one can hypothesize that the temperature (or light) may influence CB is the same manner as Chl. This would imply that if the median surface-water temperature for the growing season is higher than 15°C, one would expect a higher production of cyanobacteria. The basic regression (5.2) includes data from systems with a median surface water temperature of 17.5°C.

This means that the dimensionless moderator for temperature influences on CB (Y_{SWT}) may be given by:

$$\begin{aligned} &\text{If SWT} > 15°\text{C, then } Y_{SWT} = (0.86 + 0.63\cdot((\text{SWT}/15)^\wedge 1.5 - 1)) \\ &\qquad \text{else } Y_{SWT} = (1 + 1\cdot((\text{SWT}/15)^\wedge 3 - 1)) \end{aligned} \tag{5.4}$$

Thus, if SWT = 15°C, then $Y_{SWT} = 0.86$; if SWT = 17.5, $Y_{SWT} = 1$; if SWT = 25, $Y_{SWT} = 1.48$; SWT = 10°C, $Y_{SWT} = 0.30$, etc.

So, when the temperature is 25°C, the risks of getting high concentrations of cyanobacteria is a factor of 1.48 higher than at 17.5°C, if all else is constant, using this approach.

In many freshwater lakes, the biomass of cyanobacteria can be very high (Reynolds 1987). In brackish systems, the situation is probably slightly different. Howarth et al. (1988) found no data on N_2-fixing planktonic species in estuaries and coastal seas, except for the Baltic Sea and Pell-Harvey estuary, Australia. Also results from Marino et al. (2006) support this general lack of N_2-fixing cyanobacteria

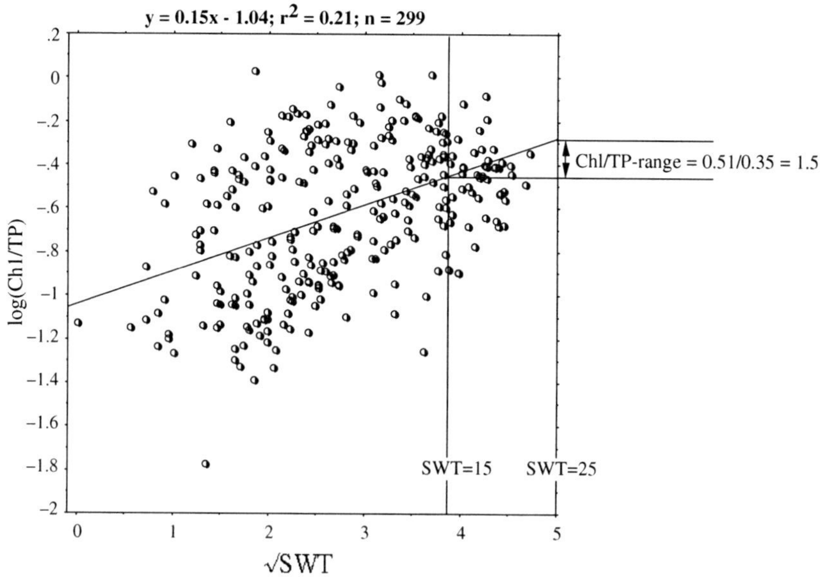

Fig. 5.15 Illustration of water temperature (surface water temperatures, SWT, in °C; the x-axis gives $\sqrt{SWT}$) influences on the ratio between chlorophyll and TP [log(Chl/TP)] using daily median values from Ringkobing Fjord in the salinity range from 5 to 10 psu

in estuaries. Common freshwater cyanobacteria species may be unable to deal with a higher concentration of Na^+, since their saline tolerance is low and they have an inefficient efflux system for Na^+ (Thomas et al. 1988). This is an indication that there may be limited amounts of cyanobacteria in brackish waters. That conclusion is not, however, supported by other studies. There are a few species of cyanobacteria tolerant to brackish water (Lehtimäki et al. 1997; Wasmund 1997; Moisander et al. 2002; Mazur-Marzec et al. 2005). A field study from the Baltic Sea (Wasmund 1997) indicates that in this brackish environment species of cyanobacteria have, interestingly, the highest biomass at 7–8 psu and that the blooms in Kattegat and Belt Sea are more frequent if the salinity is below 11.5 psu. A laboratory experiment with cyanobacteria from the Baltic Sea supports the results that the highest growth rate was at salinities in the range between 5 and 10 psu (Lehtimäki et al. 1997).

Water blooms of cyanobacteria in marine environments may not be as common as in freshwater systems but according to Sellner (1997) they can be the dominating factor in carbon and nutrient fluxes in some saline systems. In marine systems, there are just a few dominant genera. In a field study in the Pacific Ocean (Marumo and Asaoka 1974), there was no correlation between the salinity and the cyanobacteria abundance and no cyanobacteria were found in the cooler, less saline subartic waters. Those marine cyanobacteria species are mainly found in high-saline conditions. In the data discussed by Marumo and Asaokas (1974), the salinity was around 32–36 psu.

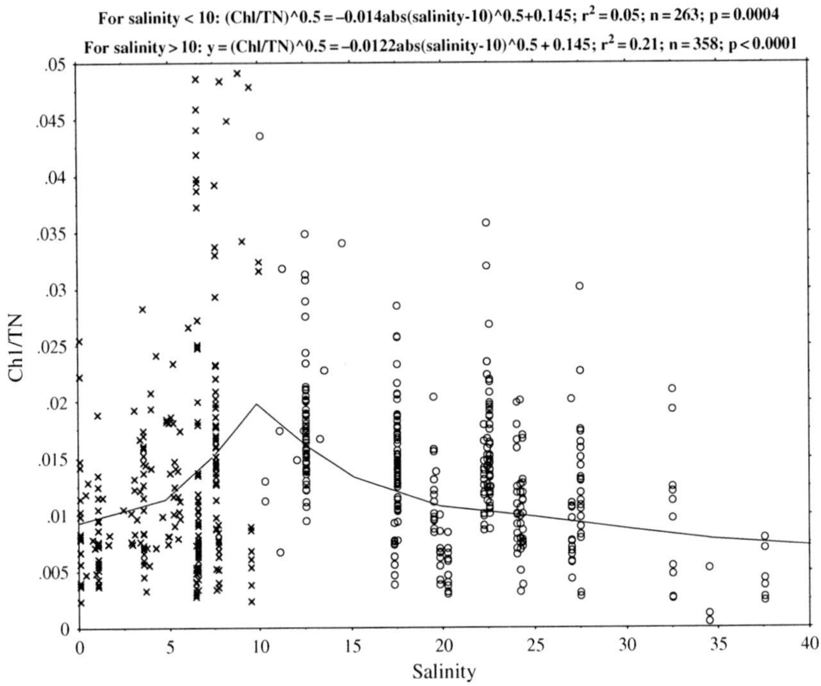

Fig. 5.16 Scatter plot of all available data relating the ratio Chl/TN to salinity (psu). The figure also gives two regressions for salinities either below (crosses) or higher than the threshold value of 10 (circles)

Figure 5.16 gives the relationship between the Chl/TN-ratio and salinity based on these data from 621 systems in the salinity range from 0 to 36. In this modeling, it has been assumed that CB would be related to salinity in the same manner as chlorophyll. This means that the salinity moderator may be derived from the information given by the two regressions in Fig. 5.16 for salinities higher and lower than the threshold value of 10 psu. The maximum expected CB-values should hence be a found if the median salinity for the growing season is 10 psu; lower values should generally be found in systems with lower and higher salinities. The two regressions in Fig. 5.16 have been transformed into an algorithm expressing how changes in salinity are likely to affect the predictions of cyanobacteria in relation to the basic regression (5.2).

$$\text{If the salinity } < 10 \text{ psu, then } Y_{sal} = (2.1 + 1.1 \cdot ((\text{salinity}/10)^2 - 1))$$
$$\text{else } Y_{sal} = (2.1 - 115 \cdot ((\text{salinity}/10)^{\wedge}0.01 - 1)) \qquad (5.5)$$

This means that at a salinity of 10, Y_{sal} is 2.1 and CB a likely factors of 2.1 higher than in freshwater systems; if the salinity is 5, Y_{sal} is 1.28; if the salinity is 20, Y_{sal} is 1.3; and if the salinity is 36, Y_{sal} is 0.62.

Model for cyanobacteria

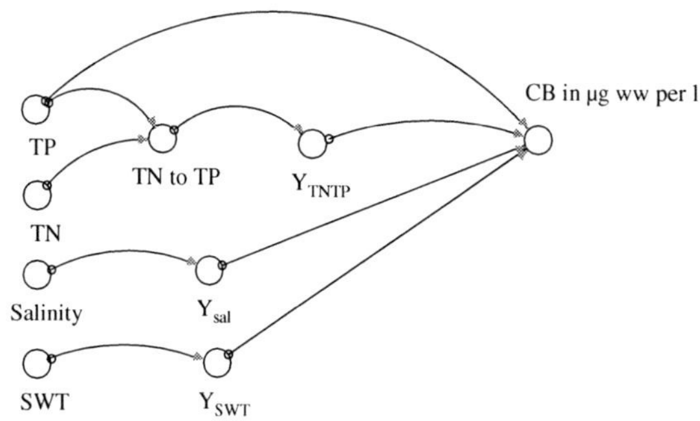

$CB = ((5.85 \cdot \log(TP) - 4.01)^4) \cdot Y_{TNTP} \cdot Y_{SWT} \cdot Y_{sal}$

CB in µg ww/l
Salinity in psu
SWT= Surface water temperature in °C
Total-N (TN) in µg/l
Total-P (TP) in µg/l

Y_{TNTP} = if TN/TP < 15 then (1-3·(TN/TP/15-1)) else 1
Y_{SWT} = if SWT > 15 then (0.86 + 0.63·((SWT/15)^1.5-1)) else (1 + 1·((SWT/15)^3-1))
Y_{sal} = if salinity < 10 then (2.1 + 1.1·((salinity/10)^2-1)) else (2.1-115·((salinity/10)^0.01-1))

Model domain: 4 < TP < 1300; 165 < TN < 6830; 0 < salinity < 40; 8 < SWT < 25

Fig. 5.17 Outline of the model for predicting median summer values of cyanobacteria in lakes and coastal areas

The CB-model is summarized in Fig. 5.17.

This model for cyanobacteria may give rather uncertain predictions for systems with high TN/TP-ratios and low temperatures. However, during such conditions, cyanobacteria generally play a small role for the N_2-fixation and as a nuisance to people and animals. Predicting high concentrations of cyanobacteria is evidently more essential than predicting low concentrations in contexts of water management, and this model serves this purpose quite well given the restrictions related to the inherent uncertainties in the available empirical data.

Due to the lack of alternatives, there is also good reason to use this approach until more data have been collected and stronger predictive models developed. Measured N_2-fixation tends to follow a similar pattern as the prevalence of cyanobacteria (Howarth et al. 1988; Tõnno 2004). Also, analyses using modern gene sequencing techniques have suggested that more organisms than is currently know may fix nitrogen in both lakes and marine systems (Zehr et al. 2003). This motivates the need for predicting general bioproduction patterns for cyanobacteria and other groups of organisms in waters with low TN/TP-ratios, rather than solely focusing on certain species. Evidently, it would be very interesting to have access to many more data on the relationship between cyanobacteria and salinity.

5.5 Macrophytes

For the determination of one of the most fundamental properties of aquatic systems, the trophic status, the basic attention is generally given to phytoplankton production or biomass. However, the macrophytes can contribute significantly to the total primary production, especially in shallow and enclosed coastal areas. Sometimes macrophyte production exceeds phytoplankton production (Wetzel 1983; Dahlgren and Kautsky 2004).

The macrophytes intercept nutrients, keep the nutrients bound for long periods and they also entrap and retain fine sediments settling out from the water phase. It is also very important to emphasize that the evolution of any aquatic system is closely connected with the overgrowing by rooted plants (Beeton and Edmondson 1972). Macrophytes also provide an important protective environment for small fish (Mann 1982; Håkanson and Boulion 2002).

Shallow coasts may be regarded as a "pantry and a nursery" for large marine areas. It has been demonstrated (see Fig. 5.18) that shallow coastal areas in the Baltic Sea can have a very high bioproduction capacity (many times higher than the most productive areas on land; see Rosenberg 1985); all three functional groups of primary producers – phytoplankton, benthic algae and macrophytes – are present in such areas and a high primary production also means high secondary production of zooplankton, zoobenthos and fish (see Duarte 1991; Håkanson and Boulion 2002). From Fig. 5.18, one can note that at water depths larger than 15 m, one can generally

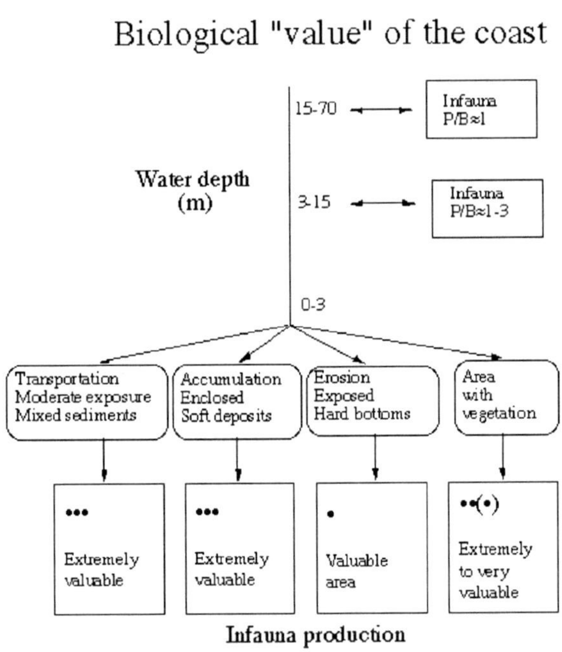

Fig. 5.18 Criteria to estimate the production potential (PR/BM; PR = production; BM = biomass; PR/BM is also called the turnover time in years) or "value" of Baltic coastal areas. At water depths > 3 m, the production potential is generally low; at water depth < 3 m, the production potential is generally high and depends on the exposure and the bottom substrate. Note that these data exemplify conditions along the Swedish coastal zone. Modified from Håkanson and Rosenberg (1985)

set the production capacity (PR/BM, PR = production in kg/year and BM = biomass in kg) of the infauna (i.e., animals > 1 mm living in the sediments) to be about 1. At water depths of 3–15 m, PR/BM is generally between 1 and 3. At water depths smaller than 3 m, it is important also to consider the sediment type, the habitat for the infauna. PR/BM is given on a relative scale of value in terms of fishery biology.

Coastal areas with a large percentage of the bottom above the Secchi depth, i.e., the water depth down to which most macrophytes may be found and where the production and biomass of benthic algae and zoobenthos may be very high, has a high bioproduction, a high "biological value" and are, hence, target areas in contexts of coastal protection. This is also exemplified by the results given in Fig. 5.19, which gives the relationship between the biomass of the mobile epifauna and water depth from a coastal area on the Swedish west coast (salinity about 22 psu).

Figure 5.20 gives a compilation of basic concepts and definitions related to bottom fauna and flora. In this figure, it is stressed that most benthic algae, macrophytes and zoobenthos appear at water depth smaller than the Secchi depth, since this area is generally well oxygenated and has a high bioproduction potential, since all three functional groups of primary producers, phytoplankton, benthic algae and macrophytes, are present in such areas. Figure 5.21 illustrates that different types of coastal areas (moderately exposed coasts, sheltered, exposed and vegetation-dominated coasts) are dominated by different sediments types, and since the sediments are the habitat for the benthic flora and fauna, different coastal areas will also be dominated by different species and functional groups (see Moen and Svensen 2004).

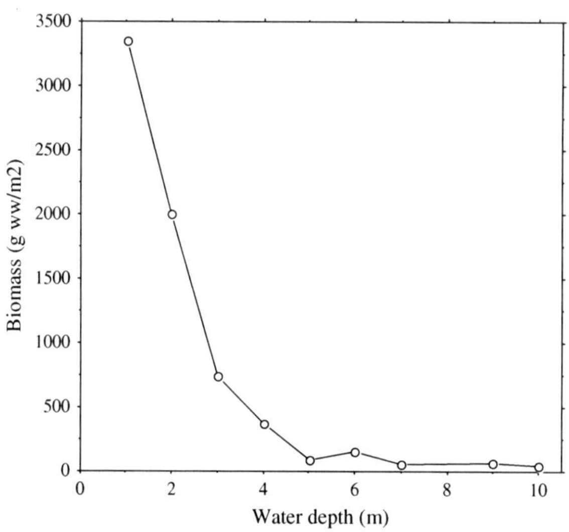

Fig. 5.19 The relationship between the biomass of the mobile epifauna (> 0.5 mm) in gram per m^2 and the water depth at Tylösand (on the Swedish west coast; data from Möller et al. 1985)

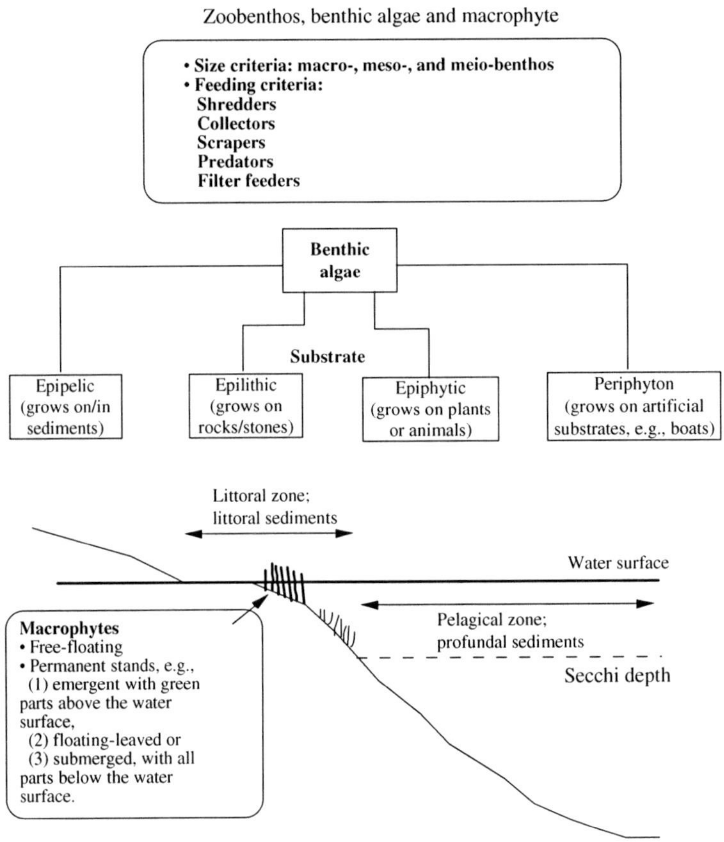

Fig. 5.20 Concepts related to zoobenthos, benthic algae and macrophytes (see Cummings 1973; Brinkhurst 1974)

To determine the relative role of macrophytes and phytoplankton in primary productivity, it is necessary to study the development of these plant groups relative to morphometric and hydro-optical properties of water bodies (see Longstaff and Dennison 1999; Menge et al. 2003; Sagert et al. 2005). Note that it may be difficult to clearly define the ratio between macrophytes plus epiphytes and benthic algae. The dominance of macrophytes or benthic algae depends on the characteristics of the bottom substratum. If it is stony, the benthic algae dominate. If it is soft, the rooted macrophytes (plus epiphytes) may shade the benthic algae. Benthic algae are also very important contributors to primary production on soft (primarily mud) bottoms.

Duarte and Kalff (1986) have demonstrated a strong relationship between littoral slope and macrophyte biomass for individual sites in lakes. It has also been shown (see, e.g., Scheffer 1990; Scheffer et al. 2000) that in a given lake with macrophytes there can be two alternative stable states, one with turbid waters, high

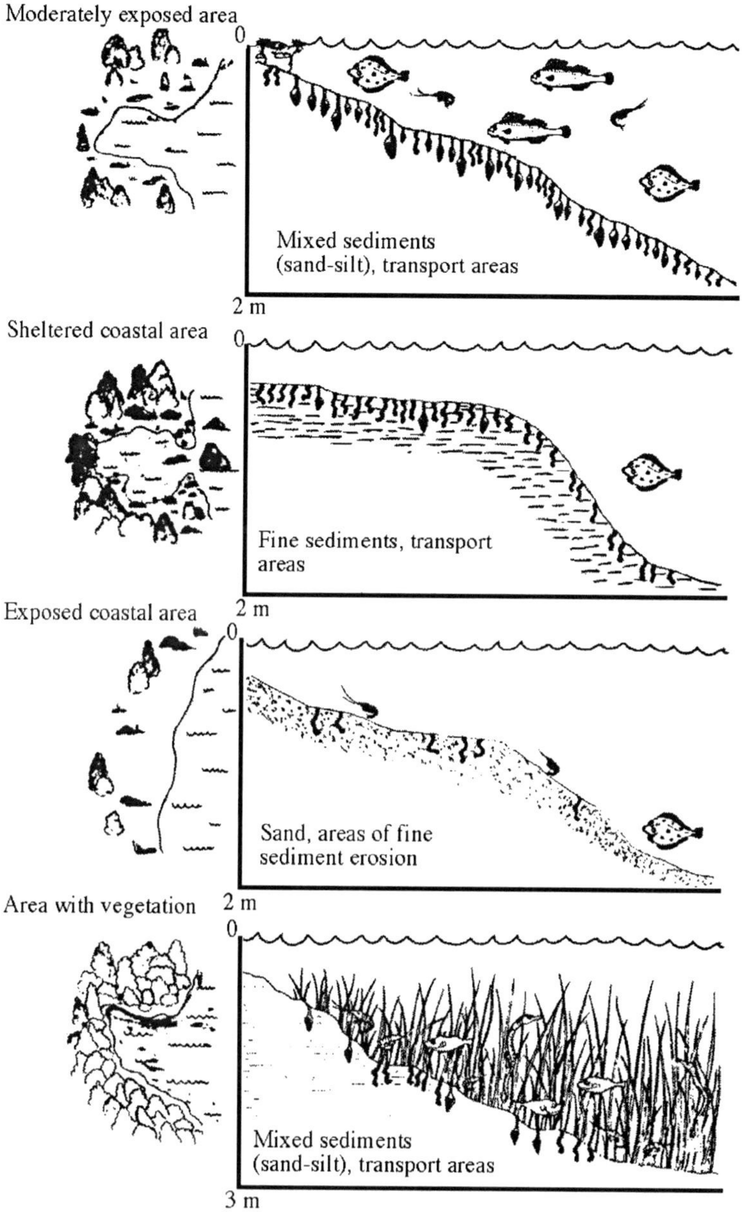

Fig. 5.21 Illustration of some typical coastal areas with a high production capacity and a high "biological value" (from Håkanson and Rosenberg 1985)

SPM-concentrations and a minimum of macrophytes, the other with clear water conditions, which maximises macrophyte production and cover.

Håkanson and Boulion (2002) have used empirical data from many lakes and statistical modeling to rank the factors influencing the variability among the lakes in macrophyte cover (MA_{cov}). Note that such data are not available to us from coastal areas.

The macrophyte database includes data on MA_{cov} from lakes covering a wide domain, large and small systems (from 0.047 to 24,000 km ^{2}), from latitudes 13 °N (Lake Tchad in Africa) to 67.5 °N, Lake Big Kharbey, Vorkuta, Russia, and deep and shallow lakes (the maximum depth varies from 1.5 to 120 m). The macrophyte cover varies from 0.29 to 100%. The light conditions are important for the macrophytes, and the database includes information on Secchi depth, which varies from 0.4 to 6.7 m. In many shallow lakes, the Secchi depth reaches its maximum value, the maximum depth. Results of stepwise multiple regressions are given in Table 5.3.

This table gives a ranking of the factors influencing the MA_{cov}-variability among the studied lakes. One can note:

- The ratio between the Secchi depth and the mean depth (Sec/D_m) is the most important factor to statistically explain the variation among these lakes regarding MA_{cov}; $r^2 = 0.52$ for n = 19 lakes.
- The next important factor for MA_{cov} is latitude, which is evidently related to water temperature. If latitude is added, 67% of the variation in MA_{cov} among the lakes can be statistically accounted for.
- The third factor is maximum depth; the deeper the system the smaller MA_{cov}.
- The fourth factor is the area of the lake above 1 m, A1; $r^2 = 0.84$.

Figure 5.22 gives a 3D-diagram relating the two most important model variables, Sec/D_m and latitude, to MA_{cov}. This diagram shows how these two model variables influence MA_{cov} when D_{max} and A1 are held constant. It is interesting to conclude the very strong influence of Secchi depth, mean depth and latitude (i.e., temperature and light) on how the macrophyte cover varies among lakes. This should also be true for brackish and marine coastal areas, but there are no data at our disposal to test that hypothesis.

Table 5.3 Stepwise multiple regression analyses (see Håkanson and Boulion 2002) to calculate the macrophyte cover (y = Mac_{cov} in % of lake area). Sec = Secchi depth (m), D_m = mean depth (m), Lat = latitude (°N), D_{max} = maximum depth (m) and A_{1a} = lake area above 1 m water depth (m^2). n = 19 lakes

Step	r^2	x-variable	Model
1	0.52	$x_1 = Sec/D_m$	$y=1.944+4.825\cdot x_1$
2	0.67	$x_2 = 90/(90\text{-Lat})$	$y=6.757+3.83\cdot x_1-1.57\cdot x_2$
3	0.74	(Days,5)^1.3	$y=8.31+2.57\cdot x_1-1.50\cdot x_2-0.286\cdot x_3$
4	0.84	$x_4 = \log(A_{1a})$	$y=10.49+1.502\cdot x_1-1.993\cdot x_2-0.422\cdot x_3+0.490\cdot x_4$

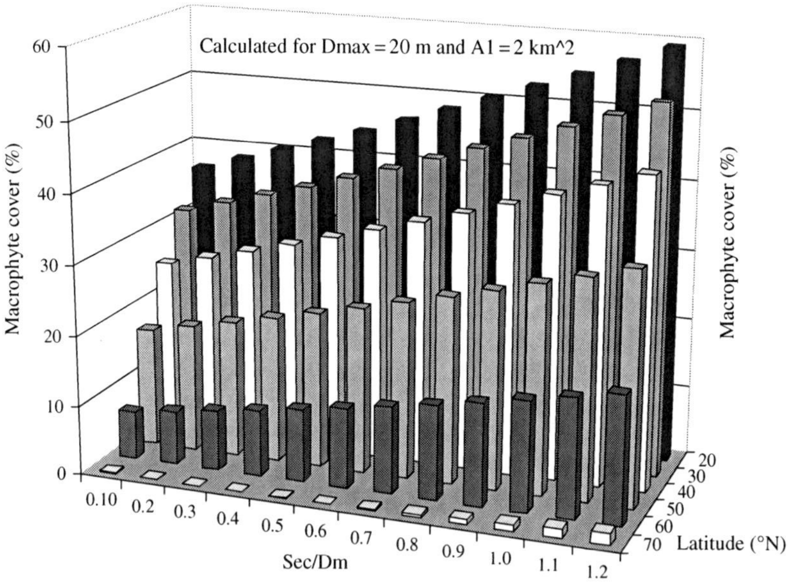

Fig. 5.22 3D-diagram illustrating how the ratio Secchi depth to mean depth (Sec/D_m) and latitude influence macrophyte cover for a lake with a maximum depth of 20 m with the 2 km^2 lake area above 1 m water depth. From Håkanson and Boulion (2002)

5.6 Index of Biological Value (IBV)

There are major differences in the bioproduction potential, or the biological "value", of different coastal areas. The aim here is to discuss criteria related to the bioproduction potential of coastal areas and to present and motivate an Index of Biological Value (IBV).

The theoretical wave base may be defined from the ETA-diagram (erosion-transport-accumulation; see Fig. 3.15). The ETA-diagram gives the relationship between the effective fetch, a measure of the free water surface over which the winds can influence the wave characteristics (speed, height, length and orbital velocity) and the bottom dynamics conditions. The ETA-diagram separates the erosion areas (dominated by coarse deposits, such as sand, gravel and rocks), from the transportation areas, with discontinuous sedimentation of fine materials (and mixed fine and coarse sediments), from the accumulation areas, with continuous sedimentation of fine suspended particles (and fine sediments). The prerequisites for the benthic production, including the oxic conditions, are generally very different between these three functional zones. The theoretical wave base (separating the T-areas from the A-areas) for open coastal areas is generally at a lower depth than the value given by the equation in Fig. 3.15 for closed lagoons or lakes. The wave base in open coastal areas may be estimated using an algorithm given in the CoastMab-model (see Sect. 9.1). If the mean effective fetch (in km) for an entire lagoon is set equal to the

$\sqrt{}$Area (Area in km^2), this approach (Fig. 3.15) gives one value for the theoretical wave base related to the water surface area of the system.

As stressed in Chap. 3, the form of the coastal areas may be described quantitatively by the volume development (= the form factor, V_d, dimensionless), which is defined by the ratio between the water volume and the volume of a cone, with a base equal to the water surface area (A in km^2) and with a height equal to the maximum depth (D_{max} in m). The form of the coast is important, e.g., for the growth of macrophytes and benthic algae and for resuspension.

From the discussion on macrophytes in Sect. 5.5, Fig. 5.23 provides one step further in the search for an index of biological value. This figure gives four relative hypsographic curves. Coast 1 has a large percentage of the bottom above the Secchi depth, which is the water depth indicating the depth of the photic zone where macrophytes may be found and where the production and biomass of benthic algae and zoobenthos may be very high. Coast 4 has only a small percentage of the bottom area above the Secchi depth, and is likely to have a lower benthic production.

From this, one can conclude (1) that coasts with a high proportion of the bottom area above the Secchi depth (A_{sec}) generally have a high potential bioproduction, a high biological value and (2) that the bioproduction potential also depends on the substrate, i.e., on the bottom dynamic conditions prevailing in the coastal area, in short on the sedimentological characteristics, which are closely linked to the exposure. Coastal areas with a high bioproduction potential should generally be more

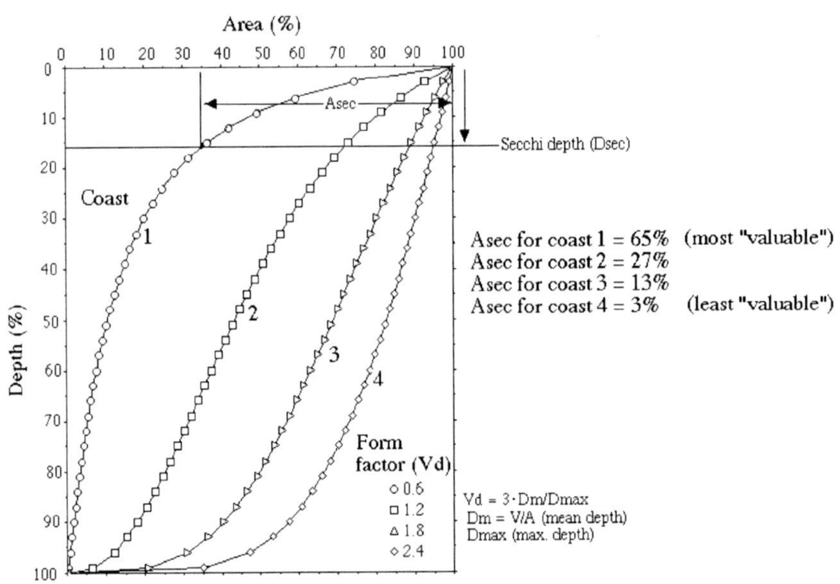

Fig. 5.23 Illustration of relative hyposographic curves (=depth-area curves) for four coasts with different forms (and form factor = volume development = Vd). The form influences the areas above the Secchi depth (A_{sec}), which indicate the production capacity and "biological value" of the coastal system

important to protect and preserve than other, less productive areas. In coastal areas which a large fraction of the bottom area above the Secchi depth, one should be particularity careful not to build marinas, harbors and/or emit contaminants (see Håkanson and Rosenberg 1985). If the exposure is very high, the coastal area would generally be dominated by coarse sediments and bare rocks, if the exposure is limited, the coastal area may have a dense macrophyte cover and be dominated by fine sediments. Such areas should have a high bioproduction potential. This may be expressed in a simple manner in by (5.6), which defines the Index of Biological Value (IBV, dimensionless; see also Fig. 5.24) from A_{sec} (dimensionless) and the exposure (Ex, dimensionless).

$$IBV = 10{\cdot}A_{sec}/(0.1 + \sqrt{Ex}) \quad (5.6)$$

The exposure generally varies between 0.0001 and 3, i.e., with a factor of 30,000 (see Fig. 3.18C, which is based on data from, 478 Baltic coastal areas; see Lindgren and Håkanson 2007); A_{sec} varies between 0.06 and 1, i.e., with a factor of 15–20. This has motivated the expression $(0.1 + \sqrt{Ex})$ in (5.6).

The latter expression will then vary between 15 and 20, which means that variations in A_{sec} and in Ex are of equal importance for the index. IBV will vary between 0.1 and 100 (extremely valuable enclosed areas with a high percentage of the bottom area above the Secchi depth). A typical Baltic Sea coastal area would have an exposure of 0.2 (see Fig. 3.18C) and an A_{sec}-value of 0.5, which gives an IBV-value of 10. According to the categories suggested in Fig. 5.24, this would indicate a coastal area at the border between "productive" and "moderately productive". The class limits for IBV, 1, 10, 25 and 50 and the categories "extremely productive, very

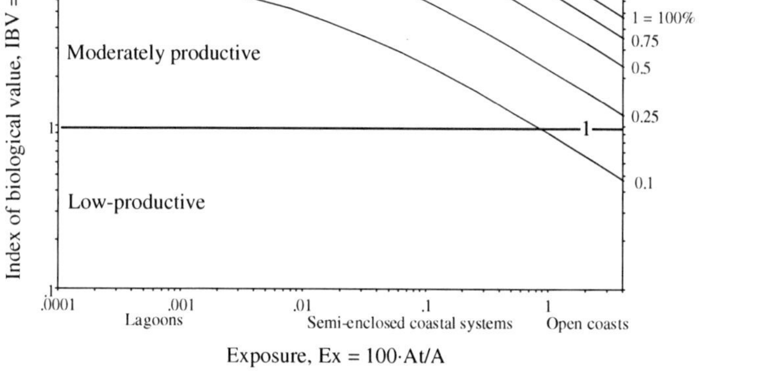

Fig. 5.24 Nomogram illustrating how the Index of Biological Value (IBV) is related to the exposure (Ex) and the area above the Secchi depth (A_{sec})

productive, productive, moderately productive and low-productive", can, of course, be discussed and this nomenclature is given here as a suggestion. However, the definition of IBV is well motivated by comprehensive empirical data. Since A_{sec} and Ex are easy to define and understand, also IBV is easy to apply in practice in coastal management. To determine A_{sec}, one would need data from standard monitoring programs on the Secchi depth as well as a bathymetric map so that the hypsographic curve can be produced. To determine the exposure, one must define the boundary lines for the coastal area by means of the topographical bottleneck approach and determine the section area (At) and the enclosed coastal area (Area). This can be done easily if digitized bathymetric information is available (see Chap. 3).

Since hypsographic curves may not be accessible for certain coastal areas, one can also estimate A_{sec} using (5.7), which was originally derived for lakes (see Håkanson 1999). Equation 5.7 gives A_{sec} (in m), as a function of the form factor, Vd, the area of the coast (Area in m^2) and the maximum depth of the coast (D_{max} in m). It would be interesting to try to derive a similar function also for open coastal areas, but that has been beyond the aim of this work. Equation 5.7 has, however, been used to calculate A_{sec} in Fig. 5.25. This also means that it would probably be possible to create a more relevant frequency distribution and obtain better statistical information regarding A_{sec} for these Baltic coastal areas, but for the present purpose these results may serve as examples illustrating the practical use of the IBV-index.

$$A_{sec} = \text{Area} - \text{Area}\cdot((D_{max} - D_{sec})/(D_{max} + D_{sec}\cdot \text{EXP}(3 - \text{Vd}^{1.5})))^{(0.5/\text{Vd})} \quad (5.7)$$

Where EXP is the exponent and D_{sec} the Secchi depth. Deep, U-formed systems generally have smaller areas above the Secchi depth.

Figure 5.25 gives frequency distributions and statistics (mean values, medians, standard deviations, minimum and maximum values) for the area above the Secchi depth (from (5.7)) and the requested Index of Biological Value calculated for 478 Baltic coastal areas.

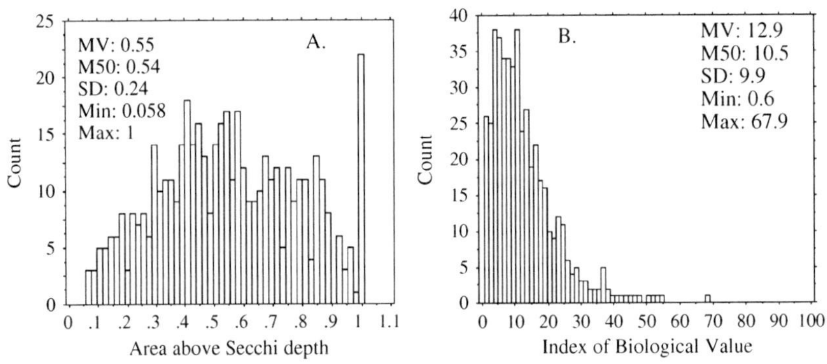

Fig. 5.25 Frequency distributions and statistics (mean values, medians, standard deviations minimum and maximum values) based on data from 478 Baltic coastal areas for the area above the Secchi depth and the Index of Biological Value

One can note that the IBV varies between 0.6 (low-productive) to 68 (extremely productive); the frequency distribution is positively skewed and the mean value is 12.9 (productive coastal area) and the median 10.5 (also indicative of productive conditions). There are 5 (1%) extremely productive coastal areas (IBV > 50), 43 (9%) very productive coastal areas (25 < IBV < 50), 209 (43.7%) productive coastal areas (10 < IBV < 25), 214 (63.0%) moderately productive coastal areas (1 < IBV < 10) and 7 (1.5%) low-productive coastal areas (IBV < 1).

Using geographical information systems (GIS) based on digitized bathymetric data, it is easy to apply these concepts and determine IBV for any given coastal area.

5.7 Summary and Conclusions

This chapter has discussed a comparative study based on a comprehensive set of empirical data from many international databases including fresh water systems, coastal brackish water areas and marine coastal areas. We have targeted on the following operational effect variables (bioindicators), which are meant to reflect key structural and functional aspects of aquatic ecosystems: Secchi depth (as a standard measure of water clarity), chlorophyll-a concentrations (a measure of primary phytoplankton biomass), the concentration of cyanobacteria (a measure of harmful algae), the oxygen saturation in the deep-water zone (an indicator reflecting sedimentation, oxygen consumption, oxygen concentrations and the habitat conditions for zoobenthos, an important functional group) and the macrophyte cover (an important variable for the bioproduction potential, including fish production, and the "biological value" of aquatic systems). For a wide range of systems, these bioindicators can be predicted using practically useful models, i.e., models based on variables that can be accessed from standard monitoring programs and maps. These bioindicators are regulated by a set of abiotic factors, such as salinity, temperature, light, nutrient concentrations (N and P), morphometry and water exchange. Empirical data ultimately form the basis for most ecological/environmental studies and this book uses maybe the most comprehensive data-set ever related to trophic level conditions and these bioindicators. We have also given compilation of empirically-based (statistical) models quantifying how the variables are interrelated and how they reflect fundamental aspects of aquatic ecosystems.

There was no simple relationship between the TN/TP-ratio and empirical chlorophyll concentrations or concentrations of cyanobacteria. Variations in TP rather than TN generally seem to be more important to predict variations among systems in chlorophyll-a and cyanobacteria. Different "bioavailable" forms of nutrients (DIN, DIP, phosphate, nitrate, etc.) have been shown to have very high coefficients of variation (CV), which means that many samples are needed to obtain reliable empirical data which are necessary in models aiming for high predictive power and practical usefulness of the operational bioindicators discussed in this chapter.

This work has motivated and introduced an index expressing the production potential or the biological "value" of coastal areas. Note that the class limits and categories defined in Fig. 5.24 should be regarded with due reservation. New data and information may motive changes in the classes and the nomenclature. Evidently, much more could be done to critically test this approach for other systems covering wider gradients in salinity and trophic status, Secchi depths and exposure. This work may be considered as a steppingstone toward such goals.

Coastal areas of different size may have similar IBV-values. From the perspective of regional, national or international sustainable management, it is, evidently, most important to try to maintain high IBV-values for the larger areas. The main threat to the bioproduction potential and the biological value, as expressed by IBV, is eutrophication, which lowers the Secchi depth and hence also A_{sec} and IBV.

Chapter 6
Case-Studies

6.1 Introduction and Aim

Process-based dynamic mass-balance models are essential tools for gaining a deeper understanding of how a number of complex processes in an aquatic system together determine the concentration of a studied substance, e.g., a nutrient or a contaminant (Håkanson 1999). Mass-balance models have long been used as a tool to study lake eutrophication (Vollenweider 1968; OECD 1982) and also used in different coastal applications (see Sect. 9.1). Mass-balance modeling makes it possible to produce predictions of what will likely happen to a system if the conditions change, e.g., a reduced discharge of a pollutant related to a remedial measure. Using dynamic ecosystem models, it is also possible to predict thresholds and points of no return before they have been reached, and hence to take action to avoid them.

Mass-balance modeling can be performed at different scales depending on the purpose of the study. A large number of coastal models do exist, all with their pros and cons. For example, the 1D-nutrient model described by Vichi et al. (2004) requires meteorological input data with a high temporal resolution, which makes forecasting for longer time periods than one week ahead very difficult. The 3D-model used by Schernewski and Neumann (2005) has a temporal resolution of 1 minute and a spatial resolution of much less than 1 m, which means that it is difficult to find reliable empirical data to run and validate the model. The model used for the case-studies in this chapter, CoastMab, is a general mass-balance model for entire coastal areas (the ecosystem scale). It is based on ordinary differential equations and monthly time steps to account for seasonal variations. It has been tested and applied for phosphorus (Håkanson and Eklund 2007b) and contaminants (Håkanson and Lindgren 2007c) with good results. For mass-balance models for coasts, data must be available both for the given coastal area and for the sea outside the given coastal area because these conditions will influence the conditions within the coastal area. This is most evident for open coasts, but also true for more enclosed systems. This explains why the topographical openness (= exposure) of coastal areas is a key attribute in the classifications system discussed in this book.

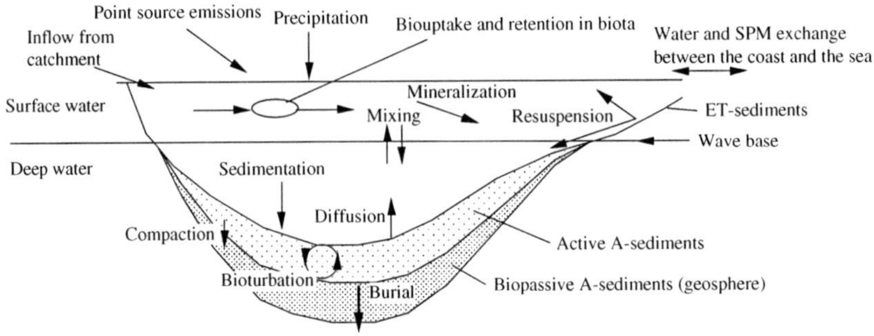

Fig. 6.1 An outline of transport processes (= fluxes) and the structure of the dynamic coastal model (CoastMab) for phosphorus and suspended particulate matter (SPM)

The transport processes (sedimentation, resuspension, diffusion, mixing, biouptake, etc.) quantified in the CoastMab-model are general and apply for all substances in all/most aquatic systems (see Fig. 6.1), but there are also substance-specific parts (mainly related to the particulate fraction and the criteria for diffusion from sediments). These processes have the same names for all systems and for all substances; so

- sedimentation is the flux from water to sediments of suspended particles and nutrients attached to such particles,
- resuspension is the advective flux from sediments back to water, mainly driven by wind/wave action and slope processes;
- diffusion in Fig. 6.1 is the flux from sediments back to water triggered by concentrations gradients, which would often be influenced by small-scale advective processes; even after long calm periods, there are currents related to the rotation of the earth, variations of low and high-pressures, temperature variations between day and night, bioturbation by benthic animals, etc.; it should be noted that it is difficult to measure flow velocities lower than 1–2 cm/sec in natural aquatic systems;
- mixing (or large-scale advective transport processes) is the transport between surface-water areas and deeper water layers;
- mineralization (and regeneration of nutrients in dissolved forms) by decomposition of organic particles by bacteria;
- primary production is creation of living suspended particles from dead matter by sunlight;
- burial is the transport of matter from the biosphere to the geosphere often of matter from the technosphere;
- outflow is the flux out of the system of water and everything dissolved and suspended in the water.

Note that CoastMab is not a model where the user should make any tuning or change model constants. The idea is to have a model based on general and mechanistically correct algorithms describing the transport processes at the ecosystem scale and to calculate the role of the different transport processes; and how a given system would react to changes in inflow related to natural variations and anthropogenic reductions of water pollutants. This chapter includes three case-studies. The first one concerns how point-source emissions of nutrients to coastal areas affect the receiving waters when all other important nutrient fluxes to, within and from the given area are accounted for. The second case-study addresses the question how to find the proper reference values related to "good" water quality to set the targets for the remedial actions. The third case-study concerns reconstruction of eutrophication.

6.2 How Important are Local Nutrient Emissions to Eutrophication in Coastal Areas Compared to Fluxes from the Outside Sea?

6.2.1 Aim of the Case-Study

This case-study addresses a key issue in coastal management: If investments are being made to reduce local nutrient emissions to coastal areas, e.g., from industries and other point sources (such as fish farms), from diffuse sources or by means of changing agricultural practices, what are the benefits for the local receiving water system? And how much of the local emissions would be transported out of the coastal system and contaminate the outside sea? To answer such questions, it is evident that one needs to quantify all major fluxes of water and nutrients/contaminants to put the reductions into the proper context. This work will use the CoastMab-model in three different forms:

1. CoastMab for salt will provide water fluxes to, within and from the given coastal area and also the basic algorithms for (a) the theoretical water retention times (which also influence the turbulence of the system and hence also sedimentation of particulate matter), (b) the mixing transport between the surface and the deep-water layers and (c) diffusive fluxes of dissolved substances (such as salt and dissolved forms of nutrients).
2. CoastMab for phosphorus will provide the requested nutrient fluxes and put the nutrient fluxes from the tributaries and from local emissions into a framework where also the exchange of nutrients between the given coastal area and the outside sea are calculated. The main difference between CoastMab for salt and CoastMab for phosphorus relates to the fact that:

    - Phosphorus may appear in two different functional forms, the particulate fraction (PF), which is subject to gravitational sedimentation, and the dissolved fraction (DF = 1 – PF), which is subject to biouptake.
    - And that the phosphorus deposited in the sediments may return to the water phase by means of advective and diffusive transport processes.

The advective fluxes are mainly caused by wind-induced wave action and slope processes and the diffusive internal loading mainly by high sedimentation of organic material leading to high oxygen consumption, low oxygen concentrations and a low redox potential in the sediments, which favors the formation of high levels of dissolved phosphorus in the sediments, which trigger a high diffusion of phosphorus from the sediments. These processes are well known and included in textbooks in recent sedimentology (see, e.g., Håkanson and Jansson 1983) and these processes are also included in the CoastMab-model.

3. CoastMab for suspended particulate matter (see Håkanson 2006), which also predicts water clarity (Secchi depth) and sedimentation of matter and how these factors relate to nutrient fluxes (how the nutrient concentrations regulate the internal production of suspended particles) and the salinity (which regulates the aggregation of suspended particles and hence also sedimentation and water clarity).

A central question in coastal management is how a given system would respond to suggested measures. How long would it take to reach a new steady state? What are the characteristic new nutrient concentrations in the water? And how would key bioindicators for eutrophication (see, e.g., Nixon 1990; Livingston 2001; Schernewski and Schiewer 2002; Schernewski and Neumann 2005; Moldan and Billharz 1997; Bortone 2005), such as chlorophyll-a concentration, concentration of cyanobacteria, oxygen concentration in the deep-water zone or Secchi depth change? In short, what is the environmental benefit related to the remedial costs? Such questions are addressed in this section using a general process-based quantitative approach, which could also be used for other coastal areas than the case-study area discussed here, the Himmerfjärden Bay in the Baltic Sea.

Eutrophication is ranked as the most severe threat to the Baltic Sea (Savage et al. 2002; Bernes 2005). The Himmerfjärden Bay has been chosen as a case-study area here because it has been investigated intensively since 1976 and long data series on nutrient levels and water quality variables are available (see Elmgren and Larsson 1997; Larsson et al. 2006). Khalili (2007) has presented a literature study on previous eutrophication research in the Himmerfjärden Bay. The research in the bay includes four large-scale nutrient regulation tests related to the discharges from a water treatment plant, Himmerfjärdsverket. The results from the Himmerfjärden Bay are often cited and used to motivate the benefits of nitrogen reductions. This strategy has been questioned and the debate has been lively (Rabalais and Nixon 2002; Rönnberg and Bonsdorff 2004; Howarth and Marino 2006; Boesch et al. 2006). Engqvist (1996) has previously modeled the salinity and the N and P cycles in the Himmerfjärden Bay, and although this section may be seen as a complement to that work, we would like to point out some major shortcomings with his model:

- Engqvist's work contains site-specific constants, which are more or less impossible to measure and have not been determined in a general way. Such modeling provides ample room for ambiguous and faulty tuning and it also makes it difficult to falsify the model in a Popperian sense (Mann 1982; Peters 1991; Bryhn and Håkanson 2007).

- His model also lacks a general algorithm to quantify sediment resuspension (and sediment internal loading), which is necessary to correctly simulate nutrient cycles, as demonstrated in this book.
- It does not contain any general algorithms for nitrogen fixation, denitrification and dry deposition of nitrogen, which are needed to predict changes in nitrogen concentrations.

The following section will present the data used in the mass-balance calculations for salt, phosphorus and suspended particulate matter (SPM). A central part of this case-study is to compare modeled data on the target variables (salinity, phosphorus concentration, Secchi depth, chlorophyll and oxygen concentrations) with empirical data. We will also present model predictions of cyanobacteria, SPM-concentrations, phosphorus concentration in sediments and sedimentation but in those cases there are no comparable reliable empirical data accessible to us. The main results concern the dynamic response of the system to reductions in nutrient loading and the analyses and interpretations of those results.

6.2.2 Information on the Himmerfjärden Bay

The Himmerfjärden Bay, situated about 60 km south of Stockholm at 59° 00' N, 17° 45', is a narrow bay divided into four sub-basins (Engqvist 1999; Boesch et al. 2006). The basins are separated by topographical thresholds, and just outside the outer basin, to the south, is the area Hållsfjärden. Hållsfjärden is commonly used as a reference area for the Himmerfjärden Bay and holds a reference station called B1. There are five sampling stations in the Himmerfjärden Bay, H2–H6. The Himmerfjärden Bay is connected to Lake Mälaren in the north but the freshwater inflow to the bay is limited to a few short periods when the water level in the lake is high (Elmgren and Larsson 1997). The Himmerfjärden Bay has been monitored since the middle of the 1970s when sewage water from the area southwest of Stockholm was redirected from Lake Mälaren to the Himmerfjärden Bay. In 1974, the treatment plant in the Himmerfjärden Bay began to remove phosphorus and 96% of the phosphorus is, on average, removed today. The treatment plant initially served about 90,000 people but the population increased rapidly, causing an increase in primarily nitrogen fluxes. Today, the plant serves 240,000 people (Boesch et al. 2006). Extensive nitrogen removal has been implemented since the late 1990s reaching about 90% in 1998 (Larsson and Elmgren 2001).

It must be stressed that it has been assumed that the emissions from the sewage treatment plant contributes with flows of nitrogen of such significance that the nitrogen reductions would have a clear effect on the eutrophication status in the Himmerfjärden Bay. This assumption was mainly based on the fact that total nitrogen concentration and inorganic concentrations of nitrogen before the spring bloom at station H4 correlated ($r^2 = 0.69$, $n = 16$) with the load from the sewage treatment plant. Changes in eutrophication status have consequently been interpreted mainly as results of treatment plant regulatory measures (Elmgren and Larsson 1997, 2001; Larsson and Elmgren 2001).

The first large-scale experiment in the Himmerfjärden Bay was performed in 1983 when the concentration of phosphorus in the water from the treatment plant was allowed to increase to about fourfold (or twofold as compared to the amount discharged annually in 1983, i.e., 31 tons). According to Elmgren and Larsson (1997), no increase in primary production was observed following this increase in phosphorus loading but a slight increase in heterocytes (the nitrogen fixing cells in cyanobacteria) was noted and this increase occurred mainly at station B1 in the reference area possibly implying that the growth of cyanobacteria in the Himmerfjärden Bay reflects the growth in the outside sea.

The second large-scale experiment started in 1985 when the treatment plant increased its capacity and began receiving sewage from Eolshälls treatment plant resulting in increased emissions of nitrogen to the Himmerfjärden Bay. The increase was followed by a successive decrease when nitrogen reduction processes were introduced and became more and more efficient reaching about 50% in 1992. As in the case with the first experiment, no clear increase in primary production occurred following increasing nitrogen inputs to the bay. Elmgren and Larsson (1997) suggested that phosphorus at this time was the main limiting nutrient in the Himmerfjärden Bay and that the excess nitrogen was exported to the outside sea causing increased eutrophication in the outside sea.

As mentioned, Elmgren and Larsson (1997) found no significant correlation between eutrophication indicators, such as chlorophyll or Secchi depth, and varying loads of nitrogen and phosphorus from the sewage treatment plant following the two first large-scale experiments. They concluded that further removal of phosphorus would not be meaningful since the emissions from the treatment plant constitute a small fraction of the total loading of phosphorus. They recommended to increase nitrogen removal from the treatment plant.

Following the recommendations from Elmgren and Larsson (1997), extensive nitrogen removal (about 90%) began in 1998. A third large-scale experiment was performed in 2001–2002 when emissions of nitrogen were deliberately doubled. As in the previous cases, no increase in chlorophyll-a levels was observed related to the increase in nitrogen from the sewage treatment plant (Boesch et al. 2006).

According to Boesch et al. (2006) both the experiment in 1983 and the two experiments with increased nitrogen emissions may have been too small and or too short to result in clear changes in primary production in the bay.

The following simulations will use morphometric data from Khalili (2007) who made an analysis of the Himmerfjärden Bay using Geographical Information Systems (GIS). Basin-specific data are compiled in Table 6.1, which gives information on, e.g., total area; volume; mean depth; maximum depth; the depth of the theoretical wave base at 19 m (see Fig. 6.2 for motivation); the volume of the surface-water layer and the deep-water layer separated by the theoretical wave base; the section area, which defines the cross-sectional area that separates the given coastal area from the outside sea (Fig. 6.2); the tributary water discharge to the bay; the discharge of water and phosphorus from the plant; the catchment area; latitude; and

Table 6.1 Data on Himmerfjärden Bay (see Khalili 2007, for more information). Bolded factors are obligatory (coastal-area typical) driving variables

Catchment area (ADA in km^2)	**1 268**
Annual precipitation (Prec, mm/yr)	**460**
Area (Area in km^2)	**234**
Area below wave base at 19 m (A_{Dwb} in km^2)	**46**
Maximum depth (D_{max} in m)	**52**
Wave base (D_{wb} in m)	**19**
Dynamic ratio (DR = $\sqrt{Area}/D_m$)	1.24
Areas of fine sediment erosion and transport (ET, dim. less)	0.80
Exposure (Ex = 100·At/Area)	0.0194
Form factor (Vd = 3·D_m/D_{max})	0.71
Land rise (LR, mm/yr)	**4**
Latitude (Lat, °N)	**59**
Mean depth (D_m=V/Area, m)	**12.3**
Water flow from plant (Q_{plant}, m^3/yr)	**35,000,000**
Water flow from rivers (Q_{trib}, m^3/yr)	**491,600,000**
Section area (At, m^2)	**45,310**
TP-emissions from plant ($F_{TPplant}$, kg/yr)	**1 632**
Volume of DW-layer (V_{SW}, km^3)	**0.236**
Volume of SW-layer (V_{DW}, km^3)	**2.642**
Total volume (V, km^3)	**2.878**

mean annual precipitation. As stressed, the theoretical wave base separates the transportation areas (T), with discontinuous sedimentation of fine materials, from the accumulation areas (A), with continuous sedimentation of fine suspended particles (see Chap. 3).

The Himmerfjärden Bay has been divided into two depth intervals:

1. The surface-water layer (SW), i.e., the water above the theoretical wave base at 19 m.
2. The deep-water layer (DW) is defined as the volume of water beneath the theoretical wave base (see Table 6.1).

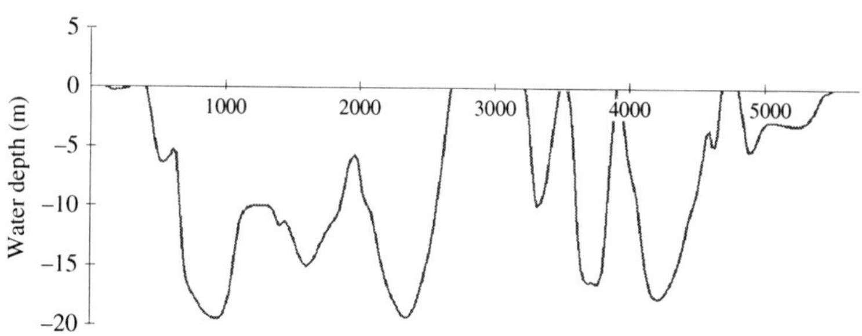

Fig. 6.2 Limiting section area profile between Askö-Torö in the Himmerfjärden Bay

It should be noted that the theoretical wave base is meant to describe average conditions. During storm events, the wave base will likely be at greater water depths (see Jönsson 2005) and during calm periods at shallower depths. Khalili (2007) has presented a hypsographic curve and a corresponding volume curve for the Himmerfjärden Bay and those curves have been used in this work to calculate the volumes given in Table 6.1. One can note that the area below the theoretical wave base (D_{wb}) is 46 km^2 and that the SW-volume is 2.6 km^3, the volume of the DW-layers is small (only 0.24 km^3) and the entire volume is 2.88 km^3. The boundary lines for the Himmerfjärden Bay used in this work are from Khalili (2007); the total section area (At), which provides a minimum value of the exposure (Ex = 100·At/A; see Fig. 6.2, for more information regarding the topographical bottleneck method to objectively define the boundary lines for coastal areas) is 45,310 m^2, which gives an exposure (Ex) of 0.0194, indicating the enclosed character of the bay.

Table 6.2 shows the driving variables used in this work (calculated from the ongoing monitoring program for the period 1997–2007). This table gives monthly mean values for the number of hours with daylight (needed to calculate chlorophyll concentration), salinity in the surface water outside the Himmerfjärden Bay in the Baltic Proper (needed in the mass-balance calculations of salt to calculate water inflow from the Baltic Proper), the Secchi depth in the area outside of the Himmerfjärden Bay (needed to calculate the inflow of SPM from the Baltic Proper), the TP-concentrations in the SW and DW-layers in the area outside of the Himmerfjärden Bay (needed to calculate the inflow of TP from the Baltic Proper), and the empirical monthly SW and DW-temperatures in the Himmerfjärden Bay (needed to quantify mixing).

Table 6.2 Monthly data on driving variables (mean monthly number of hours with daylight), surface-water (SW) salinity in the sea outside of the Himmerfjärden Bay (the Baltic Proper, BP), Secchi depth outside the bay, TP in SW and DW-water outside the bay and SW and DW-temperatures in the bay

Month	Daylight	$Salinity_{SWBP}$	$Secchi_{BP}$	TP_{SWBP}	TP_{DWBP}	$Temp_{SW}$	$Temp_{DW}$
	hr/month	psu	psu	µg/l	µg/l	°C	°C
1	9.1	7.18	8.5	30.1	31.2	0.7	1.3
2	11.8	7.12	8.5	27.2	29.3	1.1	1.2
3	14.3	7.01	7.0	22.1	24.3	3.0	2.3
4	17.1	7.01	9.0	18.2	21.2	8.0	4.0
5	18.5	7.06	7.6	18.1	22.4	12.6	5.9
6	18.1	6.96	5.3	16.6	24.5	16.1	7.5
7	15.8	6.96	5.5	17.9	24.8	18.0	8.3
8	13.0	6.93	6.7	19.3	26.0	14.6	8.0
9	10.1	6.92	8.5	23.1	27.1	10.8	8.1
10	7.3	7.00	9.5	25.9	26.9	7.6	7.1
11	6.4	7.09	7.8	30.9	33.3	1.3	2.1
12	6.4	7.09	7.8	30.9	33.3	1.3	2.1
MV	12.3	7.03	7.7	23.4	27.0	7.9	4.8
M50	12.4	7.01	7.8	22.6	26.5	7.8	4.9
SD	4.5	0.08	1.3	5.5	4.0	6.4	2.9

In the following modeling, we will compare the modeled values for the target variables mainly to the confidence intervals related to ±1 standard deviation of the mean monthly empirical data accessible to us from the bay.

6.2.3 The Dynamic CoastMab-Model Used in the Case-Study

The model consists of five compartments (see Fig. 6.1): surface water (SW), deep water (DW), erosion/transportation areas for fine sediments (ET), accumulation areas for fine sediments below the theoretical wave base (A) and also biota with short turnover times (BS; see Sect. 9.1). There are algorithms for all major internal fluxes of salt, TP and SPM (outflow, TP and SPM from land uplift, sedimentation of particulate TP and SPM, resuspension of TP and SPM, diffusion of salt and dissolved phosphorus, mixing, biouptake of dissolved phosphorus and burial of TP and SPM and mineralization of the organic fraction of SPM; see Håkanson and Eklund (2007b) and Håkanson (2006) which give and motivate all equations and model variables).

To calculate the inflow of salt, TP and SPM to the Himmerfjärden Bay (HI) from the Baltic Proper (BP), data on the concentrations in the SW and DW-layers in the Baltic Proper from Table 6.2 have been used. The inflows to the two layers from the Baltic Proper are given by the water discharges in Table 6.3 (Q_{SWBPHI} and

Table 6.3 Compilation of calculated monthly data (using the mass-balance for salt) for water transport (Q in million m^3/month) related to evaporation (Q_{eva}), surface water (SW) inflow from the Baltic Proper (BP) to the Himmerfjärden Bay (HI), deep-water (DW) inflow from BP to HI, mixing from DW to SW in HI, DW outflow from HI to BP, SW outflow from HI to BP, water flow related to direct precipitation onto the surface area of HI (Q_{prec}), inflow of water to SW and DW from the water treatment plant ($Q_{SWplant}$ and $Q_{DWplant}$) and freshwater inflow from tributaries (Q_{trib})

Month	Q_{eva}	Q_{SWBPHI}	Q_{DWBPHI}	Q_{DWSWx}	Q_{DWHIBP}	Q_{SWHIBP}	Q_{prec}	$Q_{SWplant}$	$Q_{DWplant}$	Q_{trib}
1	8.1	4050	450	278	453	4104	9.0	2.9	0.4	52.9
2	8.1	4050	450	303	453	4095	9.0	2.9	0.4	44.3
3	8.1	4050	450	248	453	4099	9.0	2.9	0.4	47.9
4	8.1	4050	450	37	453	4110	9.0	2.9	0.4	59.1
5	8.1	4050	450	18	453	4110	9.0	2.9	0.4	59.4
6	8.1	4050	450	13	453	4084	9.0	2.9	0.4	33.2
7	8.1	4050	450	10	453	4078	9.0	2.9	0.4	27.1
8	8.1	4050	450	13	453	4076	9.0	2.9	0.4	25.0
9	8.1	4050	450	151	453	4078	9.0	2.9	0.4	26.7
10	8.1	4050	450	276	453	4081	9.0	2.9	0.4	30.3
11	8.1	4050	450	243	453	4097	9.0	2.9	0.4	46.4
12	8.1	4050	450	256	453	4097	9.0	2.9	0.4	46.4
MV	8.1	4050	450	154	453	4092	9	2.9	0.4	42
M50	8.1	4050	450	197	453	4096	9	2.9	0.4	45
SD	0	0	0	125	0	13	0	0	0	13

Q_{DWBPHI}) and the given concentrations. Engqvist's (1996) calculations of the salt exchange is a factor 2.2 lower than in Table 6.3, which may be a result of different vertical resolutions and different borderlines between the bay and the Baltic Proper. The empirical Secchi depths in Table 6.2 have been recalculated into SPM-values by (5.1). So, SPM values (in mg/l) are calculated from measured Secchi depths (in m) and the salinity of the SW-layer in the area outside the Himmerfjärden Bay (Sal_{SW} in psu). The higher the salinity, the higher the aggregation and the clearer the water.

Figure 6.3A and B give comparisons between modeled salinities and measured values (the uncertainty bands related to ± 1 standard deviation). The idea is that the modeled values should lie in-between these uncertainty bands. This is one main way of controlling the model predictions, another way is shown in Table 6.4.

From extensive measurements in many coastal areas (see Håkanson et al. 1986), one can conclude that typical water velocities in limiting section areas generally range between 1 and 15 cm/sec for coastal areas in the Baltic Sea. Lower mean velocities than 1 cm/sec would be rather unrealistic on a monthly basis. The water velocity in the section area has been calculated for the total outflow (m^3/yr) divided by half the section area since there is also inflow of water to maintain a given water level ((m^3/yr)·(1/(0.5·m^2)). These calculations give an average velocity in the section area of 7.7 cm/sec, which is in the middle of the expected range.

Another way to check the modeled water fluxes between the coast and the sea is to compare these model predictions from the mass-balance for salt with data from an empirically-tested model for the theoretical surface-water retention time (T_{SW}). It has been shown (Persson et al. 1994b) that T_{SW} can be predicted very well with the regression in Table 3.2, which is based on the exposure (Ex), which, in turn is a function of section area (At) and coastal area (Area). The range of this model for T_{SW} is given by the minimum and maximum values for Ex of $0.002 < Ex < 1.3$; Ex = 0.0194 for the Himmerfjärden Bay is within this range. The model should not be used without complementary algorithms if the tidal range is >20 cm/day or for coastal areas dominated by fresh water discharges.

For open coasts, i.e., when Ex > 1.3, T_{SW} may be calculated not by this equation but from a model based on coastal currents (see Sect. 9.1 or Håkanson 2006).

From Table 6.4, one can note the good correspondence between T_{SW}-values calculated using the mass-balance for salt (mean value = 0.62 months) and with the equation in Table 3.2 (mean value = 0.59 months). One can also see from Table 6.4 that the theoretical deep-water retention time, T_{DW}, is short (0.41 months on average) because the volume of the DW-layer is small and T_{DW} is defined from the ratio between the volume of the DW-layer and the total water flux to the DW-layer. The total water residence time according to Engqvist (1996) is higher; 1.5 months or 44 days, an estimate which deviates significantly from both methods used to produce the data in Table 6.4.

Also note that there has been no calibration or tuning regarding the water fluxes given in Table 6.3 and that these fluxes are used also to calculate the TP and SPM-fluxes. The monthly data on tributary water discharge used in the modeling have been calculated from the empirical average annual value using the dimensionless moderator for this purpose (from Abrahamsson and Håkanson 1998). This

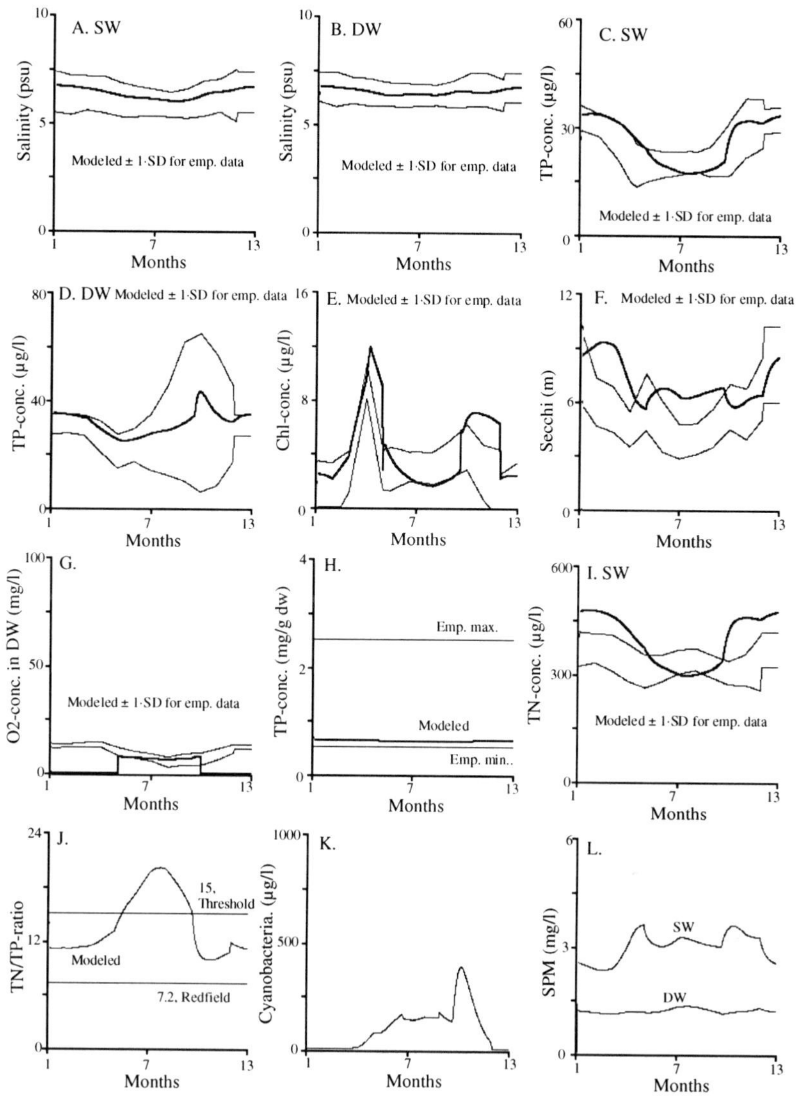

Fig. 6.3 Comparison between modeled values and uncertainty bands for the empirical mean values representing ±1 standard deviation for **A.** SW-salinity, **B.** DW-salinity, **C.** TP-concentration in SW, **D.** TP-concentration in DW, **E.** Chlorophyll, **F.** Secchi depth, **G.** O_2-concentration in DW, **H.** modeled TP-concentration in accumulation area sediments in relation to minimum and maximum reference values, **I.** modeled TN-concentration in relation to ±1 standard deviations for the empirical mean values, **J.** modeled TN/TP-ratios in relation to the Redfield ratio (7.2) and the Threshold ratio (15), **K.** modeled values of cyanobacteria, and **L.** modeled SPM-concentrations in the SW and DW-layers in the Himmerfjärden Bay

Table 6.4 Modeled monthly values the flow velocity of water in the section area, the theoretical surface-water (SW) retention time calculated from the mass-balance for salt (T_{SW}), and from the empirical morphometrical formula based on the exposure (Ex) (T_{SWEx}) and for the deep water (DW) according to the mass-balance for salt (T_{DW})

Month	Monthly flow velocity	Theor. wat. ret. time	Theor. wat. ret. time	Theor. wat. ret. time
	u_{At}, cm/sec	T_{SWEx}, months	T_{SW}, months	T_{DW}, months
1	7.66	0.59	0.60	0.32
2	7.66	0.59	0.59	0.31
3	7.66	0.59	0.60	0.34
4	7.66	0.59	0.63	0.48
5	7.66	0.59	0.63	0.50
6	7.66	0.59	0.64	0.51
7	7.66	0.59	0.64	0.51
8	7.66	0.59	0.64	0.51
9	7.66	0.59	0.62	0.39
10	7.66	0.59	0.60	0.33
11	7.66	0.59	0.60	0.34
12	7.66	0.59	0.60	0.33
MV	7.66	0.59	0.62	0.41
M50	7.66	0.59	0.62	0.39
SD	0	0	0.02	0.09

moderator is based on data on the size of the catchment area, mean annual precipitation and latitude (for data, see Table 6.1). Since we do not have access to reliable empirical monthly data on tributary water discharge for the study period (1997–2007), it should be stressed that this modeling concerns average, characteristic conditions on a monthly basis for this period of time and not the actual sequence of months.

The theoretical water retention times in the two layers from the basic mass-balance for salt (see Table 6.4) are used together with the temperature-dependent mixing rate in the mass-balance model as indicators of how the turbulent mixing influences the settling velocity for particulate phosphorus and SPM – the faster the water renewal, the more turbulence, the lower the settling velocity. The small TP-input from precipitation onto the water surface of the Himmerfjärden Bay has been estimated from the characteristic annual precipitation of 460 mm and a TP-concentration in the rain of 5 μg/l (see Håkanson and Eklund 2007b).

The internal processes are: sedimentation of particulate phosphorus from surface water to deep water (F_{TPSWDW}), sedimentation from SW to areas of erosion and transportation (F_{TPSWET}), sedimentation from DW to accumulation areas (F_{TPDWA}), resuspension (advection) from ET-areas (including TP from land uplift, F_{TPLU}) either back to the surface water (F_{TPETSW}) or to the deep water (F_{TPETDW}), diffusion of dissolved phosphorus from accumulation area sediments to the DW-layer (F_{TPADWd}), diffusion from DW-water to SW-water ($F_{TPDWSWd}$), upward and downward mixing between SW and DW ($F_{TPDWSWx}$ and $F_{TPSWDWx}$) and biouptake and elimination of phosphorus from biota (F_{TPSWBS} and F_{TPBSSW}). When there is a partitioning of a flux from one compartment to two compartments, this is handled by a distribution coefficient (DC).

1. The DCs regulating the amount of phosphorus in particulate and dissolved fractions in the SW and DW-layers. These DCs are called particulate fractions (PF). By definition, only the particulate fraction of a substance is subject to gravitational sedimentation and only the dissolved fraction (DF = 1 – PF) may be taken up by biota. Table 6.5 gives a compilation of calculated PF-values for the SW and DW-compartments. One can note that the PF-values in the SW-compartment vary between 0.2 and 0.87 depending on season of the year (and how much TP is bound in biota) and that the PF-values in the DW-compartment are low during stratified conditions (when most phosphorus appear in dissolved form). The mean PF_{SW}-value is 0.51 (see Table 6.5). This is in good correspondence with the value 0.56, which is an average values based on extensive data from many aquatic systems (see Sect. 9.1 or Håkanson and Boulion 2002).
2. The DC regulating sedimentation of particulate phosphorus either to areas of fine sediment erosion and transport (F_{TPSWET}) or to the DW-areas beneath the theoretical wave base (F_{TPSWDW}). The calculated ET-value for this bay is 0.80 (i.e., 80% of the total area of the bay are dominated by areas with fine sediment erosion and transport).
3. The DC describing the resuspension flux from ET-areas back either to the surface water (F_{TPETSW}) or to the DW-compartment (F_{TPETDW}), as regulated by the form factor (Vd, where DC = Vd/3, Vd = $3 \cdot D_m/D_{max}$, D_m = the mean depth, D_{max} = the maximum depth).
4. The DC describing how much of the TP in the water that has been resuspended (DC_{res}) and how much that has never been deposited and resuspended ($1\text{-}DC_{res}$) in the SW and DW-layers. The resuspended fraction settles out faster than the materials that have not been deposited.

Land uplift (F_{LU}) is a special case (see Sect. 9.2). Land uplift is a main contributor of TP to the Baltic Proper (Håkanson and Bryhn 2007). From the map illustrating the spatial variation in land uplift (see Fig. 6.1), one can calculate that the mean land uplift in the Himmerfjärden Bay is about 4 mm/yr and this value has been used in these calculations. Land uplift has been discussed in many contexts (Voipio 1981; Jonsson et al. 1990; Jonsson 1992) and the algorithm to quantify how land uplift influences the fluxes of TP and SPM has been given by Håkanson and Bryhn (2007). The total area above the theoretical wave base in the Himmerfjärden Bay is about 188 km^2 and the sediments in this area will be exposed to increased erosion by wind/wave action due to the land uplift. The sediments in the shallower parts, which may have been deposited more than 1000 years ago, will be more consolidated than the recent materials close to the theoretical wave base. The calculation of the TP-flux from land uplift uses (1) modeled data on the TP-concentration in the accumulation area sediments from the DW-zone, (2) a water content of the sediments exposed to increased erosion set to be 15% lower than the modeled water content of the recent sediments and (3) the total volume of sediments above the theoretical wave base lifted each year.

Table 6.5 Modeled monthly values related to accumulation area sediments 0–10 cm; bulk density, organic content (= loss on ignition), water content, sedimentation and fall velocities of suspended particulate matter and particulate phosphorus

Month	Bulk density	Organic content	Water content	Sedimentation	Sedimentation	Fall velocity	Fall velocity	Particulate fraction	Particulate fraction
	bd g ww/cm^3	IG g/g dw	W % ww	Sed_{DW} µg/cm^2·d	Sed_{DW} cm/yr	v_{SW} m/month	v_{DW} m/month	PF_{DW}	PF_{SW}
1	1.17	6.3	75	12.1	0.01	2.4	2.3	0.22	0.22
2	1.17	6.3	75	11.2	0.01	2.4	2.3	0.23	0.20
3	1.17	6.3	75	10.8	0.01	2.4	2.2	0.23	0.26
4	1.17	6.3	75	14.6	0.02	2.5	2.2	0.02	0.81
5	1.17	6.3	75	17.8	0.02	2.4	2.2	0.01	0.82
6	1.17	6.3	75	19.5	0.02	2.4	2.3	0.01	0.86
7	1.17	6.3	75	21.6	0.03	2.4	2.3	0.01	0.87
8	1.17	6.3	75	18.6	0.02	2.4	2.3	0.02	0.80
9	1.17	6.3	75	11.8	0.01	2.5	2.3	0.53	0.46
10	1.17	6.3	75	12.3	0.02	2.5	2.2	0.37	0.31
11	1.17	6.3	75	13.2	0.02	2.5	2.3	0.18	0.27
12	1.17	6.3	75	12.6	0.02	2.5	2.3	0.20	0.23
MV	1.17	6.3	75	14.7	0.018	2.4	2.3	0.17	0.51
M50	1.17	6.3	75	12.9	0.020	2.4	2.3	0.19	0.39
SD	0	0	0	3.7	0.0062	0.051	0.036	0.17	0.29

6.2.3.1 Regressions Between TP, TN and Bioindicators

In this case-study, total nitrogen (TN) concentrations have been predicted from dynamically modeled monthly TP-concentrations using the regression given in Fig. 5.6C, i.e.:

$$\log(\mathrm{TN}) = 0.70.\log(\mathrm{TP}) + 1.61 \quad (6.1)$$
$$(r^2 = 0.88; n = 58 \text{ coastal systems})$$

There are, as discussed in Sect. 4.1, several reasons why we have not done any dynamic, process-based mass-balance modeling for nitrogen.

In this work, the modeling is done on a monthly basis and in the CoastMab-model there is information on the dissolved fraction of phosphorus. This means that the basic approach for the mean conditions during the growing season (Chl_{GS} in μg/l; see Fig. 5.10) has been modified to predict the requested mean monthly chlorophyll values (Chl). These calculations use simple dimensionless moderators to account for seasonal/monthly changes in the light conditions (DayL; mean monthly number of hours with daylight in the Himmerfjärden Bay; from standard tables) and in the amount of bioavailable/dissolved phosphorus (DF_{SW}). This means the chlorophyll-a concentration are predicted from:

$$\mathrm{Chl} = (\mathrm{DayL}/12.3)\cdot(\mathrm{DF_{SW}}/0.44)\cdot\mathrm{Chl_{GS}} \quad (6.2)$$

Where the basic model between the TP-concentration in the SW-layer (TP_{SW} in μg/l, modeled), the salinity in the SW-layer (Sal_{SW}, modeled) and Chl_{GS} is shown in Fig. 5.10. (DayL/12.3) is a dimensionless moderator based on the ratio between the monthly DayL-values divided by the mean annual number of hours with daylight (12.3) at this latitude (59 °N). The modeled monthly values of the dissolved fraction of phosphorus in the SW-layer (DF_{SW}) have been transformed into a dimensionless moderator by division with the average DF-value of 0.44 for phosphorus in surface water conditions. This means that the predicted chlorophyll values are low if DF is low, the number of hours with daylight low and the modeled TP-values low. The small variations in salinity (see Fig. 6.2A and B) will not influence the predicted Chl-values very much, but such variations are also accounted for.

The empirically-based model to predict the total concentration of cyanobacteria (Håkanson et al. 2007c) is given in Fig. 5.18. The following simulations will use dynamically modeled monthly TP-concentrations in the SW-layer, empirical mean monthly SW-temperatures, dynamically modeled SW-salinities and modeled monthly TN-concentrations in the SW-layer (from (6.1)) to predict monthly values of cyanobacteria in the SW-layer. Note that there are no empirical data available to us to test the predicted values for cyanobacteria, but these values are basically predicted from an empirical approach which yielded an r^2-value of 0.78, which is close to the maximum possible predictive power for cyanobacteria because of the inherently very high coefficient of variation (CV) for cyanobacteria (see Håkanson et al. 2007c). Nitrogen fixation by different species of cyanobacteria counteracts long-term nitrogen deficits, and the N-fixation rate depends on the TP-concentration, water temperature and the TN/TP-ratio (see Fig. 5.18).

When the mean O_2-concentration is lower than about 2 mg/l, and the mean oxygen saturation (O_2Sat in %) lower than about 20%, key functional benthic groups are extinct (see Fig. 5.4 and Pearson and Rosenberg 1976). Empirical data on the amount of material deposited in deep-water sediment traps (1 m above the bottom; Sed_{DW} in g dw/m$^2\Delta$ day) were used in deriving the sub-model for oxygen used in this modeling (see Fig. 5.5, from Håkanson 2006). This empirical model for oxygen is put into the dynamic SPM-model and the empirical data on sedimentation in the deep-water zone (Sed_{DW}) will be replaced by modeled values of Sed_{DW} from the dynamic SPM-model. The values for the O_2-concentration calculated in this manner for the growing season will be compared to empirical oxygen data from the Himmerfjärden Bay.

The oxygen model was basically derived using data from coastal areas without freshwater inflow. For coastal areas with freshwater inflow, the factor Q_{FS} will account for tributary influences (Q_{trib}) in the following way:

$$\text{If } Q_{trib} = 0 \text{ then } Q_{FS} = 1 \text{ else } Q_{FS} = ((Q_{trib} + Q_{salt})/Q_{salt})^\wedge(120/(1 + T_{DW}))) \tag{6.3}$$

Where the theoretical deep-water retention time (T_{DW}) is given in days. As mentioned, dimictic coastal areas in the Baltic Sea (i.e., coastal areas which become homothermal in the spring and in the fall) rarely have longer characteristics T_{DW}-values than 120 days. Q_{salt} is the total inflow (Q_{SW} plus Q_{DW}) of saline water from the outside sea. This means that if T_{DW} is 12.3 days as it is in the Himmerfjärden Bay, if Q_{trib} is 1% of Q_{salt} (see Table 6.3), Q_{FS} is 2.2 and the predicted O_2-concentration 7.3 mg/l and not 6.5 mg/l as it would have been expected if the coastal area did not have any tributary inflow.

This oxygen model should not be used for coastal areas dominated by tides.

The following section will demonstrate how this modeling predicts the salinities in the two layers, the TP-concentrations, Secchi depths, cyanobacteria and nitrogen concentrations and also other variables of interest, such as TP-concentrations in the sediments (0–10 cm) below the theoretical wave base (the accumulation-area sediments), sedimentation in the two layers, settling velocities for particulate phosphorus (and suspended particulate matter) and the particulate fractions (PF = 1−DF). Whenever possible, the modeled values will be compared to empirical data and to the uncertainty bands related to the empirical data. All calculated TP-fluxes in Himmerfjärden Bay and all calculated TP-amounts (where would one find the TP?) will be given.

6.2.4 Results

6.2.4.1 Modeled Values Versus Empirical Data

Figure 6.3 shows modeled values compared to the standard deviations for the mean empirical monthly data for the period 1997–2007. If the model yields values close to the empirical mean value and in-between these two uncertainty bands, the predictions should be regarded as good. For all these comparisons between modeled and empirical data, it should be noted that:

- The monthly tributary inflows of water, phosphorus and SPM have not been calculated using actual monthly inflow data (since no such data are available to us). This will explain some of the differences between the modeled and the measures monthly values.
- The monthly calculations of the inflow of water, salt, SPM and phosphorus from the Baltic Proper use mean values for the SW and the DW-compartments not in the inflowing water from the area outside the Himmerfjärden Bay, which would have been more appropriate, but values from a single station meant to reflect the conditions outside the bay in the Baltic Proper. This likely also further explains some of the differences between the modeled and the measured monthly values.

With these reservations, one can note from Fig. 6.3 that:

- The TP-concentrations in the SW and DW-layer (Fig. 6.3C and D) are within the defined uncertainty bands of the empirical data.
- The modeled chlorophyll concentrations give the "twin-peak pattern" as indicated by the empirical data but the predicted values are somewhat higher than the measured data in connection with the two peaks but within the given uncertainty bands for the main part of the growing season and for the winter period. It should be stressed that the chlorophyll concentrations are predicted from a regression based on dynamically modeled TP-concentrations in the surface water, the predicted salinities in the SW-layer and a dimensionless moderator for the average light conditions at this latitude. The modeled values also account for biouptake in biota with short turnover times, but they do not include any considerations to the biouptake and retention of phosphorus in biota with long turnover times (such as fish, zoobenthos, macrophytes). Given the limitations of the modeling for chlorophyll, one can conclude that the overall correspondence is relatively good.
- The predicted Secchi depths are also quite close to the empirical values and the temporal patterns agree quite well.
- The oxygen concentrations in the DW-layer (Fig. 6.3G) are within the defined uncertainty bands of the empirical data for the growing season.
- The predicted TP-concentrations in the accumulation areas sediments (0–10 cm; 0.61 mg/g dw) below the wave base of 19 m (Fig. 6.3H) are within the minimum and maximum lines defined for Baltic Sea coastal sediments (see Håkanson and Eklund 2007b) of 0.5–2.5 mg/g dw.
- The TN-concentrations are predicted from a regression (6.1) using dynamically modeled TP-concentrations in the SW-compartment. There is a relatively good correspondence between modeled and measured TN-concentrations (Fig. 6.3I).
- The TN/TP-ratios based on modeled values are given in Fig. 6.3J in relation to the Redfield ratio of 7.2 and the threshold ratio of 15 (see Håkanson et al. 2007c). The monthly TN/TP-ratios are clearly higher than 7.2 all months, and higher than 15 most summer months. Given the situation in the Himmerfjärden Bay, as revealed by these data, a lowering of the TN/TP-ratio will imply greater risks for blooming of cyanobacteria. So, to reduce the nitrogen concentration is

generally useless when the TN/TP-ratio is higher than 15, and can be harmful when the TN/TP-ratio is lower than 15, since this would stimulated the growth of cyanobacteria, which should be avoided.

- The predicted concentrations of cyanobacteria in the Himmerfjärden Bay under default conditions are given in Fig. 6.3K. Under calm conditions, when these algae will float to the surface, the concentrations in the upper meter of water may be many times higher than the mean values for the entire SW-layer shown in Fig. 6.3K.
- The modeled SPM-concentrations in the SW and DW-compartments are shown in Fig. 6.3L. These values will, together with the modeled salinities in Fig. 6.3A and D, determine the values for the Secchi depth given in Fig. 6.3F. For the open Baltic Sea, a default value of 3 mg/l (from Pustelnikov 1977) has often been used. Håkanson and Eckhéll (2005) have presented more data on SPM-concentrations in the Baltic Proper and those data show that typical values in the SW-layer are between 0.1 and 14 mg/l with a mean value of about 2 mg/l. The predicted values in the Himmerfjärden Bay agree quite well with the empirical data from those studies.

From Fig. 6.3, one can conclude that the CoastMab-model predicts the target variables quite well given the factual limitations in the seasonal patterns in the driving variables for tributary water discharge (since this pattern in the modeling is not based on measured data for the modeled period), in the seasonal pattern for the TP and SPM-concentrations outside Himmerfjärden Bay and the fact that CoastMab is a general model. Note that there has been no tuning of the general model to achieve these predictions and that the basic model has been shown to describe the transport processes for phosphorus very well for many other coastal areas (see Sect. 9.1). This should lend credibility to the following simulations.

6.2.4.2 Fluxes and Amounts of Phosphorus

Which are the large and the small TP-fluxes? And where is the phosphorus stored? Table 6.6 gives a compilation of all monthly TP-fluxes and a ranking of the annual fluxes. One can note that the two largest fluxes are biouptake and retention (= outflow) of phosphorus to and from biota. These fluxes are significantly larger than the fluxes related to in and outflow of TP between the Baltic Proper and the Himmerfjärden Bay.

The fluxes that may be reduced by remedial measures are bolded in Table 6.6 – the SW-inflow from the Baltic Proper ($F_{TPSWBPHI}$ = 1120 tons/yr), inflow to the DW-layer from the Baltic Proper ($F_{TPDWBPHI}$ = 170 tons/yr, tributary inflow (F_{TPtrib} = 26 tons/yr) and TP from the purification plant ($F_{TPplant}$ = 19 tons/yr). From this, it is evident that smaller reductions of the TP-emissions (say a factor of 2–4) from the plant will have small effects on the conditions in the bay. The phosphorus fluxes are dominated by the exchange processes between the coast and the sea.

The diffusive flux from the accumulation areas sediment (below the theoretical wave base at 19 m) is very small (<1 tons/yr). The modeled TP-concentration in

Table 6.6 A ranking of the annual fluxes (tons/yr), as calculated using the CoastMab-model based on the monthly fluxes (tons/month) of TP to, in and out of the Himmerfjärden Bay. The key fluxes for remedial measures are bolded. F = flux, SW = surface water, DW = deep water, HI = Himmerfjärden Bay, BP = Baltic Proper, trib = tributary, d = diffusive flux, x = mixing flux, LU = land uplift, ET = erosion and transport areas for fine sediments, A = accumulation areas for fine sediments, BS = biota with short turnover times, bur = burial

Month	1	2	3	4	5	6	7	8	9	10	11	12	Annual
F_{TPBSSW}	13.6	26.3	99.2	216.2	151.9	142.9	139.9	141.8	249.1	209.8	109.7	20.0	1521
F_{TPSWBS}	12.3	28.8	109.1	67.5	151.5	141.5	140.5	142.1	260.1	197.7	96.1	20.0	1367
$F_{TPSWHIBP}$	136.3	130.0	114.2	95.8	77.9	70.2	69.2	74.2	111.5	128.4	126.5	135.1	1269
$F_{TPSWBPHI}$	**122.1**	**110.4**	**89.7**	**73.8**	**73.2**	**67.1**	**72.6**	**78.2**	**93.4**	**104.9**	**109.6**	**125.0**	**1120**
F_{TPETSW}	14.1	14.1	12.1	2.5	3.9	4.8	5.8	10.3	61.7	28.0	15.7	13.5	187
$F_{TPDWHIBP}$	**15.5**	**14.8**	**13.0**	**11.4**	**11.3**	**12.2**	**12.8**	**14.0**	**19.3**	**15.6**	**14.3**	**15.6**	**170**
$F_{TPDWBPHI}$	14.0	13.2	11.0	9.6	10.1	11.0	11.2	11.7	12.2	12.1	13.7	15.0	145
F_{TPLU}	10.4	10.3	10.3	10.3	10.3	10.3	10.2	10.2	10.2	10.3	10.4	10.4	124
F_{TPSWET}	7.5	6.7	6.8	8.0	8.5	9.6	10.3	12.8	17.8	10.7	7.0	7.7	113
$F_{TPDWSWx}$	9.5	9.9	7.1	0.9	0.5	0.3	0.3	0.4	6.4	9.4	10.1	8.8	64
F_{TPETDW}	4.4	4.4	3.8	0.8	1.2	1.5	1.8	3.2	19.1	8.7	4.9	4.2	58
$F_{TPSWDWx}$	9.0	9.2	5.7	0.5	0.2	0.1	0.1	0.1	2.2	5.7	8.1	8.2	49
$F_{TPMWSWd}$	3.8	3.4	2.3	1.5	1.5	1.9	2.2	2.9	7.5	3.9	3.1	4.0	38
F_{TPSWDW}	1.8	1.6	1.7	2.0	2.1	2.3	2.5	3.1	4.3	2.6	1.7	1.9	28
F_{TPtrib}	**2.8**	**2.3**	**2.5**	**3.1**	**3.1**	**1.8**	**1.4**	**1.3**	**1.4**	**1.6**	**2.1**	**2.5**	**26**
$F_{TPplantSW}$	**1.6**	**1.6**	**1.6**	**1.6**	**1.6**	**1.6**	**1.6**	**1.6**	**1.6**	**1.6**	**1.6**	**1.6**	**19**
F_{TPDWA}	0.7	0.8	0.7	0.0	0.0	0.0	0.0	0.1	4.6	1.8	0.8	0.7	10.3
F_{TPbur}	1.0	1.0	0.8	0.7	0.8	0.7	0.7	0.8	0.7	0.7	0.8	0.9	9.6
F_{TPprec}	0.04	0.04	0.04	0.04	0.04	0.04	0.04	0.04	0.04	0.04	0.04	0.04	0.54
$F_{TPplantDW}$	**0.02**	**0.02**	**0.02**	**0.02**	**0.02**	**0.02**	**0.02**	**0.02**	**0.02**	**0.02**	**0.02**	**0.02**	**0.20**
F_{TPADWd}	0.005	0.005	0.005	0.005	0.005	0.005	0.005	0.005	0.005	0.005	0.005	0.005	0.06

these sediments (0–10 cm) is 0.61 mg/g dw. This model provides values based on the total TP-inventory in the entire area below the theoretical wave base down to 10 cm of sediments. The biologically passive sediments below 10 cm are expected to have a TP-concentration of about 0.45 in the Baltic Proper (Jonsson et al. 1990) and this value is also used in this modeling for the Himmerfjärden Bay. This means that only a small fraction (related to the difference between 0.61 and 0.45) of the phosphorus in the accumulation area sediments could be available for diffusive transport from these sediments. But the diffusion also depends on the redox-conditions in the sediments, which depend on the sedimentation of organic matter. The average values for total sedimentation calculated by the model vary between 0.1 and 0.3 mm/yr or between 10 and 20 μg/cm^2·d on the accumulation areas. The modeled water content (W) of the accumulation area sediments (0–10 cm) is 75% ww, the modeled organic content (loss on ignition, IG) is 6.3% dw and the bulk density (d) 1.17 g/cm^3. Table 6.3 gave modeled values for the SW and DW-water inflow from the Baltic Proper ($Q_{SWBPGE} = 4050 \cdot 10^6$ m^3/month and $Q_{DWBPGR} = 450 \cdot 10^6$ m^3/month); modeled monthly tributary inflow (Q_{trib}); theoretical water retention times in the SW and DW-layers ($T_{SW} = 0.62$ months and $T_{DW} = 0.41$ months, since the DW-volume is small), as calculated from the mass-balance for salt; fall velocities for particulate phosphorus and suspended particulate matter in the SW and DW-layers (v_{SW} 2.4 m/month and v_{DW} 2.3 m/month); the particulate TP-fractions in the SW and DW-layers (PF_{SW} 0.20–0.87 and PF_{DW} 0.01–0.53; see Table 6.5).

It should also be stressed that land uplift (F_{LU}) is a rather important individual input of TP to the bay (124 tons/yr).

It is interesting to note the difference between fluxes and amounts (compare the results in Table 6.6 with the data in Table 6.7). The largest TP-fluxes are to and from biota, but the total TP-inventory in biota is only 16.8 tons TP or 6% of the total inventory. By far most TP is found in the accumulation area sediments (191 tons or 66% of the total amount).

6.2.5 Predicting the Dynamic Response of the System to Changes in Nutrient Loading

The simulations to estimate a realistic response to changes in nutrient loading will be presented in four steps. The first step concerns a two-fold increase in the direct emissions from the plant to mimic the experiment of 1983. The next example concerns a 10-fold increase in these direct emissions of TP from the plan. The third case concerns a 10% increase in the TP and SPM-concentrations in the sea outside of the Himmerfjärden Bay. The fourth test concerns a 10% reduction in TN-concentrations in the SW-layer in the bay. These four cases will be compared to the default values representing the conditions today (1997–2007). We will also show the dynamic response of the system to sudden changes in the nutrient loading. The results are compiled in Table 6.8.

The predicted values for TP-concentrations in the SW-layer, TP-concentrations in the DW-layer, Secchi depths, chlorophyll-a concentrations, concentrations of

Table 6.7 Amounts of TP (tons) in the different compartments in the Himmerfjärden Bay; accumulation areas in the DW-compartment (A-areas), in biota with short turnover times (Biota), in the DW-layer, in areas of fine sediment erosion and transport (ET) and in the SW-layer

Month	A-areas	Biota	DW	ET-areas	SW	
1	192	1.8	8.1	7.5	85.9	
2	192	3.5	7.7	6.5	80.3	
3	192	13.3	6.8	6.6	60.2	
4	191	27.3	5.9	9.0	34.0	
5	191	20.2	5.9	23.1	29.8	
6	190	19.0	6.4	36.6	26.3	
7	189	18.7	6.7	49.7	26.2	
8	189	18.9	7.3	61.1	29.2	
9	190	33.3	10.1	42.4	39.3	
10	192	27.9	8.1	14.3	55.3	
11	192	14.5	7.5	6.9	67.2	
12	192	2.7	8.1	7.7	84.5	Sum
MV	191	16.8	7.4	22.6	51.5	289
%	66	6	3	8	18	100
M50	191	18.8	7.4	11.7	47.3	
SD	1.2	3.6	7.8	6.7	82.3	

cyanobacteria and the oxygen concentrations in the deep-water layer at steady-state are given for the four cases and the results are compared to the default conditions.

1. Case 1 (a 2-fold increase of TP-emissions from the plant) would influence the system very little, e.g., the mean annual TP-concentration would increase from 25.8 to 26.2 μg/l and the Secchi depth would not change at all.
2. Case 2 (a 10-fold increase in TP-emissions from the plant) would influence the system markedly. The TP-concentrations in the SW-later would increase by 3.3 μg/l, the average chlorophyll concentration would increase by 4.4 μg/l, the maximum concentration of cyanobacteria would increase from 360 to 540 μg/l.
3. Case 3 (a 10% increase in TP and SPM-inflow from the sea) would also create substantial changes corresponding to the changes in case 2.
4. Case 4 (a 10% reduction in TN-concentrations in the bay) would only change the predicted concentration of cyanobacteria, which would increase from a maximum value of 360 to a maximum monthly value of 400 μg/l. Note that, as explained, we have not modeled the TN-concentrations dynamically.

The dynamic response of the system related to sudden 2-fold and 10-fold increases in the TP-emissions from the plant month 25 are shown in Fig. 6.4. One can see from Fig. 6.4 that the system will reach a new steady-state within a year. The main reason for this quick adjustment is related to the relatively fast water turnover of the system. The theoretical surface-water retention time is about 19 days, i.e., about 19 total water exchanges per year with the outside sea.

We also believe that these results largely explain important aspects of previous results related to the experiments carried out during the last 25 years and that, since these results are based on a general process-based modeling approach, which may be applied to most coastal areas, one would also hope that this type of analysis could be more widely used.

Table 6.8 Results related to how changes in TP loading to the Himmerfjärden Bay would influence TP-concentrations in SW, TP in DW, Secchi depth, chlorophyll, cyanobacteria and oxygen concentrations if the TP input from the plant is increased by a factor of 2, by a factor of 10, if the TP-inflow from the Baltic Proper is increased by 10% or if the TN-concentration in the bay is decreased by 10%

TP_{SW} (μg/l)					
Month	Default	2·plant	10·plant	1.1·TPsea	0.9·TN
1	33.2	33.6	36.5	36.3	33.2
2	31.7	32.1	35.0	34.7	31.7
3	27.8	28.2	31.1	30.4	27.8
4	23.2	23.6	26.5	25.4	23.2
5	18.9	19.3	22.0	20.7	18.9
6	17.2	17.5	20.3	18.8	17.2
7	17.0	17.3	20.1	18.6	17.0
8	18.2	18.5	21.3	20.0	18.2
9	27.4	27.8	31.0	29.8	27.4
10	31.5	31.9	35.1	34.2	31.5
11	30.9	31.3	34.4	33.7	30.9
12	32.9	33.3	36.3	36.0	32.9
MV	25.8	26.2	29.1	28.2	25.8
Diff.		0.4	3.3	2.4	0.0
M50	27.6	28.0	31.0	30.1	27.6
SD	6.5	6.6	6.7	7.1	6.5
TP_{DW} (μg/l)					
1	34.3	34.6	36.7	36.7	34.3
2	32.9	33.2	35.6	35.2	32.9
3	28.9	29.2	31.4	31.0	28.9
4	25.3	25.6	27.8	27.1	25.3
5	25.2	25.5	27.8	27.0	25.2
6	27.2	27.5	29.9	29.2	27.2
7	28.6	28.9	31.2	30.6	28.6

Chlorophyll (μg/l)					
Month	Default	2·plant	10·plant	1.1·TPsea	0.9·TN
1	2.0	2.0	2.2	2.2	2.0
2	3.7	3.8	4.1	4.1	3.7
3	9.4	9.5	10.5	10.2	9.4
4	2.6	2.6	3.0	2.8	2.6
5	2.7	2.7	3.1	3.0	2.7
6	1.8	1.8	2.1	2.0	1.8
7	1.5	1.5	1.8	1.6	1.5
8	2.0	2.0	2.3	2.2	2.0
9	6.4	6.5	7.3	7.0	6.4
10	6.8	6.9	7.6	7.4	6.8
11	6.1	6.2	6.8	6.7	6.1
12	2.3	2.4	2.6	2.6	2.3
MV	3.9	4.0	4.4	4.3	3.9
Diff.		0.1	0.5	0.4	0.0
M50	2.6	2.7	3.1	2.9	2.6
SD	2.6	2.6	2.9	2.8	2.6
Cyanobacteria (μg/l)					
1	0.2	0.2	0.2	0.2	0.2
2	0.5	0.5	0.7	0.7	0.6
3	7.0	7.4	10.4	9.6	8.1
4	76	81	130	109	90
5	118	125	208	153	119
6	138	147	227	183	138
7	151	160	249	201	151

Table 6.8 (continued)

TP_{SW} (µg/l)						**Chlorophyll (µg/l)**					
Month	Default	2·plant	10·plant	1.1·TPsea	0.9·TN	Month	Default	2·plant	10·plant	1.1·TPsea	0.9·TN
8	31.1	31.4	33.8	33.2	31.1	8	162	172	255	213	162
9	43.0	43.3	45.9	45.3	43.0	9	356	375	535	473	400
10	34.6	34.9	37.6	36.7	34.6	10	191	198	264	245	209
11	31.9	32.2	34.4	34.1	31.9	11	32	34	45	43	36
12	34.7	34.9	37.1	37.1	34.7	12	0.9	0.9	1.2	1.2	1.0
MV	31.5	31.8	34.1	33.6	31.5	**MV**	103	108	160	136	110
Diff.		0.3	2.6	2.1	0.0	**Diff.**		5.4	57.5	32.9	6.7
M50	31.5	31.8	34.1	33.6	31.5	**M50**	97	103	169	131	105
SD	5.0	5.0	5.1	5.1	5.0	**SD**	107	112	162	141	117
Secchi (m)						**Oxygen concentration (mg/l)**					
1	9.1	9.1	9.1	8.3	9.1	1	0	0	0	0	0
2	9.1	9.1	9.1	8.3	9.1	2	0	0	0	0	0
3	6.9	6.9	6.8	6.4	6.9	3	0	0	0	0	0
4	5.6	5.5	5.4	5.2	5.6	4	0	0	0	0	0
5	6.7	6.7	6.5	6.3	6.7	5	7.0	7.0	7.0	6.9	7.0
6	6.3	6.3	6.1	5.8	6.3	6	6.5	6.5	6.5	6.4	6.5
7	6.3	6.3	6.1	5.8	6.3	7	6.2	6.2	6.1	6.0	6.2
8	6.6	6.6	6.4	6.1	6.6	8	6.5	6.5	6.4	6.3	6.5
9	5.9	5.8	5.6	5.5	5.9	9	7.5	7.5	7.4	7.3	7.5
10	6.0	6.0	5.8	5.7	6.0	10	0	0	0	0	0
11	6.4	6.3	6.2	5.9	6.4	11	0	0	0	0	0
12	8.5	8.5	8.4	7.7	8.5	12	0	0	0	0	0
MV	6.9	6.9	6.8	6.4	6.9	**MV**	6.8	6.7	6.7	6.6	6.8
Diff.		0.0	–0.1	–0.5	0.0	**Diff.**		–0.06	–0.13	–0.23	–0.05
M50	6.5	6.5	6.3	6.0	6.5	**M50**	6.5	6.5	6.5	6.4	6.5
SD	1.2	1.3	1.3	1.1	1.2	**SD**	0.50	0.50	0.51	0.53	0.50

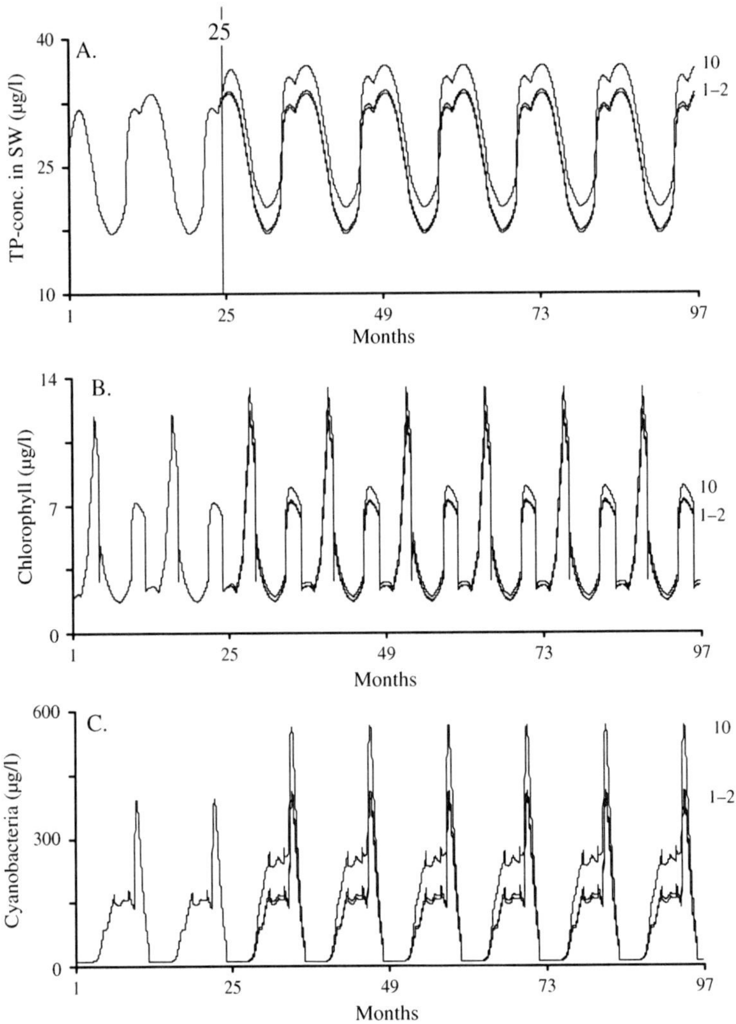

Fig. 6.4 Illustration of the dynamic response of the system (the Himmerfjärden Bay) to sudden 2-fold and 10-fold increases in phosphorus loading from the treatment plant in the bay month 25 (January) for **A.** TP-concentrations in the SW-layer, **B.** chlorophyll-a concentrations and **C.** concentrations of cyanobacteria

6.2.6 Concluding Remarks

The basic aim of this case-study has been to present a general approach to quantify how coastal systems are likely to respond to changes in nutrient loading. The conditions in many coastal areas depend on nutrients emissions from point sources, diffuse sources, river input and the exchange of nutrients and water between the given coast and the outside sea, but all these fluxes can not be of equal importance to the conditions in the given coastal area, e.g., for the water clarity, primary production and concentration of harmful algae (such as cyanobacteria). This work describes how a general process-based mass-balance model (CoastMab) has been applied for the case-study area, the Himmerfjärden Bay on the Swedish side of the Baltic Proper. These results indicate that the conditions in the Himmerfjärden Bay are dominated by the relatively quick water exchange between the bay and the outside sea. The theoretical surface-water retention time is about 19 days, as determined using the mass-balance model for salt, which is based on comprehensive and reliable empirical salinity data. This means that although this bay is quite enclosed, with an exposure of 0.0194, it is still dominated by the water exchange towards the sea. Local emissions of nutrients to the Himmerfjärden Bay are small compared to the nutrient fluxes from the sea. If the conditions in this, and many similar bays, are to be improved, it is very important to lower the nutrient concentrations in the outside sea. To do that in the best possible manner, one must apply the same process-based mass-balance principles for the larger system as discussed in this work for a coastal bay. This means that the major phosphorus fluxes to the sea, in this case the Baltic Proper, should be reduced in the most cost-efficient manner. That remedial strategy has not penetrated fully into management decisions neither for the Himmerfjärden Bay nor for the Baltic Proper or the Baltic Sea.

We have discussed a general method to calculate how a given coastal area would likely respond to changes in nutrient loading. This approach makes it possible to carry out structured analyses of the costs and environmental benefits of remedial actions designed to reduce nutrient input to coastal areas and to put such reductions or changes into a process-based holistic context where all important transport processes to, within and from the given coastal area are accounted for on a monthly basis to achieve seasonal variations, which is essential for most biological variables and key bioindicators of coastal eutrophication. The dynamic modeling also provides quantitative values of the time-dependent response of the system. The method discussed here may be applied to most coastal systems and the data necessary for this analysis have also been discussed.

6.3 An Approach to Estimate Relevant Reference Values for Key Bioindicators

6.3.1 Aim of the Case-Study

Finding relevant values representing pre-industrial or reference conditions is, evidently, an important task in most contexts dealing with water and/or sediment contamination (Håkanson and Jansson 1983; Gren et al. 2000). Very costly remedial measures are generally needed to reduce, e.g., nutrient inputs from urban areas, industries, catchments or diffuse sources (Moldan and Billharz 1997; Livingston 2001; Bortone 2005). A central question is then how a given system would respond to the suggested measures. In short, what is the environmental benefit related to the remedial costs? To address that question, two decisive issues must be dealt with:

1. A validated process-based dynamic model, which has proven to predict the target variables well, must be at hand to provide realistic values for the dynamic (time-dependent) response of a reduced nutrient loading since this can not generally be done by static empirical regression models? This case-study will use the CoastMab-model (see Sect. 9.1) for the Gulf of Riga, the selected case-study area.
2. The dynamic model must quantify all important fluxes to, within and from the system and include information on the natural load and the anthropogenic load and, preferably also, information on how much of the anthropogenic load from different sources that can, realistically, be reduced. It may be very difficult to find the appropriate reliable data to address the latter aspect but it should be possible to test various alternative options, e.g., a 25, 50 or a 95% reduction of the anthropogenic discharges from a given source and calculate how the system would respond to such changes. Pre-industrial or "natural" reference conditions may be related to what is referred to as "good" ecological status according to the Water Framework Directive (Anon 2000). General definitions of status levels are provided on the following page.

In addition, "good" ecological status implies that there have been slight (but not substantial) changes in algal biomass and in the frequency and intensity of algal blooms as a result from human activity, and it also implies that nutrient concentrations do not exceed the range beyond which the ecosystem function has been altered compared to pristine conditions (Anon 2000).

Ecological status	Definition
High	No, or only very minor, anthropogenic alterations
Good	Low distortion levels resulting from human activity, which deviate only slightly from levels associated with "high" status
Moderate	Conditions deviate moderately from those at "high" status and are significantly worse than those at "good" status
Poor	Major anthropogenic alterations compared to "high" status
Bad	Severe anthropogenic impacts under which large portions of the relevant biological communities have disappeared

The main reasons for selecting the Gulf of Riga as a case-study are that for this bay one can access the necessary data:

- First, data on the salinity inside and outside the bay and the tributary water fluxes to the bay must be available so that a reliable mass-balance for salt can be established so that the water exchange with the outside sea (the Baltic Proper) can be quantified realistically and also the theoretical water retention time in the bay, which influence important internal processes, such as sedimentation of particulate matter, stratification and mixing, resuspension and diffusion. This version of

CoastMab also includes a mass-balance model for salt structured in the same way as the sub-model for phosphorus except that the mass-balance model for phosphorus also calculates sedimentation, resuspension, burial and biouptake and retention of dissolved phosphorus in biota. By definition, total phosphorus (TP) in the water generally includes living and dead plankton (phytoplankton, bacterioplankton and herbivorous zooplankton; see Håkanson and Boulion 2002), but not larger animals such as predatory zooplankton, prey fish and predatory fish. These simulations also calculate phosphorus uptake and retention in biota with long turnover times.

- Secondly, data on the TP-concentration in the water outside the bay must be at hand so that the inflow from the sea can be calculated. Data on the TP-transport from different sources on land are also needed and such data are available for the Gulf of Riga from the HELCOM database. HELCOM also gives information on the natural background losses, the diffuse losses and the point sources discharges of TP to the Gulf of Riga and from this one can use the CoastMab-model to estimate how various reductions in the anthropogenic load would influence the system, and what the requested references values for "good ecological status" would be for the target bioindicators used in this study, the chlorophyll-a concentration, the Secchi depth and the concentration of cyanobacteria.

There are several reasons for selecting these three bioindicators in this study. Figure 6.3 shows that the Gulf of Riga generally has a very low Secchi depth, Fig. 6.7 show the relatively high chlorophyll concentrations and Figs. 6.3 and 4.4 gave the relatively high values of TP and total nitrogen (TN) in the water. Table 6.6 also shows that the topographical openness (the exposure, Ex), is only 0.0022, which is a low value compared to other coastal areas (see Lindgren and Håkanson 2007) indicating that this system is sensitive to water pollutants. The Gulf of Riga is also relatively shallow and 79% of its bottom area is dominated by resuspension fine materials (these are the erosion and transportation areas, ET; as calculated by the CoastMab-model; see Table 6.9).

6.3.2 Basic Data from the Gulf of Riga

Basin-specific data for the Gulf of Riga are compiled in Table 6.9, which gives information on total area; volume; mean depth; maximum depth; the depth of the theoretical wave base; the volume of the surface-water layer and the deep-water layer separated by the theoretical wave base; the section area, which defines the cross-sectional area that separates the given coastal area from the outside sea; the tributary water discharge to the bay (a mean value from data presented by Laznik et al. 1998; Ostmann et al. 2001 and Savchuk and Swaney 2000); the fraction of bottoms areas dominated by fine sediment erosion and transport (the ET-areas); the catchment area (from Wassman and Tamminen 2000); latitude; and mean annual precipitation (from Savchuk and Swaney 2000). The theoretical wave base is defined from the ETA-diagram (erosion-transport-accumulation; see Fig. 3.15), which

Table 6.9 Basic data for the Gulf of Riga. Bolded factors are obligatory driving variables

Area	**(km^2)**	**16,700**
Area below wave base	**(km^2)**	**3500**
Maximum depth	**(m)**	**56**
Total volume	**(km^3)**	**409,4**
Volume, SW	**(km^3)**	**391.4**
Volume, DW	**(km^3)**	**18**
Mean depth	**(m)**	**24.5**
Wave base	(m)	39
Section area	**(km^2)**	**0.36**
Exposure	(%)	0.0022
ET-areas	(%)	79
Water discharge	**(km^3/s)**	**(1149+1046+1202)/3) = 1132**
Catchment area	**(km^2)**	**135,700**
Latitude	**(°N)**	**57.7**
Precipitation	**(mm/yr)**	**590**
Land uplift	**(mm/yr)**	**0.67**

gives the relationship between the effective fetch, as an indicator of the free water surface over which the winds can influence the wave characteristics (speed, height, length and orbital velocity). The theoretical wave base separates the transportation areas (T), with discontinuous sedimentation of fine materials, from the accumulation areas (A), with continuous sedimentation of fine suspended particles. The theoretical wave base (D_{wb} in m) is at a water depth of 39 m in the Gulf of Riga. This is calculated from (6.4) (Area in km^2; see Fig. 3.15 and Håkanson and Jansson 1983):

$$D_{wb} = (45.7 \cdot \sqrt{Area})/(\sqrt{Area} + 21.4) \quad (6.4)$$

This approach gives one value for the theoretical wave base related to the area of the system. So, the Gulf of Riga has been divided into two depth intervals:

1. The surface-water layer (SW), i.e., the water above the theoretical wave base at 39 m.
2. The deep-water layer (DW) is defined as the volume of water beneath the theoretical wave base.

It should be noted that the theoretical wave base describes average conditions. The actual wave base varies around 39 m. During storm events, the wave base will be at greater water depths. Figure 6.5A gives the hypsographic curve of the Gulf of Riga and Fig. 6.5B the corresponding volume curve. The hypsographic curve in Fig. 6.5A from Håkanson and Lindgren (2007d) and has been derived using GIS and bathymetric data provided by Seifert et al. (2001). The areas and volumes calculated from these curves will be used in this modeling. One can note that the area below the theoretical wave base (D_{wb}) is 3500 km^2 and that the volumes of the SW and DW-layers are 391 and 18 km^3 and the entire volume is 409 km^3. This value for the entire volume corresponds quite well with the value of 406 given by Ostmann et al.

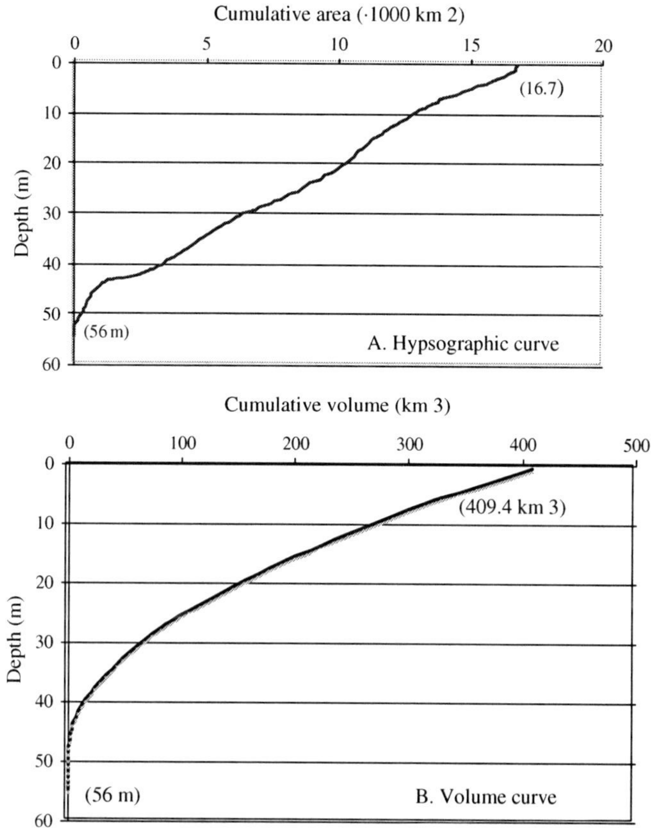

Fig. 6.5 Hypsographic curve (**A**) and volume curve (**B**) for the Gulf of Riga

(2001) but is lower than the value of 420 given by Wassman and Tamminen (2000). In this work, the boundary lines for the Gulf of Riga have been drawn according to Fig. 6.6.

Table 6.10 shows the very comprehensive set of water data from the HELCOM database. Figure 6.6 shows the geographical locations of the sample sites for Secchi depth (from Aarup 2002), which is a target variable in this work. Table 6.10 also gives monthly mean values, standard deviations and number of data for the Secchi depth, chlorophyll, TP in surface and deep water (SW and DW), total nitrogen (TN) in surface water, salinity and temperature in surface and deep water for the period from 1992 to 2005. Note that there are no corresponding data on cyanobacteria available to us. The CoastMab-model is intended to predict the measured salinities, TP, TN, Secchi depths and chlorophyll values in Table 6.10 as accurately as possible. The empirical temperature data in Table 6.10 have been used to model stratification and mixing.

The coefficients of variation (CV = SD/MV) for the monthly mean values are shown in Table 6.11. Thcses CV-values are very important in the sense that variables

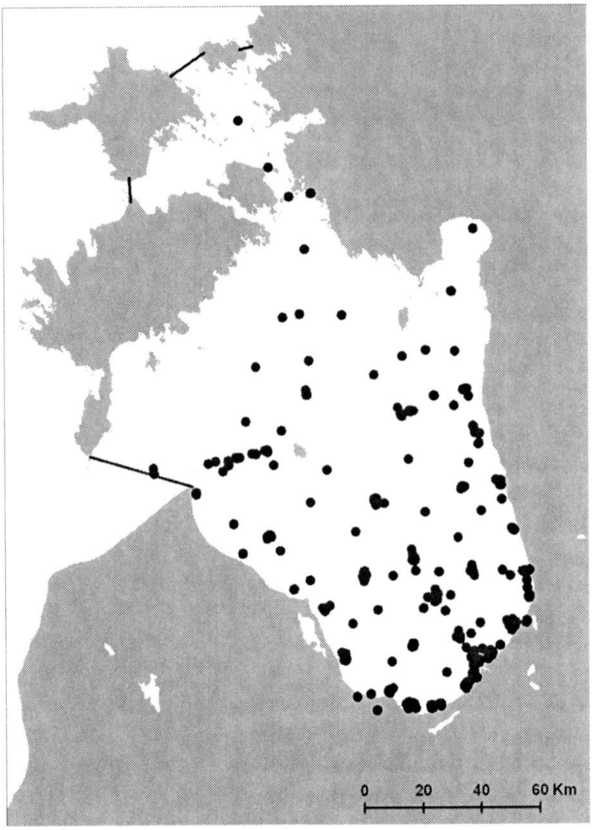

Fig. 6.6 Boundary lines (from HELCOM) and sampling sites for Secchi depth (from Aarup 2002) in the Gulf of Riga

with high inherent CV-values (such as the target variables in this study, chlorophyll and Secchi depth) cannot be expected to be predicted as well as variables with comparatively low CV-values (such as salinity) (see Chap. 4).

The Secchi depth (Sec in m) is calculated from (5.1) relating Secchi depth to the concentration of suspended particulate matter in the SW-layer (SPM_{SW} in mg/l) and the salinity of the SW-layer (Sal_{SW}). SPM_{SW}, in turn, is calculated here from dynamically modeled TP-concentrations in the SW-layer using the following regression (from Håkanson and Bryhn 2008a):

$$SPM = 0.0235 \cdot TP^{1.56} \tag{6.5}$$
$$(r^2 = 0.895;\ n = 51)$$

Equation (6.5) will translate modeled TP (in µg/l) into SPM-values (in mg/l) and together with modeled data on the salinity, translate those values into Secchi depths using (5.1). It should be noted that (6.5) is based on data from systems with

Table 6.10 Compilation of statistical information regarding monthly mean values, standard deviations (SD) and number of data (n) for measured Secchi depths, chlorophyll-a concentrations, concentrations of total phosphorus (TP) in surface water (SW) and deep water (DW), concentrations of total nitrogen (TN) in surface water, salinities and water temperatures in samples collected from 1992 to 2005 in the Gulf of Riga (data from ICES 2006; and Aarup 2002)

Month	Secchi	Chl	TP, SW	TP, DW	TN, SW	Sal., SW	Sal., DW	Temp., SW	Temp., DW
	(m)	(μg/l)	(μg/l)	(μg/l)	(μg/l)	(psu)	(psu)	(°C)	(°C)
1	3.1	0.48	37.0	39.0	509.4	6.04	5.75	0.48	2.00
2	2.8	0.97	38.2	38.5	460.3	5.84	5.87	0.93	0.86
3	2.9	3.11	41.0	42.3	582.3	5.51	5.79	0.70	0.74
4	2.8	6.90	31.3	37.3	519.6	5.67	5.91	1.92	0.96
5	2.3	6.00	27.9	32.5	513.3	5.47	6.06	4.00	1.22
6	2.7	2.76	20.9	28.8	435.1	5.43	5.86	7.81	2.04
7	2.6	4.96	21.5	35.7	480.1	5.37	6.05	10.43	1.68
8	3.1	1.25	22.1	49.9	434.3	5.53	6.02	13.08	2.83
9	3.3	2.64	24.6	65.9	464.2	5.48	6.01	12.60	4.09
10	3.5	9.12	24.3	70.7	445.9	5.52	5.88	10.15	3.81
11	3.4	3.90	29.6	43.7	445.0	5.66	5.74	6.98	7.03
12	3.3	1.59	32.3	39.9	454.5	5.58	5.65	3.14	4.03
Mean	3.0	3.6	29.2	43.7	478.7	5.6	5.9	6.0	2.6
	SD	**SD**	**SD**	**SD**	**SD**	**SD**	**SD**	**SD**	**SD**
1	0.8	0.16	2.14	7.00	34.3	0.43	0.10	0.25	0.80
2	1.1	0.58	1.97	3.03	27.3	0.08	0.09	0.38	0.40
3	1.2	1.89	3.41	6.16	142.0	0.35	0.02	0.15	0.19
4	1.9	5.06	6.42	8.62	99.9	0.43	0.11	1.57	0.47
5	0.8	5.44	10.76	8.72	131.8	0.43	0.31	2.57	0.89
6	0.9	2.65	4.18	4.78	50.1	0.28	0.06	4.80	0.59
7	0.8	2.82	5.26	14.31	40.6	0.31	0.27	6.12	0.84
8	1.0	0.91	7.60	21.21	73.7	0.25	0.21	7.11	1.54
9	1.0	1.18	8.54	35.18	50.6	0.24	0.26	3.43	2.72
10	1.1	3.82	12.79	19.34	86.1	0.23	0.11	2.49	0.67
11	1.0	3.11	5.77	31.27	88.0	0.28	0.21	0.73	0.92
12	1.2	0.16	8.15	9.85	56.4	0.18	0.12	1.06	1.16
Mean	1.1	2.3	6.4	14.1	73.4	0.3	0.2	2.6	0.9
	n	**n**	**n**	**n**	**n**	**n**	**n**	**n**	**n**
1	23	15	18	0	18	18	0	18	0
2	16	13	30	14	44	30	14	29	14
3	31	6	29	11	40	29	11	29	11
4	68	15	109	25	124	107	25	107	24
5	176	34	213	56	248	213	55	212	55
6	101	17	60	21	60	60	21	60	20
7	164	13	66	21	82	66	21	65	21
8	168	39	143	44	150	143	44	138	44
9	96	22	78	31	101	78	31	78	30
10	70	9	40	14	54	40	14	40	14
11	39	26	160	51	191	156	50	157	49
12	27	8	30	11	40	30	11	30	11

Table 6.11 Compilation of coefficients of variation (CV = SD/MV) for the studied water variables in the Gulf of Riga

Month	Secchi	Chl	TP, SW	TP, DW	TN, SW	Sal., SW	Sal., DW	Temp., SW	Temp., DW
1	0.27	0.33	0.06	0.18	0.07	0.07	0.02	0.52	0.40
2	0.39	0.60	0.05	0.08	0.06	0.01	0.02	0.41	0.47
3	0.41	0.61	0.08	0.15	0.24	0.06	0.00	0.21	0.25
4	0.68	0.73	0.21	0.23	0.19	0.07	0.02	0.81	0.49
5	0.35	0.91	0.39	0.27	0.26	0.08	0.05	0.64	0.73
6	0.35	0.96	0.20	0.17	0.12	0.05	0.01	0.61	0.29
7	0.31	0.57	0.24	0.40	0.08	0.06	0.04	0.59	0.50
8	0.31	0.73	0.34	0.42	0.17	0.05	0.03	0.54	0.54
9	0.30	0.45	0.35	0.53	0.11	0.04	0.04	0.27	0.66
10	0.30	0.42	0.53	0.27	0.19	0.04	0.02	0.25	0.17
11	0.30	0.80	0.20	0.72	0.20	0.05	0.04	0.11	0.13
12	0.35	0.10	0.25	0.25	0.12	0.03	0.02	0.34	0.29
Mean	0.36	0.60	0.24	0.31	0.15	0.05	0.03	0.44	0.41
Median	0.33	0.60	0.22	0.26	0.15	0.05	0.02	0.47	0.43

salinities = 15 psu. This means that it may provide more limited predictive power for coastal systems with salinities >15 psu.

The empirically-based model to predict the total concentration of cyanobacteria is given in Fig. 5.18. The following simulations will use dynamically modeled monthly TP-concentrations in the SW-layer, empirical mean monthly SW-temperatures, dynamically modeled SW-salinities and mean empirical monthly TN-concentrations in the SW-layer to predict monthly mean values of cyanobacteria in the SW-layer.

The following section will demonstrate how the model works and how it predicts first the TP-concentrations in the two layers and in surficial (0–10 cm) sediments below the theoretical wave base (the accumulation-area sediments), then the TN-concentrations in the SW-layer, chlorophyll-a concentrations in the SW-layer, Secchi depth and cyanobacteria. The modeled values will be compared to empirical data and to the uncertainty bands related to the empirical data (as given by the CV-values in Table 6.11).

6.3.3 Results

6.3.3.1 Modeled Values Versus Empirical Data

Figure 6.7 gives modeled values compared to the mean empirical monthly data for the period 1992–2005 and also to the 95% confidence intervals for the mean empirical values. If the model yields values close to the empirical mean value

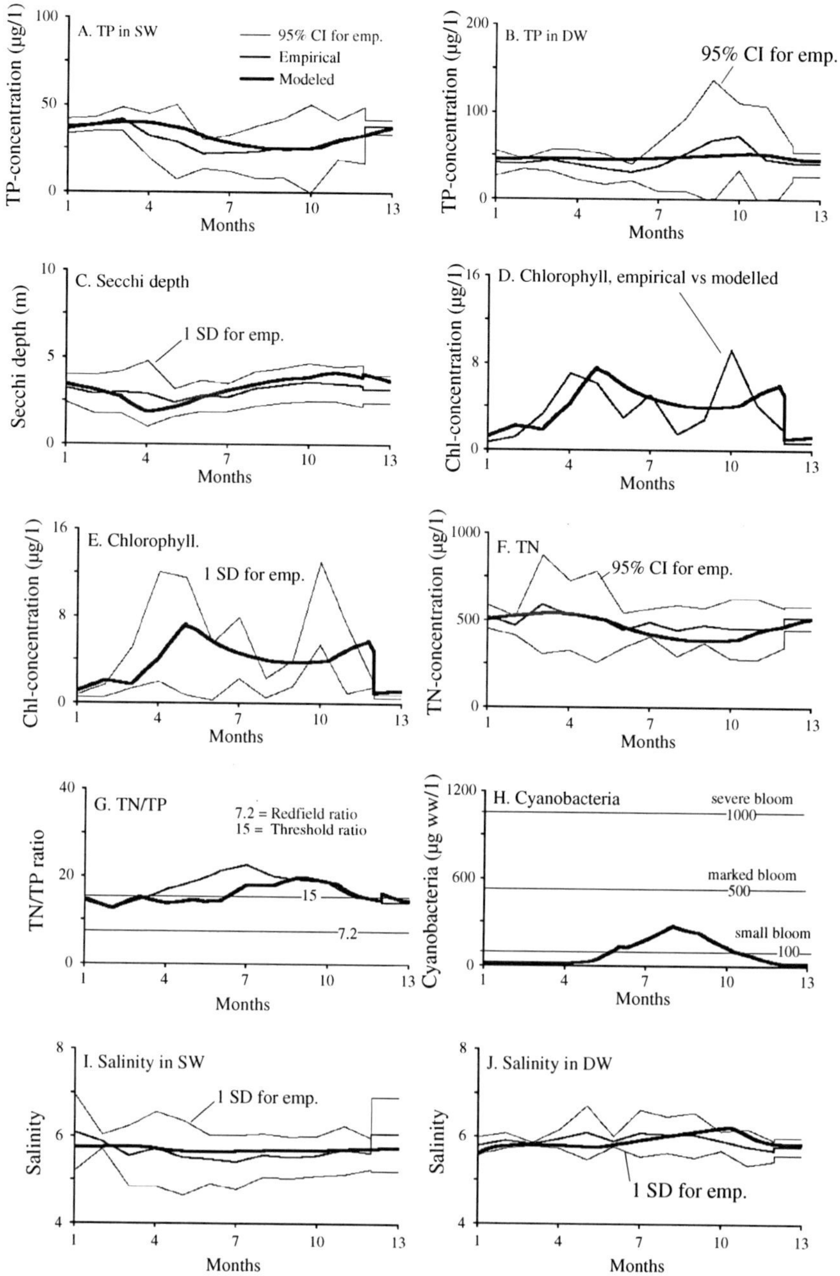

Fig. 6.7 Modeled values versus empirical data for (**A**) TP in surface water, (**B**) TP in deep water, (**C**) Secchi depth, (**D**) and (**E**) chlorophyll, (**F**) TN, (**G**) the TN/TP-ratio based on empirical data and on modeled values versus the Redfield ratio and the threshold ratio for cyanobacteria, (**H**) predicted concentrations of cyanobacteria, modeled and measured salinities in surface water (**I**) and in deep water (**J**) in the Gulf of Riga

and in-between these two uncertainty bands, the predictions should be regarded as good. For all these comparisons between modeled and empirical data, it should be noted that:

- The monthly inflows of water and phosphorus have not been calculated using data for this period of time (1992–2005) but using data from Voipio (1981). This likely explains some of the differences between the modeled and the measures monthly values.
- The monthly calculations of the inflow of water, salt and phosphorus from the Baltic Proper are based on mean values for the SW and the DW-compartments not in the inflowing water from the area just outside the Gulf of Riga, which would have been more appropriate, but for the entire Baltic Proper. The reason why the more appropriate data have not been used is simply that it has been difficult to find such data. This likely also further explains some of the differences between the modeled and the measured monthly values.

With these reservations, one can note from Fig. 6.7 that:

- (A) The TP-concentrations in the SW-layer are generally close to the empirical data and within the defined 95% uncertainty bands of the empirical data.
- (B) The average TP-concentrations in the DW-layer are also close to the empirical average value and generally within the uncertainty bands of the empirical data.
- (C) The predicted Secchi depths are close to the empirical values and within the uncertainty bands defined by one standard deviation of the empirical mean value.
- (D and E) The modeled chlorophyll concentrations do not give the "three-peak pattern" as indicated by the empirical data but a more classical "twin-peak pattern". It should be stressed that the chlorophyll concentrations are predicted from a regression based on dynamically modeled TP-concentrations in the surface water and a dimensionless moderator for the SW-temperatures. So, these predictions do not include any considerations to the biouptake of dissolved phosphorus in phytoplankton. The overall correspondence between the modeled chlorophyll values relative to the uncertainty bands of the empirical data (± 1 SD) are shown in Fig. 6.7E. Given the limitations of the modeling for chlorophyll, one can conclude that the overall correspondence is relatively good.
- (F) The TN-concentrations are also predicted from a regression using dynamically modeled TP-concentrations in the SW-compartment. There is a good correspondence between modeled and measured TN-concentrations.
- (G) The TN/TP-ratios based on empirical mean values and on modeled value are given in Fig. 6.7G in relation to the Redfield ratio of 7.2 and the threshold ratio of 15. The empirically-based monthly TN/TP-ratios are clearly higher than 7.2 all months, and higher than 15 most months, especially during the summer period. Given the situation in the Gulf of Riga, as revealed by these empirical data, a lowering of the TN/TP-ratio will imply greater risks for blooming of cyanobacteria. So, to reduce nitrogen inflow to the Gulf of Riga would be a sub-optimal approach to improve the conditions in the Gulf. The focus should instead be set on reductions of the major anthropogenic fluxes of phosphorus.

- (H) The predicted concentrations of cyanobacteria in the Gulf of Riga under default conditions are rather low.
- (I and J) The modeled salinities in the SW and DW-compartments are close to the empirical mean values and well within the uncertainty bands for one standard deviation.

From Fig. 6.7 one can conclude that the model predicts the target variables quite well given the factual limitations in the seasonal patterns in the driving variables for tributary water discharge (since this pattern in the modeling is not based on measured data for the modeled period) and in the seasonal pattern for the TP-concentrations outside the Gulf of Riga (since these data emanate not from the area outside of the gulf but from the entire Baltic Proper). Note that there has been no tuning of the general CoastMab-model to achieve these predictions. This should be noted for the following simulations, which aim to identify the reference values for "good ecological status" with respect to the target bioindicators.

6.3.3.2 Fluxes and Amounts of Phosphorus in the Gulf of Riga

Table 6.12 gives a compilation of all monthly TP-fluxes and a ranking of the annual fluxes. The two largest fluxes are biouptake and retention of phosphorus to and from biota. These fluxes are significantly larger than the fluxes related to up- and

Table 6.12 Compilation of monthly fluxes of TP (tons/month) to and from all compartments and a ranking of the annual fluxes (tons/yr). Bolded values represent external fluxes into the Gulf of Riga

	1	2	3	4	5	6	7	8	9	10	11	12	Year
F_{bioup}	748	628	1611	2512	4549	4587	4630	3982	3319	833	1370	386	29,156
F_{bioret}	850	765	983	1614	2771	3703	4173	4198	3886	2617	1960	1133	28,653
F_{DWSWx}	2001	2074	1465	1144	143	84	65	72	90	1408	1406	1331	11,284
F_{SWDWx}	1688	1801	1284	952	102	50	34	35	43	801	982	1095	8868
F_{SWGRBP}	**496**	**553**	**789**	**617**	**423**	**346**	**314**	**312**	**336**	**383**	**437**	**466**	**5471**
F_{ETSW}	559	496	355	295	44	32	30	39	57	729	523	442	3601
F_{SWET}	203	179	268	367	237	244	258	283	296	514	378	175	3401
F_{ETDW}	435	386	276	230	34	25	23	31	44	568	408	344	2804
F_{LU}	233	234	234	234	234	234	233	233	233	233	233	233	2802
F_{SWBPGR}	280	288	276	229	200	193	180	179	196	210	231	263	2725
F_{trib}	**59**	**90**	**287**	**182**	**69**	**45**	**45**	**59**	**86**	**68**	**77**	**54**	**1119**
F_{DWA}	138	147	179	128	14	9	7	8	15	151	137	141	1074
F_{bur}	85	84	82	82	84	84	85	86	86	87	86	86	1016
F_{SWDW}	54	48	71	97	63	65	68	75	78	136	100	46	902
F_{DWGRBP}	56	56	55	54	55	57	58	61	63	62	57	54	688
F_{DWBPGR}	**40**	**38**	**40**	**40**	**41**	**42**	**45**	**44**	**50**	**52**	**47**	**45**	**523**
F_{DWSWd}	0.0	0.0	0.0	0.0	6.7	9.4	11.5	12.9	13.7	3.1	0.0	0.0	57
F_{prec}	4.1	4.1	4.1	4.1	4.1	4.1	4.1	4.1	4.1	4.1	4.1	4.1	49
F_{ADWd}	1.8	1.8	1.8	1.9	1.9	1.8	1.7	1.7	1.6	1.6	1.7	1.8	21

downward advective mixing. The fluxes that may be reduced by remedial measures are bolded in Table 6.12. The SW-inflow from the Baltic Proper (F_{SWBPGR} = 5471 tons/yr), total tributary inflow (F_{trib} = 1119 tons/yr) and inflow to the DW-layer from the Baltic Proper (F_{DWBPGR} = 523 tons/yr). The diffusive flux from the accumulation areas sediment (below the theoretical wave base at 39 m) is rather small (21 tons/yr). The modeled TP-concentration in these sediments (0–10 cm) is 1.2 mg/g dw, which agrees quite well with measured values of about 2 mg/g dw in the uppermost centimeters of sediments (see Carman et al. 1996) since the TP-concentrations in the sediments generally decrease with sediment depth. This model provides values based on the total TP-inventory in the accumulation-area sediments down to 10 cm of sediments. The biologically passive sediments below 10 cm are expected to have a TP-concentration of about 0.45 in the Baltic Proper (Jonsson et al. 1990) and this value is also used in this modeling for the Gulf of Riga. This means that 61% (see Fig. 6.8) of the phosphorus in the accumulation area sediments could be available for diffusive transport from these sediments. The diffusion also depends on the redox-conditions, which depend on the sedimentation of organic matter. The average deposition (see Table 6.12) should increase from zero at the theoretical wave base to maximum values in the deepest part of the bay. The average values for total sedimentation calculated by the model vary from somewhat more than 1 mm/year to 0.1 mm/year or between 3 and 85 $\mu g/cm^2 \cdot d$ on the accumulation areas and between 23 and 65 $\mu g/cm^2 \cdot d$ on the ET-areas. The modeled water content (W) of the accumulation area sediments (0–10 cm) is 75% ww, the modeled organic content (loss on ignition, IG) is 6.3% dw and the bulk density (d) 1.17 g/cm^3. Table 6.13 also gives modeled values for the SW and DW-water inflow from the Baltic Proper (Q_{SWBPGE} = 11.3 km^3/month and Q_{DWBPGR} = 1.3 km^3/month); modeled monthly tributary inflow (Q_{trib}); theoretical water retention times in the SW and DW-layers (T_{SW} = 6.4 to 26 months and T_{DW} = 0.4 to 6.8 months, since the DW-volume is small), as calculated from the mass-balance for salt; the calculated flow velocities in the section

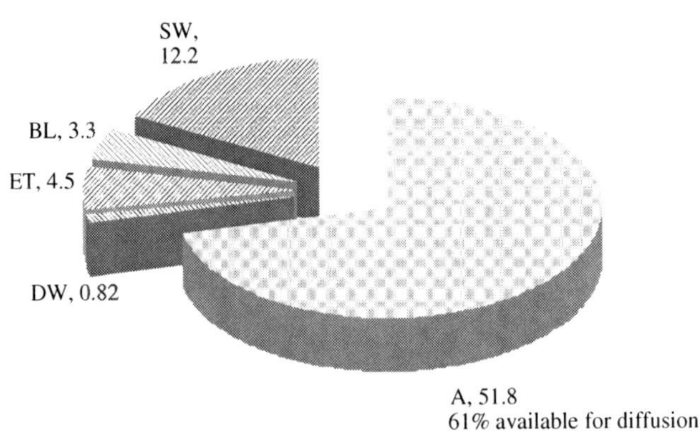

Fig. 6.8 Amounts of phosphorus in surface water (SW), deep water (DW), on ET-sediments and accumulation area sediments (A) and in biota (BL) in the Gulf of Riga; average monthly values in kt TP

Table 6.13 Compilation of calculated monthly data for water transport (Q) to surface water (SW) in the Gulf of Riga (GR) from the Baltic Proper (BP), to deep water (DW) from the Baltic Proper, from tributaries (Q_{trib}), theoretical water retention times (T_{SW} and T_{DW}), measured median TP-concentrations in the Baltic Proper at depths corresponding to the SW and DW-layers in the Gulf of Riga (using HELCOM data), calculated steady-state reference concentrations of TP in the SW and DW-layers in the Baltic Proper if 60% of all anthropogenic TP-emissions are halted, water velocities at the section area (u_{At}), settling velocities of particulate phosphorus in the SW and DW-layers (v_{SW} and v_{DW}), sedimentation on accumulation areas (Sed. A), sedimentation on ET-areas (Sed. ET), the particulate fraction of total phosphorus in the SW and DW-layers (PF_{SW} and PF_{DW}), calculated concentrations of suspended particulate matter (SPM) in the SW and DW-layers using a regression based on modeled TP-concentrations, the water content (W), organic content (IG), bulk density (d) and modeled TP-concentrations in the upper 10 cm sediment layer within below the theoretical wave base (the accumulation areas)

		1	2	3	4	5	6	7	8	9	10	11	12
Q_{SWBPGR}	km^3/month	11.3	11.3	11.3	11.3	11.3	11.3	11.3	11.3	11.3	11.3	11.3	11.3
Q_{DWBPGR}	km^3/month	1.3	1.3	1.3	1.3	1.3	1.3	1.3	1.3	1.3	1.3	1.3	1.3
Q_{trib}	km^3/month	1.9	2.9	9.2	5.8	2.2	1.4	1.4	1.9	2.7	2.2	2.4	1.7
T_{SW}	months	6.6	6.4	7.2	8.8	22.3	25.3	26.1	25.2	23.5	9.2	8.6	8.8
T_{DW}	months	0.4	0.4	0.5	0.7	4.0	5.8	6.8	6.5	5.9	0.6	0.6	0.6
TP_{SWBP}	µg/l	24.8	25.4	24.3	20.1	17.7	17.0	15.9	15.8	17.4	18.6	20.4	23.2
TP_{DWBP}	µg/l	32.1	30.1	31.9	31.8	32.4	33.5	35.5	35.1	39.7	41.5	37.0	35.7
$TP_{SWBPref}$	µg/l	16.0	17.7	19.1	20.5	21.5	21.0	18.1	15.0	12.3	10.3	11.3	13.9
$TP_{DWBPref}$	µg/l	30.4	29.9	29.5	29.7	30.3	30.9	31.6	32.3	32.7	32.2	31.6	31.1
u_{At}	cm/sec	1.6	1.7	2.4	2.0	1.6	1.5	1.5	1.6	1.7	1.6	1.6	1.5
v_{SW}	m/month	0.64	0.70	0.99	0.83	0.66	0.62	0.61	0.62	0.64	0.61	0.63	0.62
v_{DW}	m/month	0.75	0.80	1.12	0.93	0.76	0.74	0.77	0.83	0.91	0.85	0.79	0.73
Sed. A	cm/yr	0.08	0.09	0.11	0.08	0.01	0.01	0.01	0.01	0.01	0.09	0.08	0.08
Sed. A	$\mu g/cm^2 \cdot d$	65.6	70.2	85.2	60.8	6.7	4.5	3.1	4.0	7.2	71.6	65.4	67.0
Sed. ET	$\mu g/cm^2 \cdot d$	25.6	22.6	33.9	46.5	30.0	30.8	32.6	35.8	37.3	64.8	47.5	22.0
PF_{SW}	–	0.26	0.22	0.25	0.40	0.36	0.44	0.51	0.54	0.53	0.85	0.53	0.23
PF_{DW}	–	0.82	0.82	0.71	0.62	0.07	0.05	0.03	0.03	0.05	0.75	0.73	0.79
SPM_{SW}	mg/l	5.1	5.5	5.9	6.1	5.9	5.5	4.9	4.4	4.1	3.9	4.2	4.6
SPM_{DW}	mg/l	8.5	8.6	8.3	8.1	8.3	8.7	9.1	9.7	10.3	10.1	8.8	8.1
W	% ww	75	75	75	75	75	75	75	75	75	75	75	75
IG	% dw	6.3	6.3	6.3	6.3	6.3	6.3	6.3	6.3	6.3	6.3	6.3	6.3
d	g/cm^3	1.2	1.2	1.2	1.2	1.2	1.2	1.2	1.2	1.2	1.2	1.2	1.2
TP_{Ased}	mg/g dw	1.2	1.2	1.2	1.2	1.2	1.2	1.2	1.2	1.2	1.2	1.2	1.2

area are between 1.5 and 2.4 cm/s, which agree well with typical water velocities in section areas of Baltic coastal areas (Håkanson et al. 1986); fall velocities for particulate phosphorus and suspended particulate matter in the SW and DW-layers (v_{SW} 0.6–1 m/month and v_{DW} 0.85–1.1 m/month); the particulate TP-fractions in the SW and DW-layers (PF_{SW} 0.22–0.85 and PF_{DW} 0.03–0.82); and concentrations of SPM in the two layers (SPM_{SW} 3.9–6.1 mg/l and SPM_{DW} 8.1–10.3 mg/l).

Land uplift (F_{LU}) is an important individual input of TP to the system (2802 tons/yr).

Note the difference between fluxes and amounts. The largest TP-fluxes are to and from biota with long turnover times, but the total TP-inventory in biota is only

3.3 ktons (on a monthly basis) corresponding to 4.5% of the total amount of TP in the gulf (62.6 ktons on average). By far most TP is found in the accumulation area sediments (51.8 ktons or 61% of the total amount), and 61% of this is potentially available for diffusion.

6.3.3.3 Predicting "Good Ecological Status" for the Target Bioindicators

The simulations to estimate realistic reference values will be presented in two steps. The first step concerns substantial but realistic reductions of the direct anthropogenic emissions of TP to the Gulf of Riga from diffuse sources and point sources. The values given by HELCOM (2000) are 582 tons/yr and 335 tons/yr, respectively. It is not realistic to assume that all these anthropogenic emissions can be removed if one wants to sustain agricultural production in the area, and at the first step, it will be assumed that 50% of the anthropogenic emissions are eliminated.

In reality, this would require major investments and it would take a long time to implement such actions. In the following, it will be assumed that 50% of the anthropogenic emissions are suddenly removed (month 25, i.e., in January). This scenario will also demonstrate the dynamic response of the system to such a reduction. At the next step, it will be assumed that as much as 75% of all present anthropogenic emissions of TP to the bay would be suddenly removed (month 25) and also 60% of all anthropogenic TP-emission the Baltic Proper. The latter aspects have been discussed and motivated by Håkanson and Bryhn (2007) and in this simulation predicted values of TP in the Baltic Proper from such a reduction will be used. The idea is to simulate the lowest possible TP-input from anthropogenic sources that one can hope for assuming that there will be no drastic changes in the industrial or populations infrastructures in the Baltic countries.

The results are shown in Fig. 6.9:

- (A) Gives the predicted values for TP-concentrations in the SW-layer in the three cases, curves 1 give the present conditions, curves 2 when 50% of direct TP-emissions to the Gulf of Riga are removed and curves 3 the suggested reference values for "good ecological status". Note that it is possible that one cannot hope to achieve better conditions that the values in-between curves 2 and 3. Since it is difficult to follow the actual data in Fig. 6.9, Table 6.14 gives steady-state values for the monthly data in the three cases for Secchi depth, chlorophyll-a concentrations, cyanobacteria and TP-concentrations. One can note that today the mean predicted TP-concentration is 31.3 μg/l, which is in good general agreement with measured data (the mean value is 29.2 and the standard deviation 6.4; see Table 6.10). A 50% reduction of internal anthropogenic TP-fluxes would lower this value with 2.2 μg/l or with 7% to 29.1 μg/l. The reference concentration would be about 25.4 μg/l.
- (B) Gives the corresponding information for the TP-concentrations in the deep-water layer.

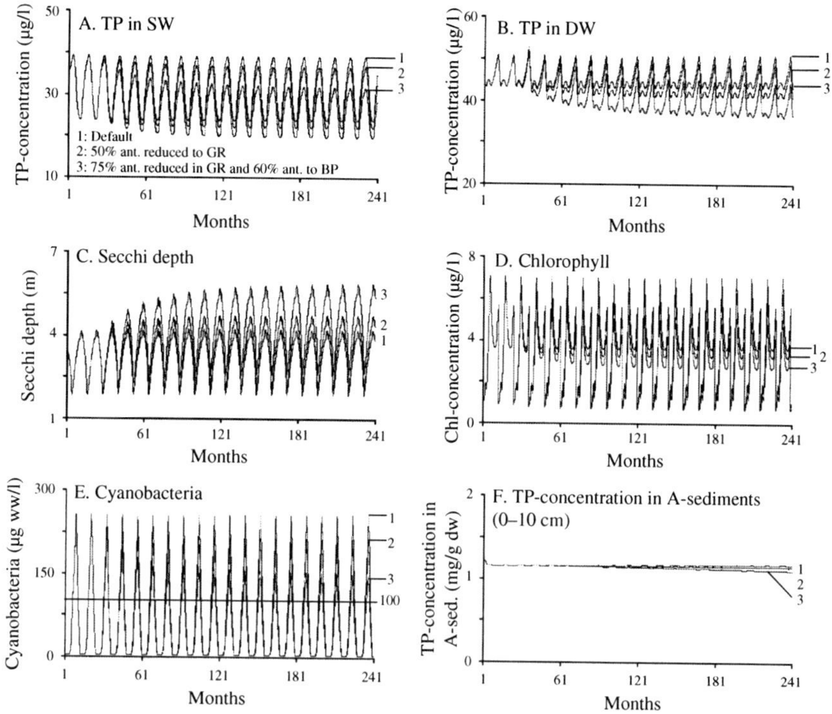

Fig. 6.9 Modeled TP-concentrations in surface water (**A**) and deep water (**B**), Secchi depths (**C**), chlorophyll-a concentrations (**D**), concentrations of cyanobacteria (**E**) and TN/TP-ratios in the Gulf of Riga under default conditions (curves 1), when 50% of the anthropogenic TP-load to the Gulf of Riga is suddenly (and hypothetically) reduced month 24 (curves 2) and when 75% of the anthropogenic load to the Gulf of Riga and 60% of the anthropogenic load to the Baltic Proper is being reduced month 24. Curves 3 represent realistic background conditions which correspond to "good" water quality

- (C) The calculated Secchi depths are shown in Fig. 6.9C. The measured mean value today is 3 m and the modeled value is 3.2 m. From these calculations, one cannot hope to get a mean Secchi depth in the Gulf of Riga higher than 4.6 m.
- (D) The present average chlorophyll concentration is about 3.6 µg/l with a standard deviation of 2.3 µg/l. The modeled value is 3.8 µg/l, and from these simulations one should not expect average chlorophyll value lower than 2.7 µg/l. This is a 29% reduction.
- (E) Predicted concentrations of cyanobacteria are shown in Fig. 6.9E. These simulations indicate that the maximum value would go down from about 250 µg/l to about 150 µg/l. Evidently, the actual situation concerning cyanobacteria in the Gulf of Riga will depend very much also on future water temperatures. If the SW-temperatures increase due to global warming, the risk of harmful algal blooms will also increase.
- (F) The TP-concentration in the accumulation area sediments will also be reduced if the TP-inflows to the system are reduced, but the response in the sediments is slow.

Table 6.14 Results for modeled monthly Secchi depths, chlorophyll-a concentrations (Chl), concentrations of cyanobacteria (CB) and TP-concentrations in the SW-layer under present (default) conditions (1), when 50% of the anthropogenic TP-load to the Gulf of Riga is hypothetically reduced month 24 (2) and when 75% of the anthropogenic load to the Gulf of Riga and 60% of the anthropogenic load to the Baltic Proper is being reduced month 24. The bolded values represent the requested "natural" background conditions

	1	2	3	4	5	6	7	8	9	10	11	12	Mean
Secchi (m), 1	3.3	2.8	1.8	2.1	2.7	3.1	3.5	3.7	3.9	4.2	3.9	3.7	3.2
Secchi (m), 2	3.7	3.1	2.1	2.4	3.0	3.5	3.9	4.2	4.4	4.7	4.4	4.2	3.6
Secchi (m), 3	4.7	4.0	2.6	3.0	3.8	4.4	4.9	5.3	5.5	5.9	5.5	5.2	**4.6**
Chl (μg/l), 1	1.9	1.5	3.8	7.0	5.5	4.4	3.8	3.5	3.6	4.7	4.6	1.0	3.8
Chl (μg/l), 2	1.7	1.4	3.4	6.3	4.9	3.9	3.3	3.2	3.2	4.2	4.1	0.9	3.4
Chl (μg/l), 3	1.3	1.1	2.7	5.0	3.9	3.2	2.7	2.6	2.7	3.5	3.3	0.7	**2.7**
CB (μg/ww/l), 1	0.4	0.1	2.8	19	113	163	256	208	112	55	7.2	0.03	78.0
CB (μg/ww/l), 2	0.3	0.1	2.1	15	80	137	211	171	92	46	5.5	0.03	63.4
CB (μg/ww/l), 3	0.2	0.1	1.4	11	58	97	148	121	66	34	4.0	0.02	**45.1**
TP_{SW} (μg/l), 1	37.4	38.8	38.4	35.8	31.0	26.9	24.4	23.5	23.8	28.2	31.6	35.5	31.3
TP_{SW} (μg/l), 2	34.8	36.2	35.7	33.3	28.8	25.0	22.7	21.8	22.1	26.3	29.4	33.0	29.1
TP_{SW} (μg/l), 2	30.1	31.3	30.9	28.9	25.1	21.8	19.9	19.3	19.6	23.2	25.7	28.6	**25.4**

6.3.4 Comments

This case-study has discussed a general method to estimate pre-industrial reference values for "good ecological status" for phosphorus and operational bioindicators of coastal water eutrophication using a process-based mass-balance approach, which makes it possible to carry out structured modifications in nutrient inflow to the given coastal system. The dynamic modeling can also provide quantitative values of the time-dependent response of the system. The CoastMab-model discussed here may be applied to most coastal systems and the data necessary for this analysis have also been discussed.

Curves 3 in Fig. 6.9 probably also describe the conditions in the Gulf of Riga as they were about 100–200 years ago.

There is also another interesting aspect of these simulations. It is well know from lake studies (see, e.g., Scheffer 1990; Scheffer et al. 2000; Carpenter 2003) that lakes can appear in two markedly different and rather stable states, one with clear water when macrophytes are abundant and one with turbid water when the macrophyte cover is small. Basically, in the first state with clear water the primary production of benthic algae and macrophytes can be significantly higher than the phytoplankton production (Mann 1982) and in the turbid state, the phytoplankton production may be higher (see Håkanson and Boulion 2002). So, for all aquatic systems, one can hypothesize that there exists a threshold value for the Secchi depth when this shift is likely to occur. Since this modeling approach can quantify the mechanisms related to changes in Secchi depth, and hence also the depth of the photic zone, this approach may also be used to clarify the conditions concerning possible regime shifts from pelagic to benthic dominance of coastal systems.

The selected case-study area, the Gulf of Riga, is sensitive to nutrient loading because of its shallowness and low openness towards the Baltic Proper. The morphometry of any coastal area, as given by the size and form parameters, influences all internal processes, such as sedimentation, resuspension, diffusion in water and from sediments to water, biouptake and retention in biota, stratification, mixing and outflow. Today, the average modeled monthly values for Secchi depth, chlorophyll (Chl), cyanobacteria (CB) and total-P (TP) are 3.2 m, 3.8 μg/l, 78 μg/l and 31.3 μg/l, respectively. If 50% of all anthropogenic sources to the Gulf of Riga via rivers, point sources and diffuse sources were to be removed, these values would be 3.6 m, 3.4 μg Chl/l, 63 μg CB/l and 29.1 μg TP/l. If 60% of the anthropogenic phosphorus fluxes to the Baltic Proper were to be omitted and as well as 75% of all direct anthropogenic sources to the Gulf of Riga, the values would be 4.6 m, 2.7 μg Chl/l, 45 μg CB/l and 25.4 μg TP/l. These values represent the reference levels and it is not realistic to expect that remedial measures would improve the conditions more than that.

6.4 Reconstruction of Eutrophication History

6.4.1 Aim of the Case-Study

The conditions in the Baltic Sea, and specifically in the Baltic Proper and the Gulf of Finland, have been discussed in many papers and books (see recent compilations by Pitkänen and Tallberg 2007; and also Ambio 2007; Schernewski and Schiewer 2002; Schernewski and Neumann 2005; Wulff et al. 2001a). The aim of this section is to focus on the conditions in the Gulf of Finland, since this is one of most heavily eutrophicated, major sub-basins in the Baltic Sea (see Fig. 5.3 and the HELCOM website), and to try to reconstruct the development that has taken place in this bay between the years 1900 and 2000. The eutrophication in the Gulf of Finland has been discussed in many papers and reports (see, e.g., Kiirikki et al. 2001; SYKE 2003, 2006; HELCOM 2006). Since the Gulf of Finland is open to the Baltic Proper, the conditions in the Baltic Proper will influence the conditions in the Gulf of Finland together with the direct discharges to the Gulf (see, e.g., Savchuk and Wulff 1999a, b).

The main objective here can be illustrated by the data on Secchi depth given in Fig. 6.10 (see also Aarup 2002). These data come from the HELCOM database and concern how the water clarity (the Secchi depth) has changed in the period from 1900 to 1991. An initial trend analysis, a linear regression between measured Secchi depths and time (month 1 is January of 1900), indicates the negative development: The Secchi depth has decreased from about 7–8 m 100 years ago to about 5 m today. One can also note the large scatter in the data from the individual sampling sites.

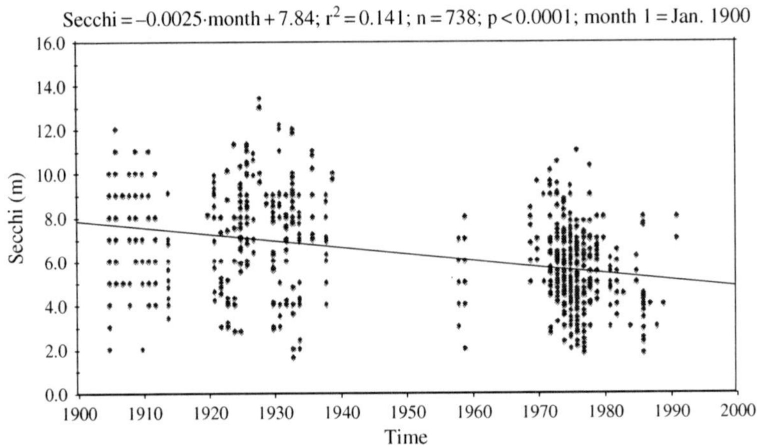

Fig. 6.10 HELCOM data on the Secchi depth in the Gulf of Finland 1990–1991 and a trend analysis; regression line, coefficient of determination (r^2), number of data (n) and statistical certainty (p)

The aim of case-study is to try to explain the development shown in Fig. 6.10 by using:

1. CoastMab for salt (see Håkanson et al. 2007d) to get realistic and reliable data on the water fluxes to, within and from the Gulf of Finland;
2. CoastMab for phosphorus (see Sect. 9.1) to quantify how the system would react to changes in nutrient loading;
3. linked to the mass-balance model for phosphorus, there are empirically-based sub-models for Secchi depth, chlorophyll-a concentration, concentration of cyanobacteria, sedimentation, total nitrogen and suspended particulate matter; and
4. linked to these models, there is also a more comprehensive dynamic food-web model for 10 functional groups (CoastWeb, see Håkanson and Gyllenhammar 2005; Håkanson and Bryhn 2008b; Håkanson and Lindgren 2007b), which have been used in this work to calculate biouptake and retention of phosphorus in biota.

The basic idea is to demonstrate (A) how this modeling works and how well modeled phosphorus concentrations, Secchi depths and chlorophyll values agree to empirical data, (B) how reductions in nutrient loading to the Gulf of Finland and to the Baltic Proper would influence the conditions for these target variables in the Gulf of Finland, and (C) how a reconstruction of the development could be made so the that the changes in the Gulf of Finland shown by the data in Fig. 6.10 may be explained and understood. This would also imply that important information to reverse the development in the Gulf of Finland and the Baltic Proper could be gained.

An important reason for selecting the Gulf of Finland as a case-study is that for this bay one can access the necessary data:

- First, data on the salinity inside and outside the bay, and the tributary water fluxes to the bay, must be available so that a reliable mass-balance for salt can be established, so that the water exchange with the outside sea (the Baltic Proper) can be quantified realistically and also the theoretical water retention time in the bay, which influence important internal processes, such as stratification, mixing and diffusion. As stressed in Sect. 6.2, the CoastMab-model includes a mass-balance model for salt structured in the same way as the model for phosphorus except that the mass-balance model for phosphorus also calculates sedimentation, resuspension, burial and biouptake of dissolved phosphorus. By definition, total phosphorus (TP) in the water generally includes living and dead plankton, but not larger animals such as zoobenthos and fish. A new aspect of this version of the CoastMab-model (as compared to the basic version in Sect. 9.1) is that this model calculates phosphorus uptake and retention in biota with short turnover times (phytoplankton, bacterioplankton, herbivorous zooplankton and benthic algae) and long turnover times (predatory zooplankton, prey and predatory fish, macrophytes, zoobenthos and jellyfish) using the CoastWeb-model.
- Second, data on the TP-concentration in the water outside the bay must be at hand so that the inflow from the Baltic Proper can be calculated. Data on the TP-transport from different sources on land are also needed and such data are available for the Gulf of Finland from the HELCOM database. HELCOM also gives information on the natural background losses, the diffuse losses and the point sources discharges of TP to the Gulf of Finland and from this one can use the CoastMab-model to estimate how various reductions in the anthropogenic load would influence the system. This is essential information for the reconstruction requested in this case-study. Even though the focus is on the changes in Secchi depth, we will also discuss changes in other bioindicators of eutrophication, mainly the chlorophyll-a concentration and the concentration of cyanobacteria.

6.4.2 Data from the Gulf of Finland and Information on the Model Structure

Figure 5.3 shows that the Gulf of Finland today generally has a relatively low Secchi depth compared to other basins in the Baltic Sea. Table 6.15 gives background data from the Gulf of Finland and the Baltic Proper and shows that the limiting section area towards the Baltic Proper is 3.74 km^2.

This means that the Gulf of Finland is quite open to the Baltic Proper. Table 6.15 also gives information on, e.g., total area, volume, mean depth, maximum depth and the depth of the theoretical wave base. Table 6.16 gives data on the volume of the surface-water layer and the middle-water layer separated by the theoretical wave base and the deep-water layer separated from the middle-water layer by the depth of the average halocline.

Table 6.15 Basic data for two main sub-basins in the Baltic Sea, the Gulf of Finland and the Baltic Proper (from Håkanson and Lindgren 2007d). Bolded factors are the obligatory driving variables

		Gulf of Finland (GF)	Baltic Proper (BP)
Area	**(km²)**	**29,600**	**211,100**
Max. depth	**(m)**	**105**	**459**
Volume	**(km³)**	**1073.2**	**13,055**
Mean depth	(m)	36.3	61.8
Wave base	(m)	41	43.8
Section area	**(km²)**	**3.74**	–
Halocline depth	**(m)**	**75**	**75**
ET-areas	(%)	63	47
Water discharge	**(km³/yr)**	**29.0**	**250**
Catchment area	**(km²)**	**421,000**	**568,973**
Latitude	**(°N)**	**60**	**58**
Precipitation	**(mm/yr)**	**593**	**750**

The annual freshwater flux to Gulf of Finland used in this work is the average annual value from Savchuk (2000; 3552 m^3/s); Myrberg (1998; 3615 m^3/s) and Stålnacke et al. (1999a, b; 3875; m^3/s). The monthly data on water discharge used in the modeling have been calculated from the average annual value using the dimensionless moderator for this purpose (from Abrahamsson and Håkanson 1998). This moderator is based on data on the size of the catchment area, mean annual precipitation and latitude (Table 6.14 gives the data). Since we do not have access to reliable monthly data on water discharge for the study period (1990–1998), it should be stressed that this modeling concerns average, characteristic conditions on a monthly base for this period of time and not the actual sequence of months. The results from 1990 to 1998 concerning the processes and fluxes will then be put into a wider time frame (1900 to 2000).

Also note that the values presented in this work relate to the definitions of the surface-water, the middle-water and the deep-water layers and the given hypsographic curves (from Håkanson and Lindgren 2007d). This means that although these data correspond quite well with other values (see, e.g., Voipio 1981; Mikulski 1985; Monitor 1988), they are not directly comparable.

Table 6.16 Data on volumes and areas (below the given depths) in the Gulf of Finland and the Baltic Proper based on hypsographic curves from Håkanson and Lindgren (2007d)

Basin	Level	Volume (km³)	Area below the given level (km²)
Gulf of Finland	SW	851.2	29,600
	MW	202.0	10,900
	DW	20.0	2400
Baltic Proper	SW	7315	211,100
	MW	3050	123,500
	DW	2690	73,000

The Gulf of Finland has been divided into three depth intervals:

1. The surface-water layer (SW), i.e., the water above the theoretical wave base at 41 m.
2. The middle-water layer (MW) between the theoretical wave base and the average depth of the halocline at 75 m.
3. The deep-water layer (DW) is defined as the volume of water beneath the average depth of the halocline at 75 m.

Note again that the theoretical wave base describes average conditions.

Håkanson and Lindgren (2007d) have presented new hypsographic curves for the Gulf of Finland calculated using GIS and bathymetric data provided by Seifert et al. (2001). One can see from Table 6.15 that the area below the theoretical wave base (D_{wb}) at 41 m is 10,900 km^2 and that the volumes of the SW, MW and DW-layers are 851.2, 202.0 and 20.0 km^3 and the entire volume is 1073.2 km^3. The Gulf of Finland is also relatively shallow and 63% of its bottom area is dominated by resuspension processes of fine materials (these are the erosion and transportation areas, ET; as calculated by the CoastMab-model; see Table 6.14).

Figure 6.11 illustrates the number of sampling sites for the Secchi depth data from the HELCOM database used in this case-study. Empirical data on salinities, Secchi depth, chlorophyll and nutrient concentrations will be given later and compared to modeled values. There are no corresponding empirical data on cyanobacteria available to us from the Gulf of Finland. The empirical temperature data from the HELCOM database (from 1990 to 1998) have been used to model stratification and mixing. The standard deviations (SD) for the monthly mean empirical values

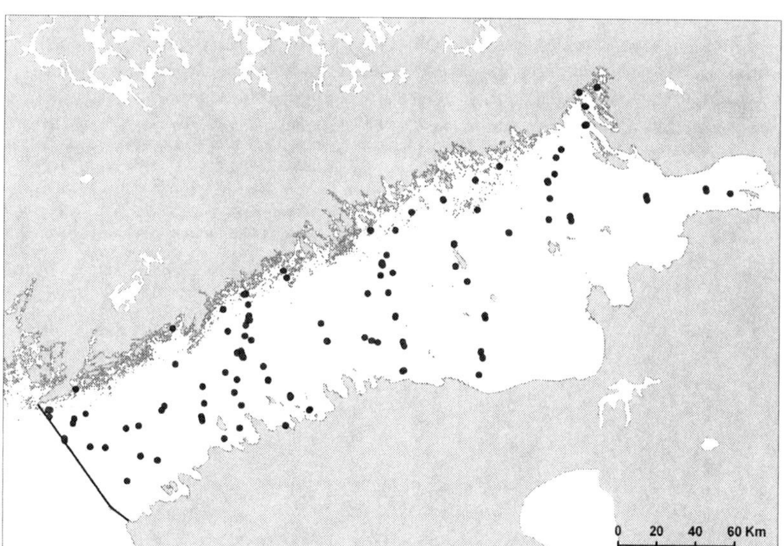

Fig. 6.11 Sampling sites for Secchi depth in the Gulf of Finland (from HELCOM)

are very important in the sense that the variables with the high inherent SD-values cannot be expected to be predicted as well as variables with comparatively low SD-values (such as salinity; see Chap. 4).

6.4.3 Dynamic Modeling

The model consists of seven compartments: surface water (SW), middle water (MW), deep water (DW), erosion/transportation areas for fine sediments (ET-areas), accumulation areas for fine sediments below the theoretical wave base (A-areas), biota with short turnover times (BS) and biota with long turnover times (BL). There are algorithms for all major internal TP-fluxes (outflow, TP from land uplift, sedimentation, resuspension, diffusion, mixing, biouptake and retention in biota and burial; see Sect. 9.1, for a more detailed model description).

To calculate the inflow of TP to the Gulf of Finland (GF) from the Baltic Proper (BP), modeled data on the TP-concentrations in the surface, middle and deep-water layers in the Baltic Proper from Håkanson and Bryhn (2007) have been used. The water fluxes between the Gulf of Riga and the Baltic Proper are calculated from the mass-balance for salt. So, the inflows to the three layers (SW, MW and DW) from the Baltic Proper are given by the water discharges in Table 6.17 (Q_{SWBPGR}, Q_{DWBPGR} and Q_{DWBPGR}) and the modeled TP-concentrations. The transport of TP from the catchment area to the Gulf of Finland uses data from HELCOM: 1191 tons TP/yr from natural background losses, 2112 tons TP/yr from diffuse losses and 2431 tons

Table 6.17 Calculated monthly values for the theoretical SW, MW and DW water retention times (T_{SW}, T_{MW} and T_{DW}), the water fluxes (Q) from the Baltic Proper (BP) to the Gulf of Finland (GF) in the three layers (SW, MW and DW), the mixing transport (abbreviated with an x) between the MW and SW and the DW and MW layers and the water velocity in the section area (u_{At}) from the mass-balance model for salt. These values are used in the mass-balance model for phosphorus

Month	T_{SW}	T_{MW}	T_{DW}	Q_{SWBPGF}	Q_{MWBPGF}	Q_{DWBPGF}	Q_{MWSWx}	Q_{DWMWx}	U_{At}
	months	months	months	km^3/month	km^3/month	km^3/month	km^3/month	km^3/month	cm/s
1	8.7	5.4	1.7	64.8	10.8	10.8	26.4	0.8	1.9
2	9.4	7.2	1.8	63.7	10.6	10.6	24.4	0.7	1.9
3	9.3	6.9	1.8	62.9	10.5	10.5	24.5	0.7	1.9
4	9.1	6.5	1.8	61.8	10.3	10.3	24.2	0.7	1.9
5	8.5	5.7	1.9	62.8	10.5	10.5	25.1	0.7	1.9
6	8.4	5.7	2.0	63.1	10.5	10.5	26.9	0.8	1.9
7	8.3	5.1	1.8	57.4	9.6	9.6	32.7	1.0	1.9
8	8.0	4.6	1.7	54.7	9.1	9.1	32.1	1.0	1.9
9	11.2	14.0	1.9	61.2	10.2	10.2	35.3	1.1	1.9
10	11.3	14.9	1.9	62.8	10.5	10.5	39.8	1.1	1.9
11	8.9	5.9	1.7	63.8	10.6	10.6	4.5	0.1	1.9
12	9.0	6.3	1.8	63.8	10.6	10.6	3.4	0.1	1.9

TP/yr from point source discharges. These annual TP-fluxes have been transformed into monthly values by division with 12 and by applying a seasonal moderator for water discharge (characteristic mean monthly Q-values divided by the annual mean water discharge).

The water velocity in the section area has been calculated for the total outflow (km^3/yr) divided by half the section area since there is also inflow of water to maintain a given water level ((km^3/yr) · (1/(0.5 · km^2). Savchuk (2006) gave a total water outflow from the Gulf of Finland of 554 km^3/yr, which is a factor of 2 lower than the value obtained in this work (990 km^3/yr) and 554 seems a less likely value since it would imply that the average water velocity in the section area would be lower than 1 cm/s. These calculations give an average velocity in the section area of 1.9 cm/s.

Some key questions for this work are: How would the system react if the diffuse losses and the point source discharges of TP to the Gulf of Finland would be reduced? How would the system respond if similar reductions to the entire Baltic Proper would also be carried out? To calculate the changes in the concentrations of TP in the SW, MW and DW-layers, the CoastMab-model has also been applied to the entire Baltic Proper (see Håkanson and Bryhn 2007).

The theoretical water retention times in the three layers (see Table 6.17) from the basic mass-balance for salt are used together with the temperature dependent mixing rate in the mass-balance model as indicators of how the turbulent mixing influences the settling velocity for particulate phosphorus – the faster the water renewal, the more turbulence, the slower the settling velocity. The small TP-input from precipitation onto the water surface of the Gulf of Finland has been estimated from the characteristic annual precipitation of 593 mm and a TP-concentration in the rain of 5 μg/l (see Håkanson and Eklund 2007b).

The internal processes are: sedimentation of particulate phosphorus from surface water to middle water and deep water (F_{SWMW} and F_{SWDW}), sedimentation from the SW-layer to areas of erosion and transportation (F_{SWET}), sedimentation from the MW and DW-layers to the respective accumulation areas (F_{MWAMW} and F_{DWADW}), resuspension (advection) from ET-areas (including TP from land uplift, F_{LU}) either back to the surface water (F_{ETSW}) or to the middle water (F_{ETMW}), diffusion of dissolved phosphorus from accumulation area sediments in the MW and DW-layers to water in the MW and DW-layers (F_{AMWMW} and F_{ADWDW}), diffusion from MW and DW-water layers to the SW and MW-layers, respectively (F_{MWSWd} and F_{DWMWd}), upward and downward mixing between the water layers (F_{SWMWx}, F_{MWSWx}, F_{DWMWx} and F_{MWDWx}) and biouptake and elimination of phosphorus from biota with short and long turnover times ($F_{bioupBS}$ and $F_{retbioBS}$ and $F_{bioupBL}$ and $F_{retbioBL}$). When there is a partitioning of a flux from one compartment to two compartments, this is handled by a distribution coefficient (DC).

From the map illustrating the spatial variation in land uplift (see Fig. 1.1), one can estimate that the mean land uplift in the Gulf of Finland is about 1.2 mm/yr and this value has been used in these calculations. The total area above the theoretical wave base in the Gulf of Finland is about 18,700 km^2 and the sediments in this area will be exposed to increased erosion by wind/wave action due to land uplift. The sediments in the shallower parts will be more consolidated than the recent materials close to the theoretical wave base. The calculation of the TP-flux from land uplift uses (1) modeled data on the TP-concentration in the accumulation area sediments from the MW-zone, (2) a water content of the sediments exposed to increased erosion set to be 15% lower than the modeled water content of the recent sediments and (3) the total volume of sediments above the theoretical wave base lifted each year.

The Secchi depth (Sec in m) is a target bioindicator in this study and it is calculated from a model illustrated in Fig. 5.1 relating Secchi depth to the concentration of suspended particulate matter in the SW-layer (SPM_{SW} in mg/l) and the salinity of the SW-layer (Sal_{SW}). One can note from Fig. 5.1. that for the Gulf of Finland with a SW-salinity of about 6 psu, one should expect a fairly rapid, non-linear improvement in Secchi depths if the SPM-concentration is lowered from 4 to 3 mg/l. The Secchi depth is important for predictions not just of water clarity and the depth of the photic zone but also of, e.g., macrophyte cover and biomass of benthic algae using the CoastWeb-model.

6.4.3.1 Modeled Values Versus Empirical Data

Table 6.18 gives modeled values for the salinity compared to the mean empirical monthly data for the period 1990–1998 and also to the standard deviations (SD) for the mean empirical values. If the model yields values in-between the uncertainty bands given by ± one standard deviation, the predictions should be regarded as good. For all these comparisons between modeled and empirical data, it should be noted that:

- The monthly inflows of water and phosphorus have not been calculated using data for this period of time (1990–1998) but using a general model for water discharge (see Abrahamsson and Håkanson 1998). This will explain some of the differences between the modeled and the measures monthly values for salinity, TP and other variables.
- The monthly calculations of the inflow of water, salt and phosphorus from the Baltic Proper use mean values for the SW and MW-compartments not in the inflowing water from the area just outside the Gulf of Finland, which would have been more appropriate, but for the entire Baltic Proper. For the inflow to the DW-compartment, a value of 70.9 μg TP/l (from Karlsson 2007) has been used. The reason why the more appropriate data have not been used is simply that it has been difficult to find such data. This likely also further explains some of the differences between the modeled and the measured monthly values.

With these reservations, one can note from Table 6.18 that the predicted monthly salinities in the SW-layer are generally close to the empirical data and within the defined uncertainty bands of the empirical data (±1SD = 0.22). The average error (the mean difference for the 12 months) is 0.02 psu. It should be noted that the available data from the MW and DW-layers are few and the values 7 and 10 for the two layers are uncertain. The CoastMab-model gives a value of 6.75 psu for the mean salinity in the MW-layer and of 10.02 psu in the DW-layer. The water

Table 6.18 A comparison between measured (empirical; data from HELCOM for 1990–1998) data and modeled values on the salinity in the SW, MW and DW-layers in the Gulf of Finland. Note that there are no reliable mean monthly data accessible from the MW and DW-layers and the data given for these layers (7.0 and 10.0) should be regarded as the "best possible" estimates based on few and scattered data from the Gulf of Finland and from the area of the Baltic Proper outside the Gulf of Finland

Month	Sal_{SW}	Sal_{SW}	Sal_{MW}	Sal_{MW}	Sal_{DW}	Sal_{DW}
	psu mod	psu emp	psu mod	psu emp	psu mod	psu emp
1	6.27	5.87	6.79	7.0	10.05	10.0
2	6.28	6.15	6.79	7.0	10.03	10.0
3	6.29	6.37	6.78	7.0	10.01	10.0
4	6.30	6.32	6.76	7.0	9.97	10.0
5	6.28	6.30	6.75	7.0	9.95	10.0
6	6.22	6.24	6.72	7.0	9.92	10.0
7	6.18	6.40	6.70	7.0	9.94	10.0
8	6.19	6.14	6.68	7.0	9.95	10.0
9	6.21	6.38	6.69	7.0	10.02	10.0
10	6.21	6.53	6.75	7.0	10.13	10.0
11	6.23	6.15	6.78	7.0	10.15	10.0
12	6.25	5.80	6.79	7.0	10.09	10.0
Mean	6.24	6.22	6.75	7.0	10.02	10.0
Median	6.24	6.27	6.76	7.0	10.02	10.0
SD for emp.		0.22		0.0		0.00
Diff. mod.-emp						
1	0.40		–0.21		0.05	
2	0.13		–0.21		0.03	
3	–0.08		–0.22		0.01	
4	–0.02		–0.24		–0.03	
5	–0.02		–0.25		–0.05	
6	–0.02		–0.28		–0.08	
7	–0.22		–0.30		–0.06	
8	0.05		–0.32		–0.05	
9	–0.17		–0.31		0.02	
10	–0.32		–0.25		0.13	
11	0.08		–0.22		0.15	
12	0.45		–0.21		0.09	
Mean diff.	0.02		–0.25		0.02	
SD for diff.	0.23		0.04		0.08	

fluxes between the Gulf of Finland and the Baltic Proper calculated from the mass-balance model for salt and the corresponding fluxes for mixing and diffusion are used without any changes also in the mass-balance calculations for phosphorus, except, of course, that phosphorus has a particulate fraction.

The main message here is that there should be no "tuning" of the mass-balance calculations and the same algorithms and values have been used for mixing, diffusion and water fluxes for salt and phosphorus. There are substance-specific parts in the CoastMab-model and they mainly concern the algorithms for the particulate fraction and for diffusion.

The results for phosphorus are given in Table 6.19.

- (A) The TP-concentrations in the SW-layer are generally close to the empirical data and (mean difference 0.5 and median difference 0.2 μg/l) well within the uncertainty bands of the empirical data (±1SD).
- (B) Also the average TP-concentrations in the MW-layer are close to the empirical average value (the mean difference between the mean values is 2.9 μg/l), which is a small difference compared to the relatively high standard deviation for the empirical data (18.6 μg/l).

Table 6.19 A comparison between empirical data (data from HELCOM from 1990 to 1998) and modeled values on TP-concentrations, Secchi depths, chlorophyll-a concentrations and total-N concentrations (TN) in the three layers (SW, MW and DW) in the Gulf of Finland. The lower part of the table gives the differences between modeled and measured values

Month	TP_{SW}	TP_{SW}	TP_{MW}	TP_{MW}	TP_{DW}	TP_{DW}	Sec	Sec	Chl	Chl	TN	TN
	μg/l mod	μg/l emp	μg/l mod	μg/l emp	μg/l mod	μg/l emp	m mod	m emp	μg/l mod	μg/l emp	μg/l mod	μg/l emp
1	23.0	34.1	37.0	37.7	90.4	71.7	6.1	6.3	3.1	0.5	366	439
2	24.0	32.2	35.0	35.1	90.5	43.2	5.6	5.1	5.9	1.3	376	422
3	24.8	30.1	32.7	33.9	89.3	31.3	5.4	5.1	7.9	2.2	386	410
4	25.3	30.1	32.0	35.6	86.9	57.2	5.3	5.4	8.3	10.5	391	394
5	25.6	22.5	32.8	32.3	85.1	47.8	4.7	5.6	7.5	7.3	394	394
6	26.5	24.0	35.0	40.9	84.7	69.2	4.2	5.5	7.2	2.4	403	458
7	26.8	16.2	36.5	39.1	85.4	69.5	4.6	5.0	6.8	3.5	407	352
8	27.0	17.5	38.0	38.2	85.0	50.6	4.6	5.8	5.9	3.0	409	329
9	25.6	16.6	38.1	39.6	87.6	68.0	5.2	5.3	7.5	5.8	394	291
10	22.7	16.9	36.2	51.5	85.3	126.1	6.4	4.8	5.1	3.6	362	294
11	21.2	24.2	38.0	40.7	79.2	47.8	7.3	3.3	1.5	0.4	345	375
12	22.1	24.2	39.0	40.7	86.2	47.8	6.7	6.3	2.3	0.5	356	447
Mean	24.5	24.1	35.8	38.8	86.3	60.9	5.5	5.3	5.7	3.4	382	384
Median	25.1	24.1	36.3	38.7	85.8	53.9	5.3	5.3	6.3	2.7	388	394
SD for emp.		±8.9		±18.6		±22.0		±1.6		±2.6		±41
Diff.												
1		−11.1		−0.7		18.7		−0.2		2.6		−73.0
2		−8.3		−0.1		47.3		0.5		4.6		−45.7
3		−5.3		−1.2		58.0		0.3		5.7		−24.0
4		−4.8		−3.7		29.7		−0.1		−2.2		−3.4
5		3.1		0.5		37.3		−0.9		0.3		0.2
6		2.5		−5.9		15.5		−1.3		4.8		−54.5
7		10.6		−2.7		15.9		−0.4		3.3		54.8
8		9.5		−0.2		34.4		−1.2		2.9		80.4
9		9.0		−1.5		19.6		−0.1		1.7		102.8
10		5.8		−15.3		−40.8		1.6		1.5		68.1
11		−3.0		−2.7		31.4		4.0		1.1		−29.8
12		−2.1		−1.7		38.4		0.4		1.8		−91.5
Mean diff.		0.5		−2.9		25.4		0.2		2.3		−1.3
Median diff.		0.2		−1.6		30.5		−0.1		2.2		13.7
SD for diff.		7.3		4.3		24.6		1.4		2.2		63.9

- (B) The average TP-concentrations in the DW-layer differ more (mean difference 25.4) but also this is close to the inherent uncertainty in the empirical data (SD = 22.0 μg/l).
- (C and D) The target variables, the two bioindicators Secchi depth and chlorophyll-a concentration in the SW-layer, are close to the empirical values (the mean error for Secchi depth is 0.2 m and for chlorophyll 2.3 μg/l) and within the uncertainty bands defined by one standard deviation of the empirical mean value. The modeled chlorophyll concentrations also give the "twin-peak pattern" as indicated by the empirical data. It should be stressed that the chlorophyll concentrations are predicted from a regression including dynamically modeled TP-concentrations in the surface water and a dimensionless moderator for the light conditions (6.2) and the calculated dissolved fraction of phosphorus. These calculations also include considerations to the biouptake of dissolved phosphorus in all types of biota (functional groups) included in the CoastWeb-model. However, the temporal patterns are calculated in the standardized pattern to reveal the most typical, characteristic condition in the Gulf of Finland, and the given presuppositions including the relatively high uncertainties in the empirical data, one cannot hope to obtain very much better predictions than these.
- (F) The TN-concentrations are predicted from a simple regression (6.1) using dynamically modeled TP-concentrations in the SW-compartment. There is a relatively good correspondence between modeled and measured TN-concentrations (the mean error is 1.3 μg/l).

The TN/TP-ratios based on modeled values, modeled salinities in the SW-layer and empirical temperature data for the SW-layer will be used to calculate the concentration of cyanobacteria.

From this, one can conclude that the model predicts the target variables quite well given the factual limitations in the seasonal patterns in the driving variables for tributary water discharge (since this pattern in the modeling is not based on measured data for the modeled period) and in the seasonal pattern for the TP-concentrations outside the Gulf of Finland (since these data for the SW and MW-layers emanate not from the area outside of the Gulf but from the entire Baltic Proper and since the empirical value for the TP-concentration in the Baltic Proper outside the Gulf of Finland is uncertain). Note that there has been no tuning of the model to achieve these predictions and that the basic model has been shown to describe the transport processes for phosphorus very well for many other coastal areas (see Sect. 9.1).

6.4.3.2 Fluxes and Amounts of Phosphorus

Table 6.20 gives a compilation of the monthly TP-fluxes and a ranking based on the annual fluxes. One can see that the two largest fluxes are biouptake and retention of phosphorus to and from biota with short turnover times (BS). These fluxes (about 500,000 tons/yr) are 25 times larger than the fluxes related to the following fluxes because the organisms in this group have very short turnover times.

Table 6.20 A ranking of the annual fluxes (tons/yr), as calculated using the CoastMab-model from the monthly fluxes (tons/month) of TP to, in and from the Gulf of Finland. The key fluxes for remedial measures are bolded. F = flux, SW = surface water, MW = middle water, DW = deep water, GF = Gulf of Finland, BP = Baltic Proper, trib = tributary, d = diffusive flux, x = mixing flux, LU = land uplift, ET = erosion and transport areas, A = accumulation areas, BS = biota with short turnover times, BL = biota with long turnover times, bur = burial

Month	1	2	3	4	5	6	7	8	9	10	11	12	Annual
$F_{TPbioupBS}$	33,010	42,102	56,886	57,493	49,979	38,693	37,909	37,807	38,434	54,519	40,518	17,391	504,740
$F_{TPbioretBS}$	31,491	41,569	54,658	56,791	49,823	39,015	37,506	37,506	38,036	52,083	41,010	16,860	496,349
$F_{TPMWAMW}$	2798	2620	2641	2461	2281	2174	2371	2240	2189	3040	1350	1832	27,998
F_{TPETSW}	2549	2464	2335	2300	2351	2840	2793	3060	3209	376	351	2941	27,568
$F_{TPSWGFBP}$	1569	1642	1718	1773	1800	1872	1958	1923	1926	1813	1606	1494	21,094
F_{TPSWET}	1694	1487	1031	1116	1412	1795	1982	2106	2120	1676	1751	2302	20,472
F_{TPLU}	1535	1535	1535	1536	1536	1537	1537	1538	1538	1538	1536	1535	18,438
$F_{TPburAMW}$	1420	1398	1379	1369	1365	1343	1326	1341	1343	1364	1410	1448	16,507
F_{TPETMW}	1344	1300	1232	1213	1240	1498	1473	1614	1693	198	185	1551	14,540
$F_{TPSWBPGF}$	**1036**	**1153**	**1238**	**1352**	**1449**	**1381**	**1303**	**1292**	**1126**	**959**	**818**	**880**	**13,987**
F_{TPSWMW}	988	867	601	650	823	1046	1155	1228	1236	977	1021	1342	11,933
$F_{TPMWSWx}$	965	920	857	831	870	1083	1134	1297	1522	172	125	1018	10,792
$F_{TPAMWMWd}$	1021	1253	1187	1040	878	754	717	762	732	976	942	488	10,748
$F_{TPDWGFBP}$	917	949	933	936	915	815	774	872	892	932	909	856	10,700
$F_{TPDWBPGF}$	**752**	**743**	**730**	**742**	**745**	**678**	**647**	**723**	**742**	**753**	**754**	**765**	**8775**
F_{TPMWDW}	790	740	746	695	644	614	670	632	618	858	381	517	7905
$F_{TPSWMWx}$	473	477	463	493	551	710	736	820	933	93	64	522	6336
F_{TPtrib}	**361**	**414**	**486**	**417**	**400**	**774**	**948**	**526**	**416**	**355**	**349**	**288**	**5735**
$F_{TPMWSWd}$	666	497	354	217	196	260	362	450	517	628	677	751	5574
$F_{TPbioupBL}$	234	331	460	511	472	379	358	354	356	445	316	116	4331
$F_{TPbioretBL}$	295	309	330	365	388	388	381	373	365	357	352	310	4212
$F_{TPDWMWd}$	313	374	381	379	359	339	337	322	307	431	373	213	4128
$F_{TPMWBPGF}$	**358**	**345**	**334**	**335**	**339**	**314**	**302**	**338**	**356**	**368**	**366**	**367**	**4123**
$F_{TPburADW}$	107	105	104	103	103	101	100	101	101	103	106	109	1245
$F_{TPDWADW}$	113	113	120	121	117	117	110	118	120	6	5	120	1180
F_{TPprec}	7.3	7.3	7.3	7.3	7.3	7.3	7.3	7.3	7.3	7.3	7.3	7.3	88
$F_{TPADWDWd}$	0.9	0.9	1.0	1.0	0.9	1.0	0.9	1.0	1.0	0.2	0.2	1.0	10

The following fluxes are: sedimentation from the MW-layer to accumulation areas within the MW-zone ($F_{TPMWAMW}$ = 28,000 tons/yr), resuspension from ET-sediments to the SW-layer (27,500 tons/yr), outflow of TP from the Gulf of Finland (GF) to the Baltic Proper (BP) in the SW-compartment (21,000 tons/yr). The fluxes that may be reduced by remedial measures are bolded in Table 6.20: the SW-inflow from the Baltic Proper (F_{SWBPGF} = 14,000 tons/yr), total tributary inflow (F_{trib} = 5735 tons/yr) and inflow to the DW and MW-layers from the Baltic Proper (F_{DWBPGR} = 8880 and F_{MWBPGR} = 4100 tons/yr, respectively). The diffusive flux from the accumulation areas sediment (below the halocline at 75 m) is rather small (10 tons/yr). The modeled TP-concentration in the accumulation area sediments (0–10 cm) in the MW-layer is 2.4 mg/g dw and in the DW-layer 0.63 mg/g dw (see Table 6.21), which agrees with measured values for the Baltic Proper (see Carman et al. 1996).

This model provides values based on the total TP-inventory in the entire area below the theoretical wave base and below the average halocline down to 10 cm of sediments. The biologically passive sediments below 10 cm are expected to have a TP-concentration of about 0.45 in the Baltic Proper (Jonsson et al. 1990) and this value is also used in this modeling for the Gulf of Finland. This means that only a small part (reflecting the difference between the calculated value of 0.63 and 0.45) of the phosphorus in the accumulation area sediments in the DW-zone could be available for diffusive transport. The diffusion rate depends on the redox-conditions in the sediments, which depend on the calculated sedimentation of matter. The average values for total sedimentation calculated by the model is about between 0.7 and 1.5 mm/yr in the MW-layer and less than 1 mm/yr in the DW-layer or between 4 and 131 $\mu g/cm^2 \cdot d$ on the accumulation areas. The predicted water content (W) of the accumulation area sediments (0–10 cm) is 75% ww, the organic content (loss on ignition, IG) 6.3% dw and the bulk density (d) 1.17 g/cm^3. Table 6.21 also gives modeled values for the dissolved fraction in the three layers and these values vary between 0.8 and 0.99 in the DW-zone, between 0.64 and 0.98 in the MW-layer and between 0.22 and 0.64 in the SW-layer.

Land uplift (F_{LU}) is also an important individual input of TP to this system (18,000 tons/yr).

The largest TP-fluxes are to and from biota with short turnover times, but the total TP-inventory in biota with short turnover times is only between 1 and 5 ktons (on a monthly basis). Most TP is found in the accumulation area sediments in the MW-zone (about 590 ktons), and a significant part of this is potentially available for diffusion (see Table 6.22).

6.4.3.3 Changes in the Gulf of Finland During the Last 100 Years

Figure 6.10 shows measured Secchi depths from the Gulf of Finland during the last 100 years and Table 6.23 gives selected results from statistical analyses of the data. One can note that:

The changes in Secchi depth are small and/or not statistically significant for the periods between 1900 to 1945 and 1980 to 19991.

Table 6.21 Compilation of monthly data for modeled TP-concentrations in accumulation area sediments (in mg/g dw) in the Gulf of Finland from the AW and DW-layers, water content (% ww) and loss on ignition (IG) of accumulation area sediments (% dw), sedimentation (Sed) in the the MW and DW-layers (cm/yr and μg/cm^2·yr), settling velocities (v) in the three layers (m/month) and dissolved fractions (DF) in the three layers (dim. less)

Month	C_{TPAMW}	C_{TPADW}	W	IG	Sed_{AMW}	Sed_{ADW}	Sed_{DW}	Sed_{MW}	v_{SW}	v_{MW}	v_{DW}	DF_{DW}	DF_{MW}	DF_{SW}
1	2.4	0.63	75	6.3	0.14	0.10	113	79	0.61	0.78	2.00	0.83	0.7	0.49
2	2.4	0.63	75	6.3	0.14	0.10	114	83	0.62	0.77	2.03	0.83	0.68	0.64
3	2.4	0.63	75	6.3	0.13	0. 10	106	84	0.62	0.76	2.04	0.82	0.68	0.63
4	2.4	0.63	75	6.3	0.12	0. 10	98	81	0.63	0.75	2.04	0.82	0.68	0.54
5	2.4	0.63	75	6.3	0.13	0. 10	94	82	0.63	0.74	2.02	0.82	0.66	0.40
6	2.4	0.63	75	6.3	0.12	0. 10	102	76	0.63	0.74	2.00	0.83	0.67	0.34
7	2.4	0.63	75	6.3	0.12	0. 10	97	82	0.63	0.74	1.99	0.81	0.67	0.33
8	2.4	0.63	75	6.3	0.16	0. 10	95	83	0.63	0.75	1.99	0.81	0.66	0.33
9	2.4	0.63	75	6.3	0.07	0.01	131	4	0.62	0.75	2.00	0.99	0.98	0.53
10	2.4	0.63	75	6.3	0.24	0.00	58	4	0.60	0.76	2.04	0.99	0.98	0.52
11	2.4	0.63	75	6.3	0.10	0. 10	79	84	0.59	0.77	2.01	0.79	0.64	0.22
12	2.4	0.63	75	6.3	0.15	0. 10	121	78	0.60	0.77	1.98	0.82	0.68	0.41

Table 6.22 Amounts of TP (1000·tons) in the different compartments in the Gulf of Finland, i.e., accumulation areas in the DW-compartment (ADW), accumulation areas in the MW-compartment (AMW), the DW-layer, areas of fine sediment erosion and transport (ET), the MW-layer, the SW-layer, in biota with long turnover times (BL) and in biota with short turnover times (BS)

Month	M_{TPADW}	M_{TPAMW}	M_{TPDW}	M_{TPET}	M_{TPMW}	M_{TPSW}	M_{TPBL}	M_{TPBS}
1	44.3	588	1.7	25.9	7.9	16.3	1.2	2.5
2	44.3	588	1.8	25.4	7.5	16.4	1.2	3.2
3	44.3	588	1.8	24.9	7.1	16.1	1.3	4.3
4	44.4	588	1.8	24.3	6.6	16.6	1.5	4.6
5	44.4	588	1.7	23.9	6.5	17.3	1.6	4.2
6	44.4	588	1.7	23.6	6.6	18.3	1.6	3.4
7	44.4	589	1.7	22.7	7.1	19.4	1.6	3.1
8	44.4	589	1.7	22.2	7.4	19.7	1.6	3.1
9	44.4	589	1.7	21.5	7.7	19.8	1.5	3.1
10	44.4	589	1.8	22.4	7.7	17.6	1.5	4.1
11	44.3	588	1.7	25.4	7.3	15.9	1.5	3.4
12	44.3	588	1.6	26.8	7.7	16.6	1.4	1.4

Table 6.23 Changes in Secchi depths in the Gulf of Finland during different periods of time

1900–1991
Sec = –0.0025·month + 7.84; r^2 = 0.141; n = 738; $p < 0.0001$
1900–1945
Sec = +0.0005·month + 7.10; r^2 = 0.0008; n = 352; p = 0.61; not significant
1945–1991
Sec = –0.0024·month + 7.67; r^2 = 0.010; n = 386; p = 0.048
1980–1991
Sec = –0.0068·month + 11.69; r^2 = 0.019; n = 60; p = 0.29; not significant
1920–1980
Sec = –0.0031·month + 8.38; r^2 = 0.141; n = 555; $p < 0.0001$

The most pronounced changes in terms of significance and slope of regression line appears for the data from 1920 to 1980.

Table 6.24 gives a statistical compilation of data (mean values, medians, standard deviations, coefficients of variation and number of data) for three interesting periods, 1900–1920, 1920–1980 and 1980–1991. One should note the high CV–values (about 0.35) and that the mean Secchi depths have decreased from 7.1 to 4.8 m. This is a significant change influencing the entire ecosystem since it influences the depth of the photic zone and the benthic production, which is highly dependent of water clarity. If there are major changes in primary phytoplankton and benthic algae production, one should also expect major changes in secondary production (of zooplankton, zoobenthos and fish). The question now is if it is possible to reconstruct the changes in Secchi depth shown in Fig. 6.10 and Tables 6.23 and 6.24.

The changes in the Baltic Proper between the years 1974 and 2006 for the chlorophyll-a concentrations in the SW-layer provide a complementary picture and

Table 6.24 Statistics of Secchi depth measurements from different periods from the Gulf of Finland

	1900–1920	1920–1980	1980–1991
Mean (MV)	7.1	6.3	4.8
Median (M50)	7.0	6.1	4.5
Standard deviation (SD)	2.45	2.2	1.6
Coefficient of variation (CV)	0.35	0.35	0.33
Number of data (n)	123	556	60

were shown in Fig. 6.9. During this period, there was a weak and continuous decline in the chlorophyll values, which demonstrates that the eutrophication is not getting worse in the Baltic Proper, but rather the opposite. It is also interesting to note that the individual data in the surface water of the Baltic Proper cover all trophic classes from oligotrophic to hypertrophic. The median chlorophyll-a value in the surface water is close to the class limit between mesotrophic and oligotrophic, i.e., at 2 μg Chl/l.

Figure 6.12 gives chlorophyll data from the Gulf of Finland relative to water depth. All these data emanate from the SW-layer, as this is defined in this work from the theoretical wave base. One can see from Fig. 6.12 that there is only a weak correlation between the measured chlorophyll values and the water depth.

This indicates that the SW-zone is relatively well mixed. That conclusion is supported by the data in Fig. 6.13 showing measured TP-concentrations in the entire water column. Some important depth intervals are also shown in this figure. The mean depth of the Gulf of Finland is 36 m, the theoretical wave base is at 41 m and the average depth of the halocline at 75 m. One can see from Fig. 6.13 and from

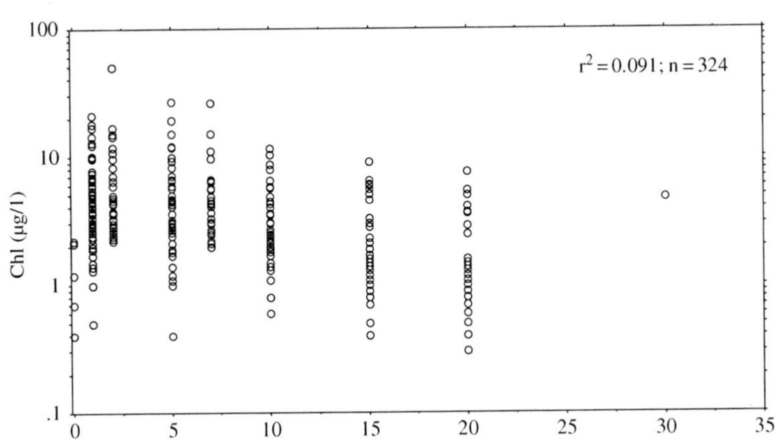

Fig. 6.12 Chlorophyll data from the Gulf of Finland (1990–1998) collected at different water depths (based on HELCOM data)

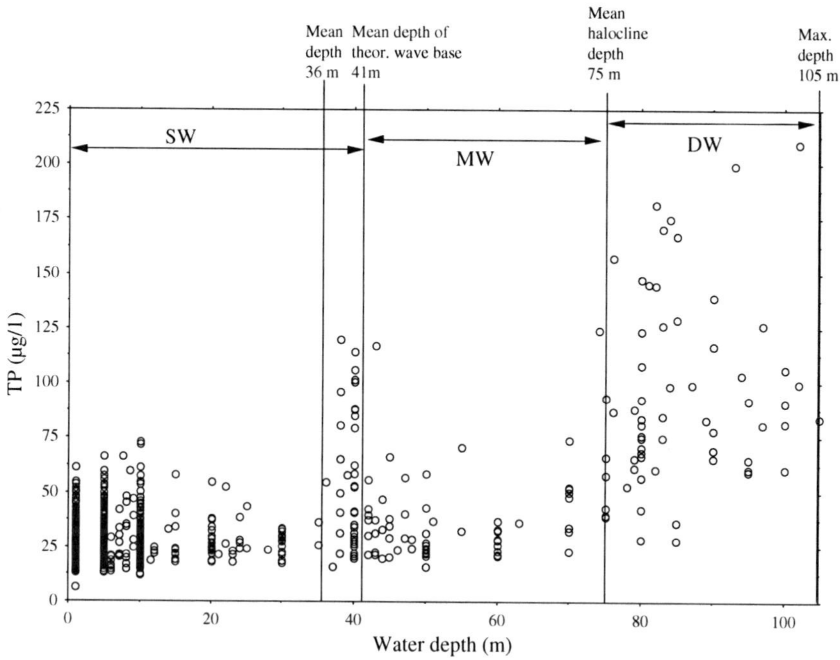

Fig. 6.13 Total phosphorus concentrations in the Gulf of Finland (1990–1998) collected at different water depths (based on HELCOM data)

Table 6.25, which gives a statistical compilation of monthly TP-data, that there are important differences between the three zones discussed in this work.

For the reconstruction of the development in the Gulf of Finland, one can conclude that (1) there were probably no or only small changes in the nutrient loading and eutrophication in the Gulf of Finland from 1900 to about 1920, (2) the most significant changes occurred between 1920 and 1980 and (3) after that the system has not changed very much (and there may even be a slight improvement, at least in the Baltic Proper which would also affect the conditions in the Gulf of Finland).

6.4.3.4 Reconstruction of the Conditions in the Gulf of Finland

The simulations to estimate the changes that have taken place during the last 100 years in the Gulf of Finland will be presented in two steps. The first step concerns substantial but realistic reductions of the direct anthropogenic emissions of TP to the Gulf of Finland from diffuse sources and point sources. The values given by HELCOM (2000) are 2112 tons/yr and 2431 tons/yr, respectively, and the natural loading is 1191 tons/yr. It is not realistic to assume that all the anthropogenic emissions can be removed, and at the first step, it will be assumed that 40% of these emissions are eliminated (i.e., 1817 tons phosphorus so that the annual loading would be reduced

Table 6.25 Statistical compilation (means values, standard deviations and number of data) of empirical monthly data on TP-concentrations in the three layers (SW, MW and DW) from the Gulf of Finland (data from HELCOM from 1990 to 1998)

Month		SW	MW	DW
1	MV	34.1	37.7	71.7
	SD	11.4	18.2	21.8
	n	138	53	7
2	MV	32.2	35.1	43.2
	SD	6.7	12.4	20.0
	n	286	91	10
3	MV	30.1	33.9	31.3
	SD	5.7	10.1	0.0
	n	117	45	1
4	MV	30.1	35.6	57.2
	SD	7.1	15.5	22.1
	n	204	77	12
5	MV	22.5	32.3	47.8
	SD	8.5	16.6	16.9
	n	149	63	11
6	MV	24.0	40.9	69.2
	SD	15.0	18.5	21.2
	n	51	22	4
7	MV	16.2	39.1	69.5
	SD	7.2	18.3	19.7
	n	377	81	12
8	MV	17.5	38.2	50.6
	SD	7.8	21.7	25.1
	n	846	168	32
9	MV	16.6	39.6	68.0
	SD	8.0	23.7	14.6
	n	79	37	6
10	MV	16.9	51.5	126.1
	SD	9.4	38.3	50.1
	n	59	26	4
11	MV	24.2	40.7	47.8
	SD	7.6	11.4	8.9
	n	282	92	14
12	MV	33.3	54.1	58.6
	SD	17.4	24.5	37.5
	n	65	30	3

from 5734 to 3917 tons TP). In reality, this would require major investments and it would also take a long time to implement such actions. In the following, it will be assumed that these emissions are suddenly removed (month 31, i.e., in July).

The aim is also to demonstrate the dynamic response of the system to such a change in nutrient loading. Next, it will be assumed that the conditions in the Baltic Proper will also be altered. An overall budget for nitrogen and phosphorus fluxes to the Baltic Proper (including the Gulf of Riga and the Gulf of Finland) was given in Table 4.1. On average, the total tributary load of phosphorus to the Baltic Proper is about 30,000 tons/yr.

Values of the proportion between natural load, load from diffuse sources and from point source emission for the nutrient to the entire Baltic Proper are given by HELCOM (2006) and here it will be assumed that in total 7200 tons TP/yr of the total transport of 30,000 tons TP/yr will be removed. There have been several tests and the following results mainly concern this particular reduction.

Steady-state results are first shown in Table 6.26. This table gives the predicted values today for the Secchi depth (mean value = 5.5 m), the monthly maximum concentration of chlorophyll-a (8.3 μg/l; since the maximum value is of great interest in contexts of algal blooms), the monthly maximum concentration of cyanobacteria in the SW-layer (86 μg/l), the mean predicted TP-concentrations in the SW, MW and DW-layers (24.5, 36 and 86 μg/l, respectively). The second column gives the corresponding steady-state values if 1871 tons TP from the direct tributary inflow to the Gulf of Finland are reduced.

Then, the Secchi depth would be 6.7 m, which is lower than the value 100 years ago (7.1 m, see Table 6.23). So, it is not enough to reduce anthropogenic TP-inflow via the rivers to the Gulf of Finland by 40%. If 7200 tons TP are removed from the present TP-inflow via rivers to the Baltic Proper (including 1817 tons to the Gulf of Finland), then the requested mean annual Secchi depth of 7.1 m in the Gulf of Finland will be reached and there are also major changes in the chlorophyll-a concentration and the maximum concentration of cyanobacteria.

Figure 6.14 shows the dynamic response of the system. In these simulations, 7200 tons of TP to the Baltic Proper (including 1871 tons to the Gulf of Finland) have been removed in month 31 and the response of the Gulf of Finland is shown for (A) the Secchi depth, (B) chlorophyll concentrations, (C) cyanobacteria, (E) TP in surface water, (E) TP in surface water in the Baltic Proper and (F) predicted SPM-concentrations in the SW, MW and DW-layers in the Gulf of Finland. Whenever possible, this figure compares the modeled values during the initial 31 months, corresponding to the conditions prevailing today, with the uncertainty bands in the empirical data. From Fig. 6.14A, one can see that during the initial period there is a very good correspondence between the modeled values and the empirical data given

Table 6.26 How would mean annual Secchi depths, maximum monthly chlorophyll-a concentrations, maximum concentrations of cyanobacteria and mean annual TP-concentrations in the three layers (SW, MW and DW) change from today if, first, 1817 tons of TP to the Gulf of Finland and, secondly a total of 7200 tons of TP from river inflow to the Baltic Proper (including 1817 tons to the Gulf of Finland) were reduced? The table gives steady-state values

	Today	1817 tons TP reduced from rivers to GF	7200 tons TP reduced to BP
Mean Secchi (m)	5.5	6.7	7.2
Max. chlorophyll-a (μg/l)	8.3	7.7	7.5
Max. cyanobacteria (μg/l)	86	63	55
Mean TP_{SW} (μg/l)	24.5	21.9	18.7
Mean TP_{MW} (μg/l)	36	32	30
Mean TP_{DW} (μg/l)	86	80	79

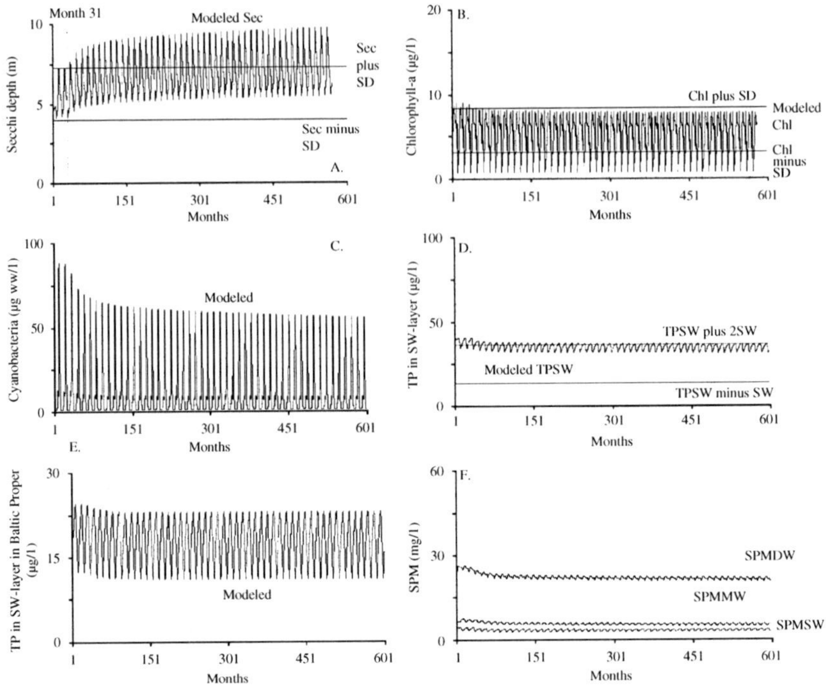

Fig. 6.14 The dynamic response of the Gulf of Finland if 7200 tons of phosphorus (including 1817 tons from tributaries to the Gulf of Finland) were hypothetically reduced month 31 for (**A**) Secchi depth, (**B**) chlorophyll, (**C**) cyanobacteria, (**D**) TP-concentrations in the SW-layer, (**E**) TP-concentrations in the SW-layer in the Baltic Proper and (**F**) concentrations of suspended particulate matter (SPM) in the SW, MW and DW-layers in the Gulf of Finland. This figure also gives uncertainty bands for the empirical data (± 1 standard deviation) valid for the initial period (the first 31 months) for Secchi depth, chlorophyll and TP in the SW-layer in the Gulf of Finland

by the uncertainty band related to ± 1 standard deviation of the measured mean Secchi depth. There is also a good correspondence between measured and modeled values for chlorophyll (Fig. 6.10B) and for TP in the SW-layer in the Gulf of Finland (Fig. 6.14D).

It is interesting to note that under these hypothetical presuppositions (that 7200 tons TP would suddenly be removed month 31), there is an initial phase with a relatively fast recovery of about 7 years and then a phase with a slow recovery related to the fact that the steady-state adjustment to changes is very slow for the sediments in the MW and DW-zones (the accumulation area sediments).

The reconstruction results are given in Fig. 6.15. The measured data on Secchi depth (from Fig. 6.10) have been compared to the modeled values, first when there were no changes in the TP-inflow to the Baltic Proper and the Gulf of Finland in the years 1900–1920 (3917 tons via tributaries to the Gulf of Finland; 22,800 via tributaries to the Baltic Proper including the tributaries to the Gulf of Finland).

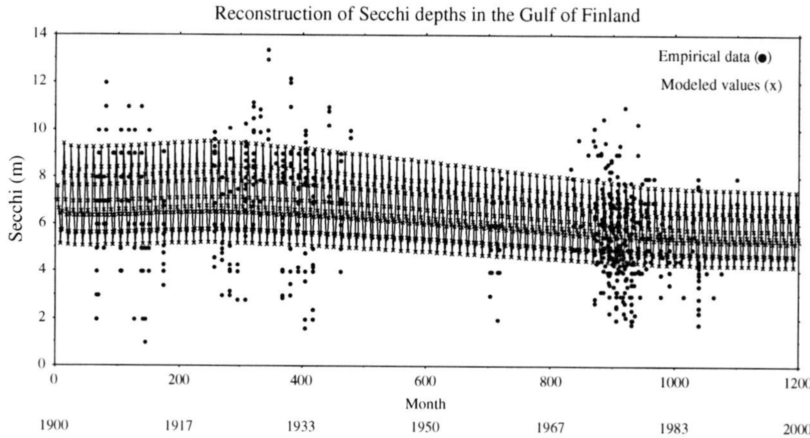

Fig. 6.15 Reconstruction of Secchi depths in the Gulf of Finland when 7200 tons of phosphorus (including 1817 tons from tributaries to the Gulf of Finland) have been reduced between the years from 1920 to 1980 compared to measured data (from HELCOM)

Then, in the period 1920–1980 7200 tons (including 1817 tons from the tributaries to the Gulf of Finland) have successively been reduced ($7200/60 \cdot 12 = 10$ tons per month). Finally, the discharges of today have been used in the period from 1980. From Fig. 6.15, one can see that this will reflect the measured Secchi depths in the Gulf of Finland quite well. There are several individual Secchi depth measurements higher and lower than the predicted mean monthly values, but the general correspondence between the measure and the modeled Secchi depths is good.

This also means that one can run this scenario in the other direction and conclude that if the present tributary TP-load could be reduced by 7200 tons, the Baltic Proper and the Gulf of Finland would return to the conditions as they were 100 years ago. If the reductions are done as in Fig. 6.15, it would take 60–70 years to get a new steady-state condition. If the reductions are implemented slower, it takes longer, and vice versa.

6.4.3.5 Sensitivity Tests

In the following sensitivity tests, one variable at a time has been changed (reduced by 50% as compared to the default situation) and all else in the model kept at the initial default conditions. The results will be presented for two target variables, the dynamically modeled TP-concentrations in the SW-layer and the Secchi depths. The first column in Table 6.27 gives the default conditions (today), the second column the results when the TP-flow from land uplift has been reduced by 50%. Then, one can note that the modeled TP-concentration would be 19.8 µg/l, which is significantly lower than the reference value (24.5 µg/l) and also the empirical mean annual value (24.1 µg/l, see Table 6.18). The Secchi depth would be 7.9 m, which is markedly higher than the empirical mean value (7.1 m) a hundred years ago.

Table 6.27 Steady state results from 8 sensitivity analyses where the influence from (1) land uplift, (2) diffusion from sediments, (3) diffusion in water, (4) biouptake and retention in biota with short turnover times and (5) in biota with long turnover times, (6) the particulate fraction of phosphorus in the deep-water zone, in (7) the middle-water zone and (8) the surface-water zone were reduced by 50%. This has been calculated for (A) TP-concentrations (μg/l) and (B) Secchi depths (m) in the Gulf of Finland

A. TP-concentrations in the surface-water layer									
Month	Today	Landup	Diff sed	Diff wat	BioS	BioL	PF_{DW}	PF_{MW}	PF_{SW}
1	23.0	18.5	21.8	21.3	23.3	23.3	22.5	23.0	24.3
2	23.9	19.0	22.7	22.4	24.2	24.2	23.4	23.9	24.9
3	24.8	19.6	23.7	23.5	25.2	25.2	24.3	24.8	25.6
4	25.2	20.0	24.1	24.1	25.7	25.7	24.7	25.2	26.1
5	25.6	20.4	24.5	24.5	26.0	26.1	25.1	25.6	26.6
6	26.4	21.1	25.3	25.3	26.8	26.9	25.9	26.4	27.6
7	26.7	21.5	25.7	25.5	27.0	27.2	26.2	26.7	27.9
8	27.0	21.5	25.9	25.7	27.2	27.4	26.4	27.0	28.1
9	25.5	20.7	24.4	24.1	25.8	26.0	25.1	25.5	26.8
10	22.6	19.2	21.4	20.8	23.2	23.3	22.3	22.6	24.0
11	21.1	18.0	19.8	19.2	21.6	21.7	20.8	21.1	22.8
12	22.0	18.1	20.9	20.2	22.3	22.4	21.6	22.1	23.6
MV	**24.5**	**19.8**	**23.4**	**23.1**	**24.9**	**24.9**	**24.0**	**24.5**	**25.7**
B. Secchi depth									
1	6.1	8.9	6.7	7.0	6.0	6.0	6.4	6.1	5.6
2	5.6	8.3	6.2	6.3	5.5	5.5	5.9	5.6	5.3
3	5.4	8.1	5.8	5.9	5.2	5.3	5.6	5.4	5.1
4	5.3	7.9	5.7	5.7	5.1	5.1	5.5	5.3	5.0
5	4.7	6.9	5.0	5.0	4.5	4.5	4.8	4.7	4.4
6	4.2	6.1	4.5	4.5	4.1	4.1	4.4	4.2	3.9
7	4.6	6.7	4.9	5.0	4.5	4.5	4.8	4.6	4.3
8	4.7	6.8	5.0	5.1	4.6	4.5	4.8	4.7	4.3
9	5.2	7.4	5.6	5.7	5.1	5.0	5.4	5.2	4.8
10	6.4	8.5	7.0	7.4	6.1	6.1	6.5	6.4	5.8
11	7.3	9.6	8.1	8.6	7.0	7.0	7.5	7.3	6.4
12	6.6	9.2	7.2	7.6	6.5	6.4	6.8	6.6	5.9
MV	**5.5**	**7.9**	**6.0**	**6.1**	**5.4**	**5.3**	**5.7**	**5.5**	**5.1**

In the next sensitivity test, the two diffusive TP-fluxes from the sediments (from accumulation area sediments in the MW and DW-layers) have been reduced by 50%. Since these diffusive fluxes are relatively small, the changes are not great: annual mean TP has been reduced from 24.5 to 23.4 μg/l, and the Secchi depth increased from 5.5 to 6 m. Also the diffuse TP-fluxes in the water have been reduced by 50%, and the results are close to the results for the diffusive sediment fluxes: mean annual TP has decreased to 23.2 μg/l and the Secchi depth increased to 6.1 m.

Since it is fairly complicated to calculate the production and biomasses of functional groups or species of organisms and since most mass-balance models for nutrients would not do this, it is also interesting to see what would happen of 50% of the biouptake to organisms with short turnover times would be changed. The results show that this would neither alter the predicted TP-concentrations very must (from 24.5 to 24.9 μg/l) nor the predicted Secchi depths (m). This is also the result if 50% of the biouptake to organisms with long turnover times are being reduced.

There are also uncertainties regarding the distribution of phosphorus in dissolved and particulate forms. Only the dissolved forms can be taken of by biota and only the particulate forms can settle out due to gravity. If one would first decrease the particulate fraction for phosphorus in the DW-zone, this could be a reflection of lower oxygen concentrations in the DW-layer and it would increase the dissolved fraction in the DW-layer, reduce sedimentation in the DW-layer, increase diffusion to the MW-layer but not influence the TP-concentration in the SW-layer much since this change would mainly influence relatively small TP-fluxes in the Gulf of Finland. If the particulate fraction in the MW-layer is reduced by 50%, it would influence the predicted TP-concentrations in the SW-layer even less and hence also the Secchi depth. If, however, the particulate fraction in the SW-layer itself is reduced by 50%, this would lower the sedimentation more and also increase the predicted TP-concentration (from 24.5 to 25.7 μg/l) and also decrease the Secchi depth (from 5.5 to 5.1 m).

6.4.4 Concluding Remarks

The Gulf of Finland is a large bay in the Baltic Sea where major changes have taken place during the last 100 years. The Secchi depth has, for example, decreased from more than 7 m to about 5 m. The basic aim of this case-study has been to try to reconstruct the development that has taken place in this bay during the last 100 years. Since the conditions in the Gulf of Finland depend very much on both the river input of nutrients directly to the bay and the exchange of nutrients and water between the bay and the Baltic Proper, this case-study has focused on such interactions. We have described how the CoastMab-model has been applied for the Baltic Proper and the Gulf of Finland. The model has previously been extensively tested and validated for phosphorus, suspended particulate matter, radionuclides and metals in several lakes and coastal areas. This is not a model where the user should make any tuning or change model constants. These results indicate that it is possible to remediate the Gulf of Finland and the Baltic Proper to the conditions that characterized the system 100 years ago. About 7000 tons of phosphorus (including 1800 tons from the tributaries to the Gulf of Finland) must then be removed on an annual basis from the present annual tributary load of about 30,000 tons to the Baltic Proper. This should be done in the most cost-efficient manner so that the largest fluxes of phosphorus are removed per euro. The trophic conditions in the Baltic Proper have varied relatively little during the last 25–30 years. The most marked changes in Secchi depth in the Gulf of Finland took place between 1920 and 1980.

6.5 Summary

This part of the book has provided three specific examples of how a general dynamic phosphorus model can be used to quantify the causes of coastal eutrophication and to evaluate the probable outcome of remedial measures in terms of reductions in phosphorus loading. In the Himmerfjärden Bay, and in many other coastal areas, local conditions depend very much on the conditions in the outside sea, so local nutrient abatement to the Himmerfjärden Bay has had, and will have, disappointingly limited impacts on the water quality. Remedial strategies should always focus on reducing the large nutrient fluxes. In the case of the Himmerfjärden Bay, this means that the large nutrient inputs to the whole of the Baltic Sea basin should be focal points for remedial actions, because of the comparatively intensive water exchange between the Baltic Proper and the Himmerfjärden Bay.

In the Gulf of Riga, it is indeed possible to achieve "good" water quality as defined in the European Water Framework Directive (Anon 2000). Since the direct anthropogenic nutrient loading to the Gulf of Riga is substantial, the best results with respect to water quality will probably be reached through a combination of an extensive reduction of anthropogenic phosphorus inputs to the Gulf itself (approximately a 75% reduction) and a substantial reduction to the Baltic Proper (about 60% of the anthropogenic phosphorus loading).

The Gulf of Finland has during the last 100 years received increasing phosphorus inputs, which has resulted in a considerably lower Secchi depth and intensified algal blooms. The situation has stabilized since the 1970s but the water quality remains poor in a historical perspective. In order for conditions to be restored, about 7200 tons of phosphorus per year must be prevented from reaching the Baltic Proper and the Gulf of Finland from their catchments. If this is done, substantially better long-term water quality can be expected. The recovery time would depend very much on how rapidly the actions can be taken. This implies that the phosphorus abatement goal set by the Ministers of Environment around the Baltic Sea in the fall of 2007, at 15,000 tons per year, may be too ambitious and very difficult to achieve.

Put in a global context, this kind of quantitative, dynamic analysis of nutrient fluxes may be a powerful tool to combat eutrophication in many coastal regions. A first step would be to collect and freely distribute relevant empirical data from areas where it can be suspected that eutrophication has occurred. The authors of this book would see it as an exciting challenge to use such data for expanding the model domain and application of the CoastMab-model, and thereby contributing to a deeper understanding of coastal eutrophication, which is indeed a global problem.

Epilogue

To create operational tools and approaches for a sustainable coastal management may be visualized as a bridge between "ecology" and "economics". The aim of this book has been to try to compile, review and present the "ecological" half of that bridge. We have discussed fundamental concepts for coastal management, such as effect-load-sensitivity analyses in Chap. 2, motivated an index of coastal area sensitivity to anthropogenous nutrient loading (eutrophication) in Chap. 3, discussed fundamental concepts related to variations and uncertainties in nutrient concentrations, predictive power of models for eutrophication, "limiting nutrients", different forms of nutrients and spurious correlations in Chap. 4, motivated and discussed operational bioindicators and an "index of biological value" in Chap. 5 and presented three important case-studies in Chap. 6; the process-based mass-balance model (CoastMab) for nutrients used in the simulations in the case-studies is given in Chap. 9.

To build the other half of the bridge is certainly very important and Table 1 is included here to stress that point.

Table 1 Costs and purification capacities for some communal water purification plats built during the 1990s. The costs to remove phosphorus per ton are based on an annual capital cost of 10% of the total costs, a time of 15 years to write off the costs and an energy cost of 0.3 million euros for plants removing less than 100 tons phosphorus per year and 1 million euros for plants removing more than 100 tons/y (from Håkanson et al. 2002)

	Costs (10^6 euro)	Reduced emissions of TP (tons/y)	Costs (10^3 euro/ton TP)
Haapsalu (Estonia)	4.7	14	77.3
Daugavpils (Latvia)	22.0	88	45.0
Liepaja (Latvia)	14.9	71	39.0
Riga (Latvia)	86.5	500	31.0
Kaunas (Lituania)	65.0	245	48.0
Klaipeda (Lituania)	21.8	67	59.0
Average cost			**49.9**

Table 1 provides a few data for discussions about the most cost-efficient remedial strategy for a given coastal area. Then, one must consider the cost to reduce 1 kg of total phosphorus (TP). Why remedial actions should target on phosphorus rather than on nitrogen is motivated in Chap. 4. Table 1 gives examples of costs for water purification plants. One can note that the average cost to remove 1 ton of TP is about 50,000 euros in the given East-Baltic example. The corresponding costs in countries like Sweden where large investments in systems to reduce nutrient loading have already been made would, on average, be a factor of 6 higher (see Håkanson et al. 2002). The point we would like to stress by this example is that much effort should be put into compilations of costs in agriculture, industry and urban areas, including the building and maintenance of water purifications plants and changes in agricultural practices at all scales and involving all aspects from governmental legislations to practical advice to individual farmers in targeted coastal catchment areas, so that it becomes possible and meaningful to carry out cost-benefit analyses. That is, how much would a given reduction cost and how would this influence the target bioindicators discussed in this book. Such analyses can utilize the ecosystem models we have used in this work.

This work demonstrates the importance of the application of a set of bioindicators. We have shown that these bioindicators fulfil key criteria of operational indicators for coastal management. Such bioindicators should be measurable, interpretable and predictable, relevant for the ecosystem function and internationally applicable. These five discussed operational bioindicators are complementary and characterize different aspects of water quality:

Chlorophyll-a:	phytoplankton biomass
Concentration of cyanobacteria:	harmful algal blooms
Secchi depth:	depth of the photic zone; depth of macrophyte and benthic algal growth
Deep water oxygen saturation:	benthic fauna; anoxic sediments; diffusion of phosphorus from sediments to water
Macrophyte cover:	fish habitat and coastal fish production; the "biological value" of the coastal zone

Different abiotic factors (nutrient concentrations, salinity, temperature, coastal morphometry and water exchange) influence the selected bioindicators in a logical and predictable manner. This is demonstrated by extensive data and models based on or tested against empirical data. Note that one would generally need a set of operational bioindicators to get an adequate framework to analyse changes in coastal ecosystem function and structure related to eutrophication, since one bioindicator cannot normally cover all relevant aspects. However, if asked if there is one candidate for a general operational bioindicator, we would argue that the Secchi depth would be our favourite for that role.

We have discussed a functional classification approach for coastal areas that is meant to be used in coastal management and science. A morphometric classification system has been presented and examples of the practical use of the system have been given both at the European and at regional and local scales. General maps of several environmental variables have been given for as large parts of the European coastal zone as possible (restricted only by the data accessibility). We have also tried to highlight the importance of presenting comparative information at different scales and we have provided several maps as examples of how this can be done using free data sources and GIS.

When using the kind of generalized maps of water variables, tides and water currents presented in this work, one must remember that the aim here has been to give an overview, a context. At the local scale, and over shorter time-scales, major variations may occur in all these water variables. Hence, more local or regional data should preferably be consulted for such modeling applications. In this coastal classification, the generalized maps were created mainly using data from the European Environment Agency (EEA) and the International Council for the Exploration of the Sea (ICES).

Many scientists and managers probably believe that information on coastal size and form is of interest mainly for descriptive purposes. This book takes a holistic ecosystem perspective and focuses on the structure and function of coastal ecosystems. From that perspective, this book demonstrates a different view. Coastal morphometry influences almost all transport processes in coastal areas, such as sedimentation, resuspension, mixing, diffusion, burial and outflow. Therefore, morphometry regulates concentrations of pollutants in water and sediments, and hence also ecosystem effect related to such concentrations. These transport processes are general and apply to all substances (nutrients, metals, organics, radionuclides, etc.). So, the morphometry regulates the nutrient concentrations in coastal systems, and hence also primary production, and secondary production of zooplankton, zoobenthos and fish.

The book has presented several databases on nutrients, different forms of nutrients and abiotic variables influencing eutrophication effects (such as salinity, water temperature and water fluxes). These variables can statistically and causally explain important variations in eutrophication effects within and among coastal areas. Many statistical/empirical models and one dynamical model, CoastMab, have also been presented and used. Hopefully, this has clarified the present state-of-the-art and also where there are gaps in data and our knowledge. Today, there are limited information in data from many types of marine coastal systems. Hopefully, the gaps may become smaller in the future. Then, it is probable that some of the results, algorithms and approaches discussed in this book need to be modified. One cannot falsify results from empirical models by new empirical data, but new empirical data may improve the description of the model domain and the boundary conditions when and where the model applies. Equations in dynamic model, on the other hand, may be falsified by new empirical data. Basically, there is only one avenue to increase the knowledge about how aquatic systems work, and that is by collecting new and better data. If this book could help revise many ongoing monitoring programs so that TN and TP (rather than DIN and DIP) and the given operational bioindicators (rather than bioindicators with high CVs and/or those requiring special competence for sampling, analyses and interpretations related to individual species) could be integral parts of all monitoring programs, that would be a positive result.

Any dynamic model can be tuned so that it describes empirical data well in a given system. However, errors in models are often – if not always – revealed when models are blind tested against independent data from new systems. So, validations are fundamental in ecosystem modeling in disclosing deficiencies in models and hence also in the modeler's understanding of how natural systems work. There are at least four basic criteria by which ecosystem models can be critically evaluated:

- By the predictive power revealed by validations.
- By the relevance of the target y-variable in disclosing fundamental ecosystem structures, functional aspects of aquatic ecosystems and threshold values related to operationally applied guidelines in water management.
- By the applicability and generality of the model, i.e., by the width of the model domain, and
- By the accessibility of the driving variables needed to make simulations.

Evidently, there exist very many models for marine systems. At a first glance, such models may look the same, but there can also be fundamental differences between seemingly similar models because the basic structures, the equations and the model constants may be different. To the best of our knowledge, no other models use the same sedimentological criteria as the CoastMab-model discussed in this book to define fundamental model structures, e.g., the surface-water compartment, the deep-water compartment, the sediment compartment for ET-areas (where there is resuspension) and the accumulation-area compartment (where there is no wind/wave-induced resuspension). This also means that key transport processes, such as sedimentation, resuspension, mixing, mineralization and outflow, are quantified differently in this modeling approach compared to most other models. All approaches to quantify these transport processes cannot be best or most relevant from a mechanistic point of view. Such a ranking of models cannot be done by arguments, only from critical validations using reliable empirical data from a wide domain of systems. We know of no dynamic models which provide seasonal variations for TP, SPM and salinity in coastal areas based on other structures than those discussed in this book that have been validated over such wide domains and given results even close to what has been reported here for the CoastMab-model.

If the scientific task to gain better understanding about how aquatic systems work, there are few more rewarding avenues than comparative studies. This book has presented many statistical/empirical models, which in themselves may reveal very little about processes. However, such models are excellent tools to rank x-variables influencing variations in target y-variables, and in this way they can provide invaluable information in building practical and operational process-oriented mass-balance models. In such models, statistical explanation may be transformed into mechanistic explanation. Then, the aim is not to account for "everything", but to try to find and quantify the most important transport processes and omit or simplify the smaller processes. This is far easier said than done. But it can be done with the help of statistical/empirical modeling approaches.

This book has a focus on the ecosystem scale. This is also a very important scale in water management, e.g., in contexts of impact assessment and when remedial measures are discussed. A very important demand for all bioindicators and models discussed in this book is that they should be practically useful. Both the empirical models and the CoastMab-model discussed in this book may be driven by readily accessible data from standard maps and monitoring programs, e.g., altitude, latitude, continentality, area, mean depth and max. depth. Meteorological data on winds and light conditions and oceanographic data on directions and speeds of currents and temperatures at individual sites have been omitted, since the models discussed in this book focus on monthly predictions at the ecosystem scale and during a period of one month, winds can blow with many speeds from many directions.

The ultimate model testing in not sensitivity or uncertainty tests, like those discussed in Sect. 9.1, but validations, i.e., blind tests against independent data. Generally, one would assume that a poor fit between modeled values and empirical data can be explained by deficiencies in the model. But empirical data are, in fact, also always uncertain. This means that it is important to compare model predictions against uncertainty bands for the empirical data. This has been stressed several times in this book. The CoastMab-model has been critically tested and demonstrated to give good predictive power in many systems. However, this does not mean that it will predict equally well for all systems. This model is meant to account for defined processes and factors in a general way so that characteristics values can be predicted of the target bioindicators. Evidently, there may be situations which are not normal, but abnormal. Then, this modeling can provide a reference value, which would express normal conditions so that the divergency from the normal can be quantified and maybe related to the factor causing the abnormal conditions.

The CoastMab-model discussed and used in this work is a powerful tool to simulate possible outcomes of remedial actions and to get realistic expectations of both positive and negative consequences of such actions.

Expected positive effects of a lower anthropogenic nutrient loading are:

- A smaller production and biomass of phytoplankton.
- An increase in water clarity and a decrease in the concentration of suspended particulate matter.
- A reduced risk of blooming of toxic algae (cyanobacteria).
- A reduced spread of laminated sediments and “dead bottoms” in fine sediment accumulation areas in the coastal zone.
- The conditions in the coastal area would become more like they were before the ongoing eutrophication started (about 100 years ago).

Expected negative effects are:

- A higher bioaccumulation of organic toxins (such as dioxins, DDTs and PCBs) and higher levels of such substances in fish consumed by man (“the clearest waters have the most toxic fish”, i.e., biological dilution) if there are no parallel reductions in the input of such toxic substances to the given system.

- A lower total production of fish, but probably a higher fraction of attractive "white fish" and predatory fish and a lower production of "less attractive" fish for professional and leisure time fishermen.
- An increased production and biomass of macrophytes, which could reduce man's access to shorelines of interest for recreation (but it would also increase coastal fish production).
- A lower pH related to the lower total primary production, which could in some areas be hazardous to certain species of clams.
- A lower carbon sequestration by phytoplankton, which implies stricter regulations on greenhouse gas emissions to prevent global warming.

References

Aarup T (2002) Transparency of the North Sea and Baltic Sea - a Secchi depth data mining study. Oceanologia 44: 323–337

Abrahamsson O, Håkanson L (1998) Modelling seasonal flow variability of European rivers. Ecol Model 114: 49–58

Abrahamsen J, Jacobsen NK, Dahl E et al (1977) Naturgeografisk regionindelning av Norden (Physical geographical delineation of the Nordic countries) NU B 1977:34, Helsingfors (in Swedish)

Aertbjerg G (ed) (2001) Eutrophication in Europe's coastal waters. European Environment Agency, Topic report 7/2002, Copenhagen

Aertebjerg G, Andersen JH, Hansen OS (eds) (2003) Nutrients and eutrophication in Danish marine waters. A challange for science and management. National Environmental Institute, Copenhagen, Denmark, 115 p

Ahlgren I, Frisk T, Kamp-Nielsen L (1988) Empirical and theoretical models of phosphorus loading, retention and concentration vs. lake trophic state. Hydrobiologia 170: 285–303

Ambio (1990) Special issue. Marine eutrophication. Ambio 19: 102–176

Ambio (2007) Special issue. Science and governance of the Baltic Sea. Ambio 2-3: 117–286

Andersson C (2000) The influence of wind-induced resuspension on sediment accumulation rates. A study of offshore and archipelago areas in the NW Baltic proper. Master Thesis, Uppsala University, Uppsala

Andersen JH, Schlüter L, Aertebjerg G (2006) Coastal eutrophication: recent developments in definitios and implications for monitoring strategies. J Plankton Res 28: 621–628

Anon (2000) Directive 2000/60/EC of the European Parliament and of the Council of 23 October 2000 establishing a framework for Community action in the field of water policy. Official Journal of the European Communities L327: 1–72

Anon (2003) Common implementation strategy for the water framework directive (2000/60/EC). Guidance document no 5. Transitional and coastal waters - typology, reference conditions and classification systems. Office for Official Publications of the European Communities, Luxembourg

Andrle R (1996) The west coast of Britain: statistical self-similarity vs. characteristic scales in the landscape. Earth Surf Proc Land 21: 955–962

Arhonditsis G, Tsirtsis G, Angelidis MO, Karydis M (2000) Quantification of the effects of non-point nutrient sources to coastal marine eutrophication: applications to a semi-enclosed gulf in the Mediterranean Sea. Ecol Model 129: 209–227

Arslan NT, Ökmen M (2006) The economical and international dimensions of the environmental problems, environmental problems in the Black Sea region and the role of the voluntary organizations. Build Environ 41: 1040–1049

Baden SP, Loo L-O, Pihl L, Rosenberg R (1990) Effects of eutrophication on benthic communities including fish: Swedish West coast. Ambio 19: 113–122

Bailey RG, Zoltan S, Wiken EB (1985) Ecological regionalization in Canada and the United States. Geoforum 16: 265–275

Balls PW (1989) The partition of trace metals between dissolved and particulate phases in European coastal waters: A compilation of field data and comparison with laboratory studies. Neth J Sea Res 23: 7–14

Bakan G, Büyükgüngör H (2000) The Black Sea. Mar Pollut Bull 41: 24–43

Beck MB, Van Straten G (eds) (1983) Uncertainty, system identification and the prediction of water quality. Springer, Heidelberg, 387 p

Beeton AM, Edmondson, WT (1972) The eutrophication problem. J Fish Res Can 29: 673–682

Benoit G (1995) Evidence of the particle concentration effect for lead and other metals in fresh waters based on ultraclean technique analyses. Geochim Cosmochim Ac 59: 2677–2687

Berges JA (1997) Ratios, regression statistics, and "spurious" correlations. Limnol Oceanogr 42: 1006–1007

Bernes C (2005) Förändringar under ytan - Sveriges havsmiljö granskad på djupet (Changes under the surface - a survey on Sweden's marine environment). Monitor 19, Swedish EPA, Stockholm (in Swedish)

Bierman VJ Jr (1980) A comparison of models developed for phosphorus management in the Great Lakes. In: Loehr C, Martin CS, Rast W (eds) Phosphorus management strategies for lakes. Ann Arbor Science Publishers, Ann Arbor, pp 235–255

Bird ECF (1984) Coasts: an introduction to coastal geomorphology. Basil Blackwell Publisher Limited, Oxford

Bird ECF (2000) Coastal geomorphology: an introduction. John Wiley & Sons, West Sussex

Bloom NS, Effler SW (1990) Seasonal variability in the mercury speciation of Onondaga lake (New York). Water Air Soil Poll 53: 251–265

Boers PCM, Cappenberg TE, van Raaphorst W (1993) Proceeding of the Third International Workshop on Phosphorus in Sediments. Hydrobiologia 253

Boesch DF, Brinsfield RB, Magnien RE (2000) Chesapeake Bay Eutrophication: Scientific Understanding, Ecosystem Restoration, and Challenges for Agriculture. J Environ Qual 30: 303–320

Boesch D, Hecky R, O'Melia C, Schindler D, Seitzinger S (2006) Eutrophication of Swedish Seas. Report 5509, Swedish EPA, Stockholm

Bonham-Carter GF (1994) Geographical information systems for geoscientists: modeling with GIS. Elsevier Science, Kidlington

Bortone SA (ed) (2005) Estuarine indicators. CRC Press, Boca Raton

Boynton WR, Kemp WM, Keefe CW (1982) A comparative analysis of nutrients and other factors influencing estuarine phytoplankton production. In: Kennedy, V.S (ed) Estuarine comparisons. Academic Press, London, pp 69–90

Box GEP, Cox DR (1964) An analysis of transformations. J Roy Stat Soc B 26: 211–252

Breman J (ed) (2002) Marine geography: GIS for the oceans and seas. ESRI Press, Redlands, California

Brinkhurst RO (1974) The benthos of lakes. Macmillan, London

Bryhn AC, Håkanson L (2007) A comparison of predictive phosphorus load-concentration models for lakes. Ecosystems 10: 1084–1099

Bryhn AC, Håkanson L, Eklund J (2007) Variabilities and uncertainties in management-related coastal water variables. Manuscript, Uppsala Univ

Burban P-Y, Lick W, Lick J (1989) The flocculation of fine-grained sediments in estuarine waters. J Geophys Res 94: 8223–8330

Burban P-Y, Xu Y-J, McNeil J, Lick W (1990) Settling speeds of flocs in fresh water and seawater. J Geophys Res 95: 18213–18220

Cairns J Jr, Pratt JR (1987) Ecotoxicological effect indices: A rapidly evolving system. Water Sci Technol 19: 1–12

Capone DG (2001) Marine nitrogen fixation: what's the fuss? Curr Opin Microbiol 4: 341–348

Carlson RE (1977) A trophic state index for lakes. Limnol Oceanogr22: 361–369

Carlson RE (1980) More complications in the chlorophyll - Secchi disk relationship. Limnol Oceanogr25: 379–382

Carman R, Aigars J, Larsen B (1996) Carbon and nutrient geochemistry of the surface sediments of the Gulf of Riga, Baltic Sea. Mar Geol 134: 57–76

Carroll J, Harms IH (1999) Uncertainty analysis of partition coefficients in a radionuclide transport model. Wat Res 33: 2617–2626

Carpenter SC (2003) Regime shift in lake ecosystems: pattern and variation (see http://limnology.wisc.edu/regime)

Casazza G, Silvestri C, Spada E (2003) Classification of coastal waters according to the new Italian water legislation and comparison with the European Water Directive. J Coast Conserv 9: 65–72

Chapra SC (1980) Application of the phosphorus loading concept to the Great Lakes. In: Loehr C, Martin CS, Rast W (eds) Phosphorus management strategies for lakes. Ann Arbor Science Publishers, Ann Arbor, pp 135–152

Chapra SC, Reckhow K (1979) Expressing the phosphorus loading concept in probabalistic terms. J Fish Res Bd Can 36:225–229

Chesapeake Bay Program (2007) Water quality database, http://www.chesapeakebay.net/data/index.cfm

Christiansen C, Gertz F, Laima MJC et al (1997) Nutrient (P, N) dynamics in the southwestern Kattegat, Scandinavia: sedimentation and resuspension effects. Environ Geol 29: 66–77

Clark RB (2001) Marine Pollution, 5th ed. Oxford University Press, Oxford

Cooper SR, Brush GS (1993) A 2,500-year history of anoxia and eutrophication in Chesapeake Bay. Estuaries 16: 617–626

Cooper JAG, McLaughlin S (1997) Contemporary multidisciplinary approaches to coastal classification and environmental risk analysis. J Coast Res 14: 512–524

Cornett RJ, Rigler FH (1979) Hypolimnetic oxygen deficits: their prediction and interpretation. Science 205: 580–581

Cox DC, Baybutt P (1981) Methods for uncertainty analysis: a comparative survey. Risk Analysis 1: 251–258

Cummings KW (1973) Trophic relations in aquatic insects. Annu Rev Entomol 18: 183–206

Dal Cin R, Simeoni U (1994) A model for determining the classification, vulnerability and risk in the Southern coastal zone of the Marche (Italy). J Coast Res 10: 18–29

Davies JL (1964) A morphogenic approach to world shorelines. Z Geomorphol Sonderh 8: 127–142

Davis JL (1972) Geographical variation in coastal development. Oliver and Boyd, Edinburgh

Davies JL (1980) Geographical variation in coastal development, 2nd ed. Longman Group, London

Davis RA (1996) The evolving coast. Scientific American Library, New York

Dahlgren S, Kautsky L (2004) Can different vegetative states in shallow coastal bays of the Baltic Sea be linked to internal nutrient levels and external nutrient load? Hydrobiologia 514: 1–3

de Jonge VN (2000) Importance of temporal and spatial scales in applying biological and physical process knowledge in coastal management, an example for the Ems estuary. Cont Shelf Res 20: 1655–1686

Diaz RJ, Solan M, Valente RM (2004) A review of approaches for classifying benthic habitats and evaluating habitat quality. J Environ Manage 73: 165–181

Dillon PJ, Rigler FH (1974) The phosphorus-chlorophyll relationship in lakes. Limnol Oceanogr 19: 767–773

Dillon PJ, Rigler FH (1975) A simple method for predicting the capacity of a lake for development based on lake trophic status. J Fish Res Board Can 32:1519771–1531

Di Stefano J, Fidler F, Cumming G (2005) Effect size estimates and confidence intervals: an alternative focus for the presentation and interpretation of ecological data. In: Burk AR (ed), New trends in ecology research, Nova Science Publishers, New York, pp 71–102

Dodds WK (2003) Misuse of inorganic N and soluble reactive P concentrations to indicate nutrient status of surface waters. J N Am Benthol Soc 22: 171–181

Downing JA (1997) Marine nitrogen: phosphorus stoichiometry and the global N:P cycle. Biogeochemistry 37: 237–252

Downing JA, Watson SB, McCauley E (2001) Predicting Cyanobacteria dominance in lakes. Can J Fish Aquat Sci 58: 1905–1908
Draper NR, Smith H (1966) Applied Regression Analysis. Wiley, New York
Duarte CM (1991) Seagrass depth limits. Aquat Bot 40: 363–377
Duarte CM, Kalff J (1989) The influence of catchment geology and lake depth on phytoplankton biomass. Arch Hydrobiol 115: 27–40
Dyer JL (1972) Estuarine hydrography and sedimentation. Cambridge Univ Press, Cambridge
Eckhéll J, Jonsson P, Meili M, Carman R (2000) Storm influence on the accumulation and lamination of sediments in deep areas of the northwestern Baltic proper. Ambio 29: 238–245
Edler L (1979) Phytoplankton succession in the Baltic Sea, Acta Bot Fenn 110: 75–78
Elmgren R (1989) Trophic dynamics in the enclosed brackish Baltic Sea. Rapp P-v Cons Int Explor Mer 183: 152–169
Elmgren R (2001) Understanding human impact on the Baltic Ecosystem: Changing views in recent decades. Ambio 30: 4–5
Elmgren R, Larsson U (eds) (1997) Himmerfjärden. Förändringar i ett näringsbelastat kustekosystem i Östersjön. Swedish EPA, Stockholm
Elmgren R, Larsson U (2001) Nitrogen and the Baltic Sea: Managing nitrogen in relation to phosphorus. TheScientificWorld 1: 371–377
Engqvist A (1996) Long-term nutrient balances in the eutrophication of the Himmerfjärden estuary. Estuar Coast Shelf S 42: 483–507
Engqvist A (1999) Estimated retention times for a selection of coupled coastal embayments on the Swedish west, east and north coasts. Swedish EPA, Report 4910, Stockholm
Evans MS, Arts MT, Robarts RD (1996) Algal productivity, algal biomass, and zooplankton biomass in a phosphorus-rich, saline lake: deviations from regression model predictions. Can J Fish Aquat Sci 53: 1048–1060
Ferreira JG, Nobre AM, Simas TC et al (2006) A methodology for defining homogeneous water bodies in estuaries - Application to the transitional systems of the EU Water Framework Directive. Estuar Coast Shelf S 66: 468–482
Fisher TR, Peele ER, Ammerman JW, Harding LW (1992) Nutrient limitation of phytoplankton in Chesapeake Bay. Mar Ecol-Progr Ser 82: 51–63
Floderus S (1989) The effect of sediment resuspension on nitrogen cycling in the Kattegatt - variability in organic matter transport. Dr Thesis, Uppsala Univ, UNGI Report 71
Floderus S, Håkanson L (1989) Resuspension, ephermal mud blankets and nitrogen cycling in Laholmsbukten, south east Kattegat. Hydrobiologia 176/177: 61–75
France RL, Peters RH (1992) Temporal variance function for total phosphorus. Can J Fish Aquat Sci 49: 975–977
FRP (1978) Havet; naturförhållanden och utnyttjande (The sea; natural conditions and use). Fysisk riksplanering (FRP), Bostadsdepartementet, Nr 7, Stockholm (in Swedish)
Geider RJ, La Roche J (2002) Redfield revisited: variability of C:N:P in marine microalgae and its biochemical basis. Eur J Phycol 37: 1–17
Glasby GP, Szefer P, Geldon J, Warzocha J (2004) Heavy-metal pollution of sediments from Szczecin Lagoon and the Gdansk Basin, Poland. Sci Total Environ 330: 249–269
Gren I-M, Turner K, Wulff F (eds) (2000) Managing a sea. Earthscan, London
Groffman PM, Baron JS, Blett T et al (2006) Ecological thresholds: The key to successful environmental management or an important concept with no practical application. Ecosystems 9: 1–13
Gruber N, Sarmiento JL (1997) Global patterns of marine nitrogen fixation and denitrification. Global Biogeochem Cy 11: 235–266
Guildford SJ, Hecky RE (2000) Total nitrogen, total phosphorus, and nutrient limitation in lakes and oceans: Is there a common relationship? Limnol Oceanogr 45: 1213–1223
Håkanson L (1977) The influence of wind, fetch, and water depth on the distribution of sediments in Lake Vänern, Sweden. Can J Fish Aquat Sci 14: 397–412
Håkanson L (1984) Sediment sampling in different aquatic environments: statistical aspects. Water Resour Res 20: 41–46

Håkanson L (1991) Physical Geography of the Baltic. The Baltic University. Session 1. Uppsala University, Uppsala, ISBN 91-506-0876-2

Håkanson L (1999) Water pollution - methods and criteria to rank, model and remediate chemical threats to aquatic ecosystems. Backhuys Publishers, Leiden

Håkanson L (2000) Modelling radiocesium in lakes and coastal areas - new approaches for ecosystem modellers. A textbook with Internet support. Kluwer Academic Publishers, Dordrecht

Håkanson L (2004) Lakes - Form and function. The Blackburn Press, New Jersey

Håkanson L (2006) Suspended particulate matter in lakes, rivers and marine systems. The Blackburn Press, New Jersey

Håkanson L, Borg H, Uhrberg R (1990) Reliability of analyses of Hg, Fe, Ca, K, P, pH, alkalinity, conductivity, hardness and colour from lakes. Int Rev Hydrobiol 75: 79–94

Håkanson L, Boulion V (2002) The Lake Foodweb - modelling predation and abiotic/biotic interactions. Backhuys Publishers, Leiden

Håkanson L, Bryhn AC (2007) A process-based mass-balance model for phosphorus/eutrophication including a climate change scenario, as exemplified for the Baltic Proper. Manuscript, Dept of Earth Sci, Uppsala Univ, Uppsala

Håkanson L, Bryhn AC (2008a) Goals and remedial strategies for water quality and wildlife management in a coastal lagoon - a case-study of Ringkobing Fjord, Denmark. J Environ Manage 86:498–519

Håkanson L, Bryhn AC (2008b) Modeling the foodweb in coastal areas: a case study of Ringkøbing Fjord, Denmark. Ecol Res (in press)

Håkanson L, Bryhn AC, Blenckner T (2007a) Operational effect variables and functional coastal ecosystem classifications - a review on empirical models and a case study for Ringkobing fjord, Denmark. Int Rev Hydrobiol 92:326–357

Håkanson L, Bryhn AC, Eklund JM (2007b) Modelling phosphorus and suspended partuculate matter in Ringkobing Fjord to understand regime shifts. J Marine Syst 68:65–90

Håkanson L, Bryhn AC, Hytteborn JK (2007c) On the issue of limiting nutrient and predictions of cyanobacteria in aquatic systems. Sci Total Environ 379: 89–108

Håkanson L, Duarte CM (2007) Data variability and uncertainty limits the capacity to identify and predict critical changes in coastal systems - a review of key concepts. Manuscript, Dept of Earth Sci, Uppsala Univ, Uppsala

Håkanson L, Eckhéll J (2005) Suspended particulate matter (SPM) in the Baltic - new empirical data and models. Ecol Model 189: 130–150

Håkanson L, Eklund JM (2007a) Relationships between chlorophyll, salinity, phosphorus and nitrogen in lakes and marine areas. Manuscript, Dept of Earth Sci, Uppsala Univ, Uppsala

Håkanson L, Eklund JM (2007b) A dynamic mass-balance model for phosphorus fluxes and concentrations in coastal areas. Ecol Res 22: 296–320

Håkanson L, Floderus S, Wallin M (1989) Sediment trap assemblages - a methodological description. Hydrobiologia 176/177: 481–490

Håkanson L, Gallego E, Rios-Insua S (2000) The application of the lake ecosystem index in multi-attribute decision analysis in radioecology. J Environ Radioactiv 49: 319–344

Håkanson L, Gyllenhammar A (2005) Setting fish quotas based on holistic ecosystem modelling including environmental factors and foodweb interactions - a new approach. Aquat Ecol 39: 325–351

Håkanson L, Gyllenhammar A, Brolin A (2004) A dynamic model to predict sedimentation and suspended particulate matter in coastal areas. Ecol Model 175: 353–384

Håkanson L, Gyllenhammar A, Karlsson M (2002) Östersjön - hur läget är, hur det borde vara och hur man kommer dit! (The Baltic - present status, what it should be and how to get there). LUVA 02–01, Geotryckeriet, Uppsala

Håkanson L, Jansson M (1983) Principles of lake sedimentology. Springer, Berlin

Håkanson L, Jonsson P, Jonsson B, Martinsen K (1988) Distribution of chlorinated organic substances from paper and pulp industries. Wat Sci Tech 20: 25–36

Håkanson L, Karlsson M (2004) A dynamic model to predict phosphorus fluxes, concentrations and eutrophication effects in Baltic coastal areas. In: Karlsson, M., 2004. Predictive modelling - a tool for aquatic environmental management. Lic Thesis, Dept of Earth Sci, Uppsala Univ, Uppsala

Håkanson L, Kulinski I, Kvarnäs H (1984) Water dynamics and bottom dynamics in coastal areas (in Swedish, Vattendynamik och bottendynamik i kustzonen). SNV PM 1905, Solna, 228 p

Håkanson L, Kvarnäs H, Karlsson B (1986) Coastal morphometry as regulator of water exchange - a Swedish example. Est Coast Shelf Sci 23: 1–15

Håkanson L, Lindgren DAR (2007a) On regime shifts in the Baltic Proper - evaluations based on extensive data between 1974 and 2005. Manuscript, Dept of Earth Sci, Uppsala Univ, Uppsala

Håkanson L, Lindgren DAR (2007b) CoastWeb, a foodweb model based on functional groups for coastal areas including a mass-balance model for phosphorus. Manuscript, Dept of Earth Sci, Uppsala Univ, Uppsala

Håkanson L, Lindgren DAR (2007c) and application of a general process-based dynamic coastal mass-balance model for contaminants. Manuscript, Dept of Earth Sci, Uppsala Univ, Uppsala

Håkanson L, Lindgren DAR (2007d) Water transport and water retention in five connected sub-basins in the Baltic Sea - simulations using a general mass-balance modeling approach for salt and substances. Manuscript, Dept of Earth Sci, Uppsala Univ, Uppsala

Håkanson L, Lindgren DAR, Omstedt A (2007d) Water fluxes to, within and from a coastal system using a general mass-balance model for salt as exemplified using data for the Baltic Sea. Manuscript, Dept of Earth Sci, Uppsala Univ, Uppsala

Håkanson L, Peters RH (1995) Predictive limnology - methods for predictive modeling. SPB Academic Publishers, Amsterdam

Håkanson L, Parparov A, Ostapenia A, Boulion V (2000) Development of a system of water quality as a tool for management. Final report to INTAS, 2000-11-07, Dept of Earth Sci, Uppsala Univ, Uppsala

Håkanson L, Rosenberg R (1985) Praktisk kustekologi (Practical coastal ecology). SNV PM 1987, Solna (in Swedish)

Hansson M (2006) Cyanobakterieblomningar i Östersjön, resultat från satellitövervakning 1997–2005 (Blooms of cyanobacteria in the Baltic Sea, results from satellite monitoring 1997–2005). SMHI Reports Oceanography No 82, Norrköping (in Swedish)

Hansson S, Rudstam LG (1990) Eutrophication and the Baltic fish communities. Ambio 19: 123–125

Hassett RPB, Cardinale B, Stabler LB, Elser JJ (1997) Ecological stoichiometry of N and P in pelagic ecosystems: Comparison of lakes and oceans with emphasis on the zooplankton-phytoplankton interaction. Limnol Oceanogr 42: 648–662

Havens KE, James RT, East TL, Smith VH (2003) N:P ratios, light limitation, and cyanobacterial dominance in a subtropical lake impacted by non-point source nutrient pollution. Environ Pollut 122: 379–390

Hecky RE, Kilham P (1988) Nutrient limitation of phytoplankton in freshwater and marine Limnol Oceanogr 33: 796–822

HELCOM (1986) Water balance of the Baltic Sea. A Regional Cooperation Project of the Baltic Sea States. International Summary Report. Baltic Sea Environment Proceedings 16

HELCOM (1990) Second periodic assessment of the state of the marine environment of the Baltic Sea, 1984–1988; Background document. Baltic Sea Environment Proceedings 35B

HELCOM (2000) Baltic Sea Environment Proceedings 100

HELCOM (2003) The Baltic Marine Environment 1999–2002. Baltic Sea Environment Proceedings 87

HELCOM (2006) Development of tools for assessment of eutrophication in the Baltic Sea. Baltic Sea Environment Proceedings 104

Hellström T (1991) The effect of resuspension on algal production in a shallow lake. Hydrobiologia 213: 183–190

Himmerfjärden (2007) www2.ecology.su.se/dbHFJ/index.htm, 2007-09-04

Howarth RW (1988) Nutrient limitation of net primary production in marine ecosystems. Annu Rev Ecol Syst 19: 89–110

Howarth RW, Cole JJ (1985) Molybdenum availability, nitrogen limitation, and phypoplankton growth in natural waters. Science 229: 653–655

Howarth RW, Marino R (2006) Nitrogen as the limiting nutrient for eutrophication in coastal marine ecosystems: Evolving views over three decades. Limnol Oceanogr 51: 364–376

Howarth RW, Marino R, Lane J, Cole JJ (1988) Nitrogen fixation in freshwater, estuarine, and marine ecosystems. 1. Rates and importance. Limnol Oceanogr 33: 669–687

Højerslev NK (1978) Daylight measurements appropriate for photosynthetic studies in natural sea waters. J Cons Int Explor Mer 38: 131–146

Huang B, Hong H (1999) Alkaline phosphatase activity and utilization of dissolved organic phosphorus by algae in subtropical coastal waters. Mar Pollut Bull 39: 205–211

IAEA (2000) Modelling of the transfer of radiocaesium from deposition to lake ecosystems. The VAMP Aquatic Working Group. International Atomic Energy Agency, Vienna, IAEA-TECDOC-1143

ICES (2006a) ICES Oceanographic data. http://www.ices.dk/ocean/dotnet/HydChem/HydChem.aspx

ICES (2006b) ICES. HELCOM data. http://www.ices.dk/ocean/asp/helcom/helcom.asp?Mode=1

ICES (2007) www.ices.dk, 2007–09–04

Inman DL, Nordstrom CE (1971) On the tectonic and morphologic classification of coasts. J Geol 79: 1–21

Istvanovics V, Somlyody L (2001) Factors influencing lake recovery from eutrophication - the case of Basin 1 of Lake Balaton. Water Res 35: 729–735

Irvine K (2004) Classifying ecological status under the European Water Framework Directive: the need for monitoring to account for natural variability. Aquat Conserv 14: 107–112

Jackson DA, Harvey HH, Somers KM (1990) Ratios in aquatic sciences: Statistical shortcomings with mean depth and the morpoedaphic index. Can J Fish Aquat Sci 47: 1788–1795

Jeffrey SW, Humphrey, GF (1975) New spectrophometric equations for determination chlorophylls a, b, c_1 and c_2 in higher plants, algae and natural phytoplankton. Biochem Physiol Pfl 167: 191–194

Jeppesen E, Søndergaard M, Jensen SP et al (2005) Lake responses to reduced nutrient loading - an analysis of contemporary log-term data from 35 case studies. Freshwater Biol 50: 1747–1771

Johansson H, Lindström M, Håkanson L (2001) On the modelling of pariculate and dissolved distributions of substances in aquatic ecosystems - sedimentological and ecological interactions. Ecol Model 137: 225–240

Johnson DW (1919) Shore processes and shoreline development. Wiley & Sons, New York

Jonsson P (1992) Large-scale changes of contaminants in Baltic Sea sediments during the twentieth century. Dr Thesis, Uppsala Univ, Uppsala.

Jonsson P, Carman R, Wulff F (1990) Laminated sedments in the Baltic - a tool for evaluating nutrient mass balances. Ambio 19: 152–158

Jönsson A (2005) Model studies of surface waves and sediment resuspension in the Baltic Sea. Dr thesis No 332, Linköping Univ, Linköping

Jørgensen SE, Johnsen J (1989) Principles of environmental science and technology (2nd edition). Studies in environmental science, 33. Elsevier, Amsterdam

Kalff J (2002) Limnology. Prentice Hall, New Jersey

Karlsson M (2007) Dynamisk massbalansmodellering av fosfor i Östersjön (Dynamic mass-balance modeling of phosphorus in the Baltic Sea). Master Thesis, Dept of Earth Sci, Uppsala Univ, Uppsala (in Swedish)

Karlsson M, Håkanson L (2001) Miljökonsekvensanalys av Korsnäsverkens fosforutsläpp till Gävle Yttre fjärd (Environmental consequence analysis of phosphorus emissions from the Korsnäs Factories to the Gävle Outer Bay. of Earth Sci, Uppsala Univ, Uppsala

Kautsky L (1991) Life in the Baltic Sea. Uppsala Univ, Uppsala, ISBN 91-506-0877-0

Kautsky L, Kautsky H (1989) Algal species diversity and dominance along gradients of stress and disturbance in marine environments. Vegetatio 83: 259–267

Kenney BC (1982) Beware of spurious self-correlations! Water Resour Res 18: 1041–1048

Khalili M (2007) Salt, water and nutrient fluxes to Himmerfjärden bay. Master thesis, Dept of Earth Sci, Uppsala Univ, Uppsala

Kiefer DA, Austin RW (1974) The effect of varying phytoplankton concentration on submarine transmission in the Gulf of California. Limnol Oceanogr 19: 55–64

Kiirikki M, Inkala A, Kuosa H, Pitkänen H, Kuusisto M, Sarkkula J (2001) Evaluating the effects of nutrient load reductions on the biomass of toxic nitrogen-fixing cyanobacteria in the Gulf of Finland, Baltic Sea. Boreal Environ Res 6: 1–16

Kjerfve B (ed) (1994) Coastal lagoon processes. Elsevier. Amsterdam

Knowton MF, Hoyer MV, Jones JR (1984) Sources of variability in phosphorus and chlorophyll and their effects on the use of lake survey data. Water Resour Bull 20: 397–407

Knowlton MF, Jones JR (2006a) Temporal variation and assessment in trophic state indicators in Missouri reservoirs: Implication for lake monitoring and management. Lake Reserv Manage 22: 261–271

Knowlton MF, Jones JR (2006b) Natural variability in lakes and reservoirs should be recognized in setting nutrient criteria. Lake Reserv Manage 22: 161–166

Konopka A, Brock TD (1978) Effect of temperature on blue-green algae (Cyanobacteria) in Lake Mendota. Applied Environ Microb 36: 572–576

Krambeck H-J (1995) Application and abuse of statistical methods in mathematical modelling in limnology. Ecol Model 78: 7–15

Kranck K (1973) Flocculation of suspended sediment in the sea. Nature 246: 348–350

Kranck K (1979) Particle matter grain-size characteristics and flocculation in a partially mixed estuary. Sedimentology 28: 107–114

Kraufvelin P, Sinisalo B, Leppäkoski E et al (2001) Changes in zoobenthic community structure after pollution abatement from fish farms in the Archipelago Sea (N. Baltic Sea). Marine Environ Res 51: 229–245

Kristensen P, Søndergaard M, Jeppesen E (1992) Resuspension in a shallow eutrophic lake. Hydrobiologia 228: 101–109

Labry G, Herbland A, Delamas D (2002) The role of phosphorus on planktonic production of the Gironde plume waters in the Bay of Biscay. J Plankton Res 24: 97–117

Larsson U, Elmgren R (2001) Eutrophication in the Baltic Sea area. Integrated coastal management issues. Sci Integr Coast Manage pp 15–35

Larsson U, Hajdu S, Walve J, Elmgren R, (2001) Baltic Sea nitrogen fixation estimated from the summer increase in upper mixed layer total nitrogen. Limnol Oceanogr 46: 811–820

Larsson U, Hajdu S, Walve J, Andersson A, Larsson P, Edler L (2006) Bedömningsgrunder för kust och hav. Växtplankton, näringsämnen, klorofyll och siktdjup (Evaluation criteria for the coast and sea. Phytoplankton, nutrients, chlorophyll and Secchi depth). Swedish EPA report No 4914, Stockholm (in Swedish)

Laursen MB, Gertz F, Hansen JW et al (2004) Marine områder - miljøtilstand i fjordområder: Ringkøbing Fjord og Nissum Fjord (Marine areas - the state of the environment in fjord areas: Ringkobing Fjord and Nissum Fjord). County of Ringkjøbing, Ringkøbing (in Danish)

Laznik M, Stålnacke P, Grimvall A, Wittgren HB (1998) Riverine input of nutrients to the Gulf of Riga - temporal and spatial variation. J Marine Syst 23: 11–25

Lehtimäki J, Sivonen K, Luukainen R, Niemelä SI (1994) The effects of incubation time, temperature, light, salinity, and phosphorus on growth and hepatotoxin production by *Nodularia* strains. Arch Hydrobiol 130: 269–282

Lehtimäki J, Moisander P, Sivonen K, Kononen K (1997) Growth, nitrogen fixation, and nodularin production by two Baltic Sea cyanbacteria. Appl Environ Microb 63: 1647–1656

Le Pape O, Menesguen A (1997) Hydrodynamic prevention of eutrophication in the Bay of Brest (France), a modeling approach. J Marine Syst 12: 171–186

Lick W, Lick J, Ziegler, CK (1992) Flocculation and its effect on the vertical transport of fine-grained sediments. Hydrobiologia 235/236: 1–16

Lindgren D, Håkanson L (2007) Functional classification of coastal areas as a tool in ecosystem modeling and management. Manuscript, Dept of Earth Sci, Uppsala Univ, Uppsala

Liu H, Jezek KC (2004) A complete high-resolution coastline of Antarctica extracted from orthorectified Radarsat SAR imagery. Photogramm Eng Rem S 70: 605–616

Livingston RJ (2001) Eutrophication processes in coastal systems. CRC Press, Boca Raton

LOICZ (2007) LOICZ/Hexacoral Environmental Typology Database. http://hercules.kgs.ku.edu/hexacoral/envirodata/hex_modfilt_firststep3dev1.cfm

Longstaff BJ, Dennison WC (1999) Seagrass survival during pulsed turbidity events: the effects of light deprivation on the seagrasses Halodule pinifolia and Halophila ovalis. Aquat Bot 65: 105–121

Lukatelich RJ, McComb AJ (1986) Nutrient levels and the development of diatoms and blue-green algal blooms in a shallow Australian estuary. J Plankton Res 8: 597–618

Lundin L-C (ed) (1999) Water in Society. Sustainable water management in the Baltic Sea basin, The Baltic Univ Progr, Uppsala Univ, Uppsala

Lundin L-C (ed) (2000a) The Waterscape. Sustainable water management in the Baltic Sea basin, The Baltic Univ Progr, Uppsala Univ, Uppsala

Lundin L-C (ed) (2000b) River Basin Management. Sustainable water management in the Baltic Sea basin, The Baltic Univ Progr, Uppsala Univ, Uppsala

Mann KH (1982) Ecology of coastal waters. A systems approach. Blackwell Scientific Publications, Oxford

Marino R, Chan F, Howarth RW et al (2006) Ecological constraints on planktonic nitrogen fixation in saline estuaries. I. Nitrogen and trophical controls. Mar Ecol-Progr Ser 309: 25–39

Marumo R, Asaoka O (1974) Distribution of pelagic blue-green algae in the Northen Pacific Ocean. J Oceanogr Soc Jpn 30: 77–85

Mason RP, Reinfelder JR, Morel FMM (1995) Bioaccumulation of mercury and methylmercury. Water Air Soil Poll 80: 915–921

Mazur-Marzec H, Zeglinska L, Plinski M (2005) The effect of salinity on the growth, toxin production, and morphology of *Nodularia spumigena* isolated from the Gulf of Gedansk, southern Baltic Sea. J Appl Phycol 17: 171–179

McCave IN (1981) Location of coastal accumulations of fine sediments around the southern North Sea. Rapp P v Reun Int Explor Mer 181: 15–27

McQueen DJ, Lean DRS (1987) Influence of water temperature and nitrogen to phosphorus ratios on the dominance of blue-green algae in Lake St. George, Ontario. Can J Fish Aquat Sci 44: 598–604

MEDAR Group (2002) MEDATLAS/2002 database. Mediterranean and Black Sea database of temperature salinity and bio-chemical parameters. Climatological Atlas. IFREMER Edition (4 Cdroms)

Meeuwig JJ, Kauppila P, Pitkänen H (2000) Predicting coastal eutrophication in the Baltic: a limnological approach. Can J Fish Aquat Sci 57: 844–855

Menge BA, Allison G, Freidenburg T (2003) Local to coastal-scale macrophyte community structure: Surprising patterns and possible mechanisms. J Phycol 39: 1–42

Mikulski Z (1985) Water Balance of the Baltic Sea. Baltic Sea Environment, Proceedings 16, Helsinki Commission, Helsinki

Moisander PH, McClinton E, Paerl HW (2002) Salinity effects on growth, photosynthetic parameters, and nitrogenase activity in estuarine planktonic cyanobacteria. Microbial Ecology 43: 432–442

Moen FE, Svensen E (2004) Marine fish and invertebrates. AquaPress, Essex

Moldan B, Billharz S (eds) (1997) Sustainability indicators. Wiley, see http://www.icsu-scope.org/downloadpubs/scope58/.

Möller P, Pihl L, Rosenberg R (1985) Fisk och bottendjur i grundområden i Bohuslän och Halland - en biologisk värdering (Fish and zoobenthos in shallow areas of Bohuslän and Halland - a biological estimate). SNV PM 1911: 7–95 (in Swedish)

Monitor (1988) Sweden's Marine Environment - Ecosystem under Pressure. Swedish Environmental Protection Agency, Stockholm

Monte L (1997) A collective model for predicting the long-term behaviour of radionuclides in rivers. Sci Total Environ 201: 17–29

Monte L, Håkanson L, Perianez R et al (2006) Experiences from a case study of multi-model application to assess the behaviour of pollutants in the Dnieper-Bug Estuary. Ecol Model 195: 247–263

Mosteller F, Tukey JW (1977) Data Analysis and Regression: a second course in statistics. Addison-Wesley Publ. Reading, Massachusetts

Muir Wood AM (1969) Coastal hydraulics. Macmillan, London

Myrberg K (1998) Analysing and modelling the physical processes of the Gulf of Finland in the Baltic Sea. Monographs of the Boreal Environment Research 10, Helsinki

Neumann T, Schernewski G (2005) An ecological model evaluation of two nutrient abatement strategies for the Baltic Sea. J Marine Syst 56: 195–206

Newman MC (1993) Regression analysis of log-transformed data: statistical bias and its correction. Environ Toxicol Chem 12: 1129–1133

Newton A, Icely JD, Falcao M (2003) Evaluation of eutrophication in the Ria Formosa coastal lagoon, Portugal. Cont Shelf Res 23: 1945–1961

Nissling A, Johansson U, Jacobsson M (2006) Effects of salinity and temperature conditions on the reproductive success of turbot (Scophthalmus maximus) in the Baltic Sea. Fish Res 80: 230–238

Nixon SW (1990) Marine eutrophication: a growing international problem. Ambio 3: 101

Nixon SW, Pilson I (1983) Nitrogen in estuarine and coastal marine ecosystems. In: Carpenter EJ, Capone, DG (eds) Nitrogen in the marine environment. Academic Press, New York, pp 565–648

Nordvarg L (2001) Predictive models and eutrophication effects of fish farms. Dr Thesis, Uppsala Univ, Uppsala

Nordvarg L, Håkanson L (2002) Predicting the environmental response of fish farming in coastal areas of the Åland archipelago (Baltic Sea) using management models for coastal water planning. Aquaculture 206: 217–243

Nürnberg GK, Shaw M (1998) Productivity of clear and humic lakes: nutrients, phytoplankton, bacteria. Hydrobiologia 382: 97–112

OECD (1982) Eutrophication of waters. Monitoring, assessment and control. OECD, Paris

Olsson V (1966) Kustens fåglar (Coastal birds). Gebers, Stockholm

Omstedt A, Axell LB (2003) Modeling the variations of salinity and temperature in the large Gulfs of the Baltic Sea. Cont Shelf Res 23: 265–294

O'Neill RV, Gardner RH, Barnthouse LW et al (1982) Ecosystem risk analysis: A new methodology. Environ Toxicol Chem 1: 167–177

Ostmann M, Suursaar Ü, Kullas T (2001) The oscillatory nature of the flows in the system of straits and small semienclosed basins of the Baltic Sea. Cont Shelf Res 21: 1577–1603

Ottosson F, Abrahamsson O (1998) Presentation and analysis of a model simulating epilimnetic and hypolimnetic temperatures in lakes. Ecol Model 110: 223–253

Panagiotopoulos C, Sempere R (2005) The molecular distribution of combined aldoses in sinking particles in various oceanic conditions. Mar Chem 95: 31–49

Pearson K (1897) On a form of spurious correlation which may arise when indices are used in the measurements of organs. Proc R Soc Lon 60: 489–502

Pearson TH, Rosenberg R (1976) A comparative study on the effects on the marine environment of wastes from cellulose industries in Scotland and Sweden. Ambio 5: 77–79

Pedersen OB, Christiansen C, Laursen MB (1995) Wind-induced long term increase and short term fluctuations of shallow water suspended matter and nutrient concentrations, Ringkobing Fjord, Denmark. Ophelia 41: 273–287

Penna N, Capellacci S, Ricci F (2004) The influence of the Po River discharge on phytoplankton bloom dynamics along the coastline of Pesaro (Italy) in the Adriatic Sea. Mar Pollut Bull 48: 321–326

Persson J, Håkanson L (1996) A simple empirical model to predict deepwater turnover time in coastal waters. Can J Fish Aquat Sci 53: 1236–1245

Persson J, Håkanson L, Wallin M (1994a) Ett geografiskt informationssystem för kustvattenplanering baserad på sjökortsinformation (A geographical information system for coastal water planning based on chart information). Nordic Council, TemaNord 1994:667, Copenhagen (in Swedish)

Persson J, Håkanson L, Pilesjö P (1994b) Prediction of surface water turnover time in coastal waters using digital bathymetric information. Environmetrics, 5: 433–449

Persson G, Jansson M (1988) Phosphorus in freshwater ecosystems. Kluwer Academic Publishing, Dordrecht

Peters RH (1986) The role of prediction in limnology. Limnol Oceanogr 31: 1143–1159

Peters RH (1991) A critique for ecology. Cambridge University Press, Cambridge

Petersen JK, Hansen JW, Conley D (2004) Strukturskift i Ringkøbing Fjord (Structural change in Ringkøbing Fjord). Manuscript, Ringkjøbing County (in Danish)

Petersen JK, Hansen JW, Laursen MB, Conley D (2006) Regime shift in a marine coastal ecosystem. Manuscript, National Environmental Research Institute & County of Ringkjøbing

Pfaffenberger RC, Patterson JH (1987) Statistical methods. Irwin, Illinois

Pinet PR (2003) Invitation to oceanography (3rd ed). Jones and Bartlett Publishers, Sudbury, Massachusetts

Pilesjö P, Persson J, Håkanson L (1991) Digital sjökortsinformation för beräkning av kustmorfometriska parametrar och ytvattnets utbytestid (Digital bathymetric information for calculations of morphometrical parameters and surface water retention time for coastal areas). Swedish EPA (SNV) Report No. 3916, Solna (in Swedish)

Pitkänen H, Tallberg P (eds) (2007) Searching efficient protection strategies for the wutrophied Gulf of Finland: the integrated use of experimental and modeling tools (SEGUE), Finnish Environment Institute, Helsinki, 15

Postma H (1982) Sediment transport and sedimentation. In: Olausson E, Cato, I (eds) Chemistry and biogeochemistry of estuaries. Wiley & Sons, Chichester

Prairie Y (1996) Evaluating the predictive power of regression models. Can J Fish Aquat Sci, 53: 490–492

Preisendorfer RW (1986) Secchi disk science: Visual optics of natural waters. Limnol Oceanogr 31: 909–926

Pustelnikov OS (1977) Geochemical features of suspended matter in connection with recent processes in the Baltic Sea. Ambio 5: 157–162

Rabalais NN, Nixon SW (2002) Dedicated issue: Nutrient over-enrichment in coastal waters: global patterns of cause and effect. Estuaries 25: 639–900

Rahm L, Jönsson A, Wulff F (2000) Nitrogen fixation in the Baltic proper: an empirical study. J Marine Syst 25: 239–248

Rasmussen B, Josefson AB (2002) Consistent estimates for the residence time of micro-tidal estuaries. Estuar Coast Shelf S 54: 65–73

Redfield AC (1958) The biological control of chemical factors in the environment. Am Sci 46: 205–222

Redfield AC, Ketchum BH, Richards FA (1963) The influence of organisms on the composition of sea-water. In: Hill N (ed) The Sea 2. Interscience, New York, pp 26–77

Reed JL (1921) On the correlation between any two functions and its application to the general case of spurious correlation. J Wash Acad Sci 11: 449–455

Remane A (1934) Die Brackwasserfauna (Brackish water fauna). Verh Deutsch Zool Ges, 36: 34–74 (in German)

Reynolds CS (1987) Cyanobacterial water-blooms. Adv Bot Res 13: 67–143

Reynolds CS, Walsby AE (1975) Water-blooms. Biol Rev 50: 437–481

Riley ET, Prepas EE (1985) Comparison of the phosphorus-chlorophyll relationships in mixed and stratified lakes. Can J Fish Aquat Sci 42: 831–835

Robarts RD, Zohary T (1987) Temperature effects on phytosynthetic capacity, respiration, and growth rates of bloom-forming cyanobacteria. New Zeal J Mar Fresh 21: 391 - 399

Rönnberg C, Bonsdorff E (2004) Baltic sea eutrophication: area-specific ecological consequences. Hydrobiologia 514: 224–241

Rose KA, McLean RI, Summers JK (1989) Development and Monte Carlo analysis of an oyster bioaccumulation model applied to biomonitoroing. Ecol Mod 45:111–132

Rosenberg R (1985) Eutrophication - the future marine coastal nuisance? Mar Pollut Bull 16: 227–231

Rubio VC, Sánchez-Vázques FJ, Madrid JA (2005) Effects of salinity on food intake and macronutrient selection in European sea bass. Physiol Behav 85: 333–339

Ryther JH, Dunstan WM (1971) Nitrogen, phosphorus, and eutrophication in the coastal marine environment. Science 171: 1008–1013

Sagert S, Krause Jensen D, Henriksen P et al (2005) Integrated ecological assessment of Danish Baltic Sea coastal areas by means of phytoplankton and macrophytobenthos. Estuar Coast Shelf S 63: 109–118

Salomons W, Förstner U (1984) Metals in the Hydrocycle. Springer, Heidelberg

Sandberg J, Elmgren R, Wulff F (2000) Carbon flows in Baltic Sea food webs - a re-evaluation using a mass balance approach. J Marine Syst 25: 249–260

Santschi PH, Honeyman BD (1991) Radioisotopes as tracers for the interactions between trace elements, colloids and particles in natural waters. In: Vernet, JP (ed) Heavy metals in the environment, Elsevier, Amsterdam, pp 229–246

Savage C, Elmgren R, Larsson U (2002) Effects of sewage-derived nutrients on an estuarine macrobenthic community. Mar Ecol-Progr Ser 243: 67–82

Savchuk OP (2000) Studies of the assimilation capacity and effects of nutrient load reductions in the eastern Gulf of Finland with a biogeochemical model. Boreal Environ Res 5: 147–163

Savchuk, OP (2006) SANBaLTS - Simple as Necessary Long-Term large-Scale simulation model of the nitrogen and phosphorus biogeochemical cycles in the Baltic Sea, version 3. http://www.mare.su.se/nest/docs/SANBalTS_QAv3.pdf (2007–01–15)

Savchuk OP, Swaney DP (2000) Water and Nutrient Budget of the Gulf of Riga. http://data.ecology.su.se/mnode/Europe/Gulf%20of%20Riga/rigabud.htm (2006–04–19)

Savchuk OP, Wulff F (1999) Modeling the Baltic Sea eutrophication in a decision su8pport system. Ambio 2–3: 141–148

Savchuk OP, Wulff F (1999) Modelling regional and large-scale response of Baltic Sea ecosystems to nutrient reductions. Hydrobiologia 393: 35–43

Savchuk OP, Wulff F (2007) Modeling the Baltic Sea eutrophication in a decision support system. Ambio 36:141–148

Scheffer M (1990) Multiplicity of stable states in freshwater systems. Hydrobiologia 200/201: 475–486

Scheffer M, Brock W, Westley F (2000) Socioeconomic mechanisms preventing optimum use of ecosystem services: An interdisciplinary theoretical analysis. Ecosystems 3: 451–4571

Schernewski G, Dolch T (eds) (2004) The Oder estuary - against the background of the European Water Framework Directive. Marine Science Reports (Meereswissenschaftliche Berichte) 57, ISSN: 0939-396X

Schernewski G, Neumann T (2005) The trophic state of the Baltic Sea a century ago: a model simulation study. J Marine Syst 53: 109–124

Schernewski G, Schiewer U (eds) (2002) Baltic coastal ecosystems. Springer, Berlin

Schernewski G, Wielgat M (2004) A Baltic Sea typology according to the EC-Water Framework Directive: Integration of national typologies and the water body concept. Coastline Reports 4, ISSN 0928-2734, 1–26

Schindler DW (1977) Evolution of phosphorus limitation in lakes. Science 195: 260–262

Schindler DW (1978) Factors regulating phytoplankton production and standing crop in the world's freshwaters. Limnol Oceanogr 23: 478–486

Schindler DW (2006) Recent advances in the understanding and management of eutrophication. Limnol Oceanogr 51: 356–363

Secchi A (1866) Relazione della esperienze fatta a bordo della pontificia pirocorvetta L'Immacolata Concezione per determinare la transparenza del mare (Reports on experiments made on board the papal steam sloop L'Irnmacolata Concezione to determine the transparency of the sea). In: Cialdi A (ed) Sul moto ondoso del mare e su le correnti di esso specialmente su quelle littorali, Rome, pp 258–288 (in Italian)

Seibold E, Berger WH (1982) The sea floor. Springer-Verlag, Heidelberg

Seifert T, Tauber F, Kayser B (2001) A high resolution spherical grid topography of the Baltic Sea - 2nd edition. Baltic Sea Science Congress, Stockholm, 25–29 November 2001, Poster #147, http://www.io-warnemuende.de/iowtopo

Seitzinger SP, Sanders RW (1999) Atmospheric Inputs of Dissolved Organic Nitrogen Stimulate Estuarine Bacteria and Phytoplankton. Limnol Oceanogr 44: 721–730

Sellner KG (1997) Physiology, ecology and toxic properties of marine cyanobacteria blooms. Limnol Oceanogr 42: 1089–1104

SEPA (2002) The Future for Scotland's Waters. Guiding principles on the technical requirements of the Water Framework Directive. Scottish EPA, Stirling

Seritti A, Petrosini A, Ferrara R, Barghigiani C (1980) A contribution to the determination of reactive and total mercury in seawater. Environ Technol Lett 1: 50–57

SLU (2007) www.ma.slu.se, 2007–09–04

Sly PG (1978) Sedimentary Processes in Lakes. In: Lerman A (ed) Lakes: Chemistry, Geology, Physics. Springer, Berlin, pp 65–89

SMHI (1994) Havsområdesregister 1993 (Marine areas 1993). Swedish Meteorological and Hydrological Institute (SMHI), Oceanografi nr 60 (in Swedish)

SMHI (2003) Djupdata för havsområden 2003 (Depth data for marine areas). Swedish Meteorological and Hydrological Institute (SMHI), Oceanografi nr 73 (in Swedish)

SMHI (2007) www.smhi.se. 2007–09–04

Smith VH (1979) Nutrient dependence of primary productivity in lakes. Limnol Oceanogr 24: 1051–1064

Smith VH (1985) Predictive models for the biomass of blue-green algae in lakes. Water Resour Bull 21: 433–439

Smith VH (1990) Nitrogen, phosphorus, and nitrogen fixation in lacustrine and estuarine ecosystems. Limnol Oceanogr 35: 1852–1859

Smith VH (2003) Eutrophication of freshwater and coastal marine ecosystems: a global problem. Environ Sci and Pollut R 10:126–139

Smith VH (2006) Responses of estuarine and coastal marine phytoplankton to nitrogen and phosphorus enrichment. Limnol Oceanogr 51: 377–384

Smith VH, Joye SB, Howarth RW (2006) Eutrophication of freshwater and marine ecosystems. Limnol Oceanogr 51: 351–355

Smith VH (1979) Nutrient dependence of primary productivity in lakes. Limnol Oceanogr 24: 1051–1064

Smith VH, Joye SB, Howarth RW (2006) Eutrophication of freshwater and marine ecosystems. Limnol Oceanogr 51: 351–355

SNV (1993) Eutrofiering av mark, sötvatten och hav (Eutrophication of land, freshwater and the Sea). Swedish EPA, Report 4134, Stockholm

Stanley DJ, Swift DJ (eds) (1976) Marine sediment transport and environmental management. Wiley & Sons, New York

Stålnacke P, Grimvall A, Sundblad K, Wilander A (1999a) Trends in nitrogen transport in Swedish rivers. Environ Monit Assess 59: 47–72

Stålnacke P, Grimvall A, Sundblad K, Tonderski A (1999b) Estimation of riverine loads of nitrogen and phosphorus to the Baltic Sea, 1970–1993. Environ Monit Assess 58: 173–200

Stålnacke P, Grimvall A, Libiseller C (et al) (2003) Trends in nutrient concentrations in Latvian rivers and the response to the dramatic change in agriculture. J Hydrol 283: 184–205

Stålnacke P, Vandsemb SM, Vassiljev A (et al) (2004) Changes in nutrient levels in some Eastern European rivers in response to large-scale changes in agriculture. Water Sci Technol 49: 28–36

Swedish EPA (2000) Environmental quality criteria. Coasts and seas. Swedish EPA report 5052, Stockholm

Swedish Environmental Advisory Council (2005) A strategy for ending eutrophication of seas and coasts. Swedish Government Official Reports, Environmental Advisory Council, Jo 1968: A, Stockholm

SYKE (2006) Syrebristen i Finska viken exceptionellt omfattande, tillståndet i havsbottnen sämre än tidigare på 2000-talet (Oxygen depletion in the Gulf of Finland exceptionally widespread, seafloor conditions worse than before in the 21st century). Finnish Environmental Center (SYKE), press release, 2006–08–17 (in Swedish)

SYKE (2003) Intern belastning reglerar kraftigt algblomningen i Finska viken (Internal load strongly regulates the algal bloom in the Gulf of Finland). Environmental Center (SYKE), press release 2003–06–11 (in Swedish)

Thomas RL (1972) The distribution of mercury in the sediments of Lake Ontario. Can J Fish Aquat Sci 9: 636–651

Thomas RL, Kemp ALW, Lewis CFM (1972) Distribution, composition and characteristics of the surficial sediments of Lake Ontaio. J Sediment Petrol 42: 66–84

Thomas RL, Jacquet JM, Kemp ALW, Lewis CFM (1976) Surficial sediments in Lake Erie. J Fish Res Board Can 33: 385–403

Thomas J, Apte SK, Reddy BR (1988) Sodium metabolism in cyanobacteria nitrogen fixation and salt tolerance. In: Bothe H, Bruijn FJ, Newton WE (eds) Nitrogen fixation: hundred years after. Stuttgart, New York, pp 195–201

Thresholds IP (2006) Thresholds of Environmental Sustainability IP. http://www.thresholds-eu.org/

Tilman D, Kiesling RL (1984) Freshwater algal ecology: taxonomic trade-offs in temperature dependence of nutrient competitive abilities. In: Klug MJ, Reddy CA (eds) Current perspectives in microbial ecology. proceedings of the third international symposium on microbial ecology. American Society of Microbiology, Washington, D.C, pp 314–319

Tilzer MM (1988) Secchi disk - chlorophyll relationships in a lake with highly variable phytoplankton biomass. Hydrobiologia 162: 163–171

Tiwari JL, Hobbie JE (1976) Random differential equations as models of ecosystems. Monte Carlo simulation approach. Math Biosci 28:25–44

Tõnno I (2004) The impact of nitrogen and phosphorus concentration and N/P ratio on cyanobacterial dominance and N_2 fixation in some Estonian lakes. Dr thesis, Tartu University Press, Tartu

Turner A (1996) Trace-metal partitioning in estuaries: importance of salinity and particle concentration. Mar Chem 54: 27–39

Turner A, Hyde TL, Rawling MC (1999) Transport and retention of hydrophobic organic micropollutants in estuaries: Implications of the particle concentration effect. Estuar Coast Shelf Sci 49: 733–746

Turner RE, Rabalais NN, Justic D, Dortch Q (2003) Global patterns in dissolved N, P and Si in large rivers. Biogeochemistry 64: 297–317

Tyrrell T (1999) The relative influences of nitrogen and phosphorus on oceanic primary production. Nature 688: 525–531

Ursella L, Gacic M (2001) Use of the Acoustic Doppler Current Profiler (ADCP) in the study of the circulation of the Adriatic Sea. Ann Geophys 19: 1183–1193

Vahtera E, Conley DJ, Gustafsson BG et al (2007) Internal ecosystem feedbacks enhance nitrogen-fixing Cyanobacteria blooms and complicate management in the Baltic Sea. Ambio 2–3: 186–194

Valentin H (1952) Die Kusten der Erde (The coasts of the Earth). Petem Geogr Mitt Erg H 246, 2 Aufl (in German)

Valentin H (1979) Ein system der Zonalen Kustenmorphologie (A system of zonal coastal morphology). Z Geomorphol N F 23: 113–131 (in German)

Vichi M, Ruardij P, Baretta JW (2004) Link or sink: a modelling interpretation of the open Baltic biogeochemistry. Biogeosciences Discuss 1:79–100

Vidal M, Duarte CM, Agusti S (1999) Dissolved Organic Nitrogen and Phosphorus Pools and Fluxes in the Central Atlantic Ocean. Limnol Oceanogr 44: 106–115

Voipio A (ed) (1981) The Baltic Sea. Elsevier Oceanographic Series, Amsterdam

Vollenweider RA (1958) Sichttiefe und Produktion (Secchi depths and production). Verh Int Ver Limnol 13: 142–143 (in German)

Vollenweider RA (1960) Beitrage zür Kenntnis optischer Eigenschaften der Gewässer und Primärproduktion (Contribution to the knowledge of optical characteristics of waters and primary production). Mem Ist Ital Idrobiol 12: 201–244 (in German)

Vollenweider RA (1968) The scientific basis of lake eutrophication, with particular reference to phosphorus and nitrogen as eutrophication factors. Tech Rep DAS/DSI/68.27, OECD, Paris

Vollenweider RA (1976) Advances in defining critical loading levels for phosphorus in lake eutrophication. Mem Ist Ital Idrobiol 33:53–83

Vollenweider RA, Giovanardi F, Montanari G, Rinaldi A (1998) Characterization of the trophic conditions of marine coastal waters with a special reference to the NW Adriatic Sea: Proposal for a trophic scale, turbidity and generalized water quality index. Environmetrics 9: 329–357

Vorobev GA (1977) Landshaftnye tipy zarastaniya ozyor Vologodskogo poozer'ya (Landscape types of lake overgrowing in the Vologda lake area). In: Prirodnye usloviya i resursy Severa Yevropeyskoy chasti SSSR. BGPI, Vologda, pp 48–60 (in Russian)

Wallin M, Håkanson L (1991) Morphometry and sedimentation as regulating factors for nutrient recycling in shallow coastal waters. Hydrobiologia 75: 33–46

Wallin M, Håkanson L, Persson J (1992) Belastningsmodeller för närsaltsutsäpp i kustvatten, speciellt fiskodlingars miljöpåverkan (Load models for nutrients in coastal areas, especially from fish farms). Nordic Council 1992:502, Copenhagen (in Swedish with English summary)

Wasmund N (1997) Occurrence of cyanobacteria blooms in the Baltic Sea in relation to environmental conditions. Int Rev Hydrobiol 82: 169–184

Wasmund N, Nausch G, Schneider B et al (2005) Comparison of nitrogen fixation rates determined with different methods: a study in the Baltic Proper. Mar Ecol-Progr Ser 297: 23–31

WassmannP, Tamminen T (2000) Pelagic eutrophication and sedimentation in the Gulf of Riga: an introduction. J Marine Syst 23: 1–10

Weston K, Jickells TD, Fernand L, Parker ER (2004) Nitrogen cycling in the southern North Sea: consequences for total nitrogen transport. Estuar Coast Shelf S 59: 559–573

Wetzel RG (1983) Attached algal-substrata interactions: fact or myth, and when and how? In: Wetzel RG (ed) Periphyton of freshwater ecosystems. Dr. W. Junk Publisher, The Hague, pp 207–215

Wetzel RG (2001) Limnology. Academic Press, London

Worley BA (1987) Deterministic uncertainty analysis. Oak Ridge National Laboratory Report ORNL-6428, Oak Ridge, U.S.A., 53 p

Wright DJ, Bartlett DJ (eds) (2000) Marine and coastal geographic information systems. Taylor & Francis, London

Wulff F (2006) Östersjön är inte som andra hav (The Baltic Sea is not like other seas). In: Johansson B, (ed) Östersjön - hot och hopp, Formas Fokuserar, Stockholm, pp 51–63 (in Swedish)

Wulff F, Rahm L, Larsson P (eds) (2001) A systems analysis of the Baltic Sea. Ecological studies 148, Springer-Verlag, Berlin

Wulff F, Rahm L, Hallin AK, Sandberg J (2001) A nutrient budget model of the Baltic Sea. In: Wulff F, Rahm L, Larsson P (eds) A Systems Analysis of the Baltic Sea, Ecological studies 148, Springer-Verlag, Berlin, pp 353–372

Zehr JP, Jenkins BD, Short SM, Steward GF (2003) Nitrogenase gene diversity and microbial community structure: a cross-system comparison. Environ Microbiol 5: 539–554

Zhou JL, Fileman TW, Evans S et al (1999) The partition of fluoranthene and pyrene between suspended particles and dissolved phase in the Humber Estuary: a study of the controlling factors. Sci Total Environ 244: 305–321

Appendix

A.1 The Process-Based Mass-Balance Model, CoastMab

A.1.1 Introduction and Aim

During the last 10 years, there has been something of "a revolution" in predictive aquatic ecosystem modeling. A major reason for this development is the Chernobyl accident. To follow the pulse of radionuclides through ecosystem pathways has meant that important transport routes have been revealed and algorithms to quantify them developed and tested (Håkanson 2000). It is important to stress that many of those equations are valid not just for radionuclides, but for most types of contaminants, e.g., metals, nutrients and organics in most types of aquatic environments (coastal areas, rivers and lakes).

There exist several different kinds of models for phosphorus in lakes (Vollenweider 1968, 1976; Dillon and Rigler 1974, 1975; Chapra and Reckhow 1979, 1983; OECD 1982; Bryhn and Håkanson 2007). The most commonly used model today for lakes is probably the OECD-model. The model presented here is, to the best of our knowledge, the first *general* dynamic mass-balance model for phosphorus applicable for entire coastal areas (the ecosystem scale). However, for radionuclides in coastal areas several distributed 2D or 3D-models have been presented and those models are often based on partial differential equations (e.g., Monte et al. 2006). Unlike the model presented here, those distributed models are mainly designed to handle short-term (hours to days) spatial variations and they are driven by online meteorological data (winds, temperature and precipitation) and cannot be used for predictions over longer periods than 2–3 days since it is not possible to make reliable weather forecasts for longer periods that that. Such models may be excellent tools in science and may give descriptive power rather than long-term predictive power at the ecosystem scale. There are also different types of ecosystem-oriented models for nutrients and organic toxins in coastal areas (see, e.g., Wulff et al. 2001a; Savchuk and Wulff 2007). However, there are major differences between the model discussed here and other models related to differences in target variables (from conditions at individual sites to mean values over larger areas), modeling scales (daily to annual predictions), modeling structures (from using empirical/regression models to the use of ordinary or partial differential equations) and driving variables (whether accessed from standard monitoring programs, climatological measurements or specific studies). To make meaningful model comparisons is not a simple matter, and this is not the focus here.

The aim with this section is to present a model (CoastMab, from Håkanson and Eklund 2007b) that is process-based in the sense that it should handle all the important factors regulating the concentration of the target variable (phosphorus) in a defined coastal area. CoastMab accounts for, e.g., point source emissions, freshwater input, surface and deep-water exchange processes, land uplift, internal loading and mixing in a general manner designed to achieve practical utility and monthly variations. Also the fundamental unit, the defined coastal area, is determined in a way that, to the best of our knowledge, is not used by other groups in dynamic coastal modeling. This approach (the topographical bottleneck method, see Chap. 3 and Pilesjö et al. 1991) also makes it possible to estimate the theoretical surface-water and deep-water retention times (which are fundamental components in coastal mass-balance modeling) in bays and semi-enclosed coastal areas from morphometric information received from bathymetric maps.

In this section, we will first briefly present the studied areas and the utilized data, and then go through the set-up of the modeling (and sub-models connected to it). The different parts of the dynamic model will be described and all the fluxes explained. We will demonstrate how well the model works when tested for the studied coastal areas, in terms of predicting the target variable (the TP-concentration in the water) and associated bioindicators for coastal eutrophication. There will also be an emphasis on the different fluxes included in the model, which have been ranked and investigated by means of sensitivity analyses.

A.1.2 Data and Methods

A.1.2.1 Studied Areas and Utilized Data

Data from 21 areas located in three different archipelagos in the Baltic Sea have been used. Five of the areas are at the Swedish east coast (SE), seven areas in the south of Sweden (SS), six areas south west of Finland (F), one is a harbour area of Åland (Ål), and two areas are located in the Bothnian Sea (NS). Table A.1 gives a compilation of the data the areas: area code, latitude, land uplift, coastal area, maximum depth, mean depth, section area (between the defined coast and the adjacent sea), chlorophyll-a concentration, salinity, fish farm production in the given coastal area, sedimentation in deep water and in surface water, TP-concentration and Secchi depth in the sea.

Water samples were generally taken at about 3 sites in all sub-areas. In large and/or irregular areas, the number of sites was 3–6. Assemblages of sediment traps

Table A.1 Data for the 21 studied Baltic coastal areas (most data come from Wallin et al. 1992)

Area	Code	Latitude	Land uplift	Water area	D_{max}	D_m	At	Chl	Salinity	Fish prod.	Sed_{DW}	Sed_{SW}	C_{TPsea}	Sec_{sea}
		(°N)	(mm/y)	(km^2)	(m)	(m)	(km^2)	(µg/l)		(t/y)	(g dw/m^2·d)		(µg/l)	(m)
Matvik	SS1	56	0	3.1	14.3	5.2	0.0067	1.4	6.5	135	12.7	4.2	20	5
Bokö-fjärd	SS2	56	0	6.8	21.6	7.1	0.0141	1.4	7.2	70	6.7	1.6	20	5
Tärnö	SS3	56	0	1.5	11.1	5.1	0.0062	1.6	7.2	50	7.6	3.1	20	5
Guavik	SS4	56	0	2.7	22.8	5.2	0.0074	2.0	7.3	50	6.4	1.9	21	5
Järnavik	SS5	56	0	3.4	18.6	5.7	0.0081	1.3	7.3	10	8.1	1.5	21	5
Spjutsö	SS6	56	0	3.5	15.6	5.8	0.0188	0.9	7.4	50	8.7	4.4	21	5
Ronneby (†)	SS7	56	0	11.0	17.6	4.3	0.0176	2.1	6.5	100	20.2	4.1	21	4
Lilla Rimmö	SE1	58	2	2.5	17.6	8.3	0.0172	2.3	6.4	41	20.2	4.2	20	3
Eknön	SE2	58	2	14.0	19.5	8.5	0.0168	3.5	5.4	32	5.3	2.0	22	2
Lagnöstr.	SE3	58	2	5.3	20.1	3.8	0.0032	4.6	6.4	125	22.7	11.1	22	2
Gräsmarö	SE4	58	2	13.8	46.9	13.8	0.0825	2.6	6.6	200	18.3	1.1	18	3
Ålön	SE5	58	2	6.2	35.2	8.0	0.0162	2.1	6.6	300	9.5	3.7	18	4
Gävle (†)	NS1	61	6	17.1	17.0	8.4	0.0063	3.5	4.2	(1)	–	–	17	3
Gårdfj. (†)	NS2	61	7.5	0.9	18.0	6.0	0.0010	1.6	3.9	(2)	–	–	15.9	3
Svibyviken (†)	Ål1	60	7.5	3.2	22.2	7.8	0.0066	4.3(*)	6.0	(3)	–	–	20	3.2
Särkensalmi	F1	61	5	3.7	38.0	11.3	0.0145	4.2	5.2	78	67.8	15.5	24	2

Table A.1 (continued)

Area	Code	Latitude	Land uplift	Water area	D_{max}	D_m	At	Chl	Salinity	Fish prod.	Sed_{DW}	Sed_{SW}	C_{TPsea}	Sec_{sea}
		(°N)	(mm/y)	(km^2)	(m)	(m)	(km^2)	(µg/l)		(t/y)	(g $dw/m^2 \cdot d$)		(µg/l)	(m)
Käldö	F2	61	5	3.0	16.7	7.6	0.0040	2.7	6.5	50	20.0	11.0	24	2
Haverö	F3	61	5	2.3	22.5	8.6	0.0172	2.1	6.5	22	38.7	9.2	24	2
Hämmäröns.	F4	61	5	2.2	19.3	7.9	0.0114	2.2	6	381	13.4	9.1	24	3
Laitsalmi	F5	61	5	4.2	18.5	7.6	0.0080	2.7	6	85	56.4	17.6	24	1
Kaukolanlahti	F6	61	5	1.4	13.3	4.8	0.0006	9.6(*)	6.5	35	26.3	15.3	24	1
Min.		56	0	0.9	11.1	3.8	0.0006	0.9	3.9	10	5.3	1.1	15.9	1
Max		61	7.5	17.1	46.9	13.8	0.0825	4.6	7.4	381	67.8	17.6	24	5.5
Mean (MV)		58.6	2.9	5.3	21.3	7.2	0.0135	2.4	6.4	100.8	20.5	6.7	21.0	3.2

(1) 50.5 t/y TP from point sources (Karlsson and Håkanson 2001).
(2) 31.5 t/y TP from point sources (Håkanson and Karlsson 2004).
(3) 0.89 t/y TP from point source.
(†) these coastal areas are estuaries with a tributary inflow of phosphorus and SPM accounted for in the calculations.
(*) these empirical data are too high by a factor of 2 as compared to the modeled values which relate to the median conditions in the entire surface-water volume during the growing season.

(generally 2–3) were deployed in all areas at the beginning of every sampling period and collected at the end of the sampling periods (one week periods). Water samples for nitrogen, phosphorus, chlorophyll-a and oxygen analyses were collected 2–3 times per sampling period. Simultaneously, measurements of temperature, salinity and Secchi depth were done. Water samples were collected from the surface water (3 m depth) and from the deep water (1 m above the bottom). The thermocline was generally at about 10 m depth. Water for nitrogen and phosphorus analyses were preserved by freezing and thawed rapidly before making standard N and P analyses in the laboratory. Water for the chlorophyll-a analyses were preserved with $MgCO_3$ and filtrated aboard the ship on Whatman GF/C-filters. Chlorophyll-a was later analyzed in the laboratory after extraction with acetone by spectrophotometric methods (Jeffrey and Humphrey 1975). The Secchi depth was measured on the shady side of the boat with a white-painted disk with a diameter of 25 cm. Most data in Table A.1 emanate from sampling carried out during the summers of 1986, 1987 and 1988.

The data from Åland (the harbour area of Mariehamn) emanate from monitoring done during 1997, 1999–2000 and 2002–2003. Samples were generally taken from surface water at 2 stations within the bay. The frequency of the data differs between and within years. Generally, data are missing from December to April. From the two latter years, there are between 4 and 5 data per station from each of the remaining months, whereas from the earlier years the monitoring was less frequent with between 1 and 3 data from each month.

The data on sedimentation come from a study where sediment traps where placed in the deep water (Sed_{DW}) and the surface water (Sed_{SW}) (Wallin et al. 1992; Wallin and Håkanson 1991). Sediment trap techniques are described in Håkanson (1984a) and Håkanson et al. (1989). Each trap constitutes of two cylinders with a diameter of 5 cm and a height of 30 cm. The cylinders were placed in a mooring system that keeps the cylinders in a stable, vertical position throughout the collection period. One sediment trap was always placed 3 m below the water surface and one 1 m above the bottom. Depending on the stratification of the water mass, more sediment traps were sometimes used between these two depths. At each sampling event, trap water and settled material were poured from the cylinders into wide-neck 1-liter plastic bottles. In the laboratory, about 10 ml aliquotes were taken from the diluted, well-agitated water sample and sucked trough pre-combusted, pre-weighted 24 mm GF/C Whatman glass fiber filters. After drying at 80oC and weighting, the filters were combusted in a Carlo Erba 1106 CNH Elementary analyzer for analyses of carbon and nitrogen content. The data used here are the mean values for July, August and September.

Several of the areas included in this study were monitored during a project where the impact of fish cage farming emissions was studied (Wallin et al. 1992). "Fish production" in Table A.1 is the annual fish production in each area. This information is needed to calculate the point source emissions of total phosphorus from the farms, the only point source for nutrients in these areas. The studied areas are all located in the Baltic Sea, but they differ in, e.g., morphometry, local discharges of nutrients, salinity and land uplift. In this modeling, the purpose is to describe general conditions and therefore, instead of using time series for the variables of interest, we will use one representative median value for the growing season for each coastal area calculated from all available data.

A.1.2.2 Basic Concepts and Regressions in this Modeling

The basic structure of the CoastMab-model for TP comes from the CoastMab-model for SPM (suspended particulate matter) presented by Håkanson (2006; which is a modified version of the coastal model for SPM presented by Håkanson et al. 2004). The SPM-model has been calibrated and validated earlier with good results, which is shown in Fig. A.1. This figure is meant to demonstrate that the modeling is general and that it has been tested and that it works very well. The r^2-value when empirical data on sedimentation (from sediment traps) were compared to modeled values in 17

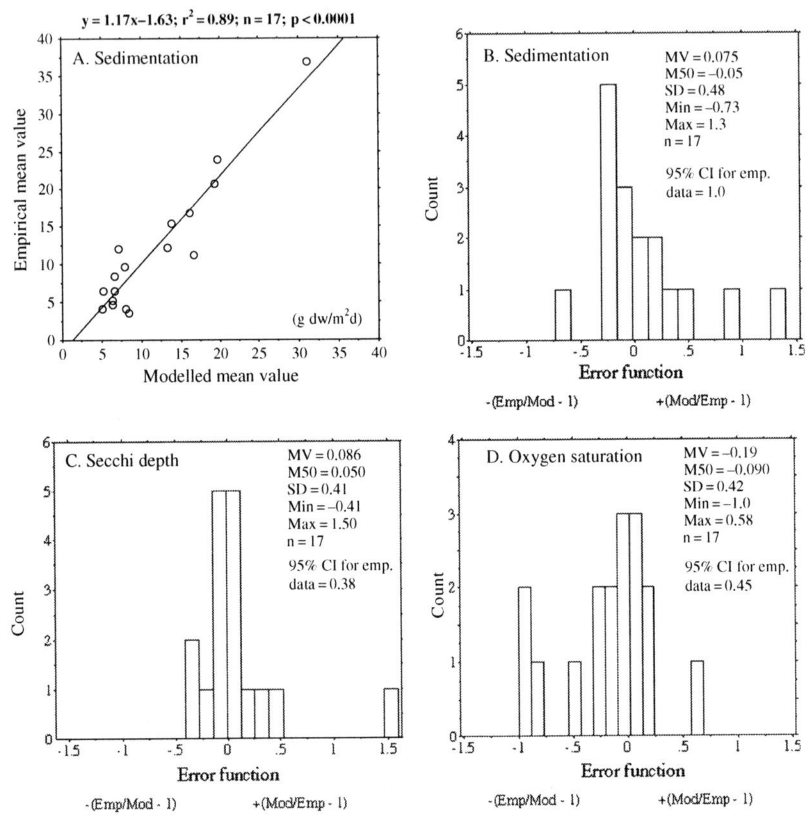

Fig. A.1 Compilation of validation results for (**A**) Sedimentation (the figure gives the regression between empirical data and modeled values; regression line, r^2, n and p); (**B**) the error function and statistics for sedimentation; (**C**) the error function and statistics for Secchi depth; and (**D**) the error function and statistics for the oxygen saturation in the deep-water zone (modified from Håkanson 2006)

Baltic coastal areas was 0.89 (Fig. A.1A) and it is not really possible to obtain much better predictions due to limitations and uncertainties in the available empirical data.

The error function is shown in Fig. A.1B; the mean error (MV) is 0.075 and the median error (M50) of –0.05; the standard deviation is 0.48, which should be compared to the 95% confidence interval for the uncertainty in the empirical data used to measure sedimentation, which is 1.0. The SPM-model also calculates the Secchi depth (as a measure of water clarity; see Chap. 5) and the oxygen saturation in the deep water (O_2Sat). These results are shown in Fig. A.1C and D. It should, however, be noted that O_2Sat is not predicted dynamically (causally) but by means of regression models (see Table A.2).

The SPM-model has been calibrated and validated with data from several coastal areas and is meant to have a wide range of applicability. This is also requested of the TP-model. Table A.1 gives information about several parameters from the different studied areas and provides information about the domain for the TP-model. These parameters are also vital for the setup of the CoastMab-model for TP. Several parts of the CoastMab-model are general and apply for all substances and have been described and used in other contexts. These parts will not be discussed in detail here, but they concern:

Latitude, which is used to calculate surface and deep-water temperatures. The temperature sub-model has been presented by Ottosson and Abrahamsson (1998). It is well known that water (and air) temperature is governed by many complicated climatological relationships but in this approach only information on latitude needs to be supplied. Then, the monthly variability of both the surface and deep-water temperatures is predicted. Evidently, the modeled temperature values could be replaced by measured data, but in all the following calculations modeled data have been used. The temperature data are utilized to quantify stratification and mixing between the surface and the deep-water volumes.

In the Baltic Sea, land uplift is a major contributor of matter and nutrients. This has been discussed in many contexts (Voipio 1981) and the algorithm to quantify how land uplift influences the concentrations of SPM and TP have been presented by Håkanson et al. (2004), see Sect. A.2.

In mass-balance modeling, it is important to define the coastal area and it is crucial to use a technique that provides an ecologically meaningful and practically

Table A.2 Results of the stepwise multiple regression for the oxygen saturation in the deep-water zone (mean O_2Sat in the deep water during the growing season in %; data from Wallin et al. 1992). n = 23 Baltic coastal areas. y = log(101 – O_2Sat). Sed_{DW} = sedimentation in sediment traps placed in the deepest part of the coastal area (g dw/m^2·d); ET = the fraction of ET-areas; T_{DW} = the theoretical deep-water retention time (days); D_m = the mean depth (m)

Step	r^2	Model variable	Model
1	0.43	$x_1 = \log(Sed_{DW})$	$y = 0.925 \cdot x_1 + 0.132$
2	0.64	$x_2 = \sqrt{ET}$	$y = 0.974 \cdot x_1 - 0.185 \cdot x_2 + 1.72$
3	0.74	$x_3 = \log(1+T_{DW})$	$y = 0.866 \cdot x_1 - 0.151 \cdot x_2 + 0.244 \cdot x_3 + 1.39$
4	0.80	$X_4 = \sqrt{D_m}$	$y = 0.643 \cdot x_1 - 0.118 \cdot x_2 + 0.301 \cdot x_3 + 0.323 \cdot x_4 + 0.470$

useful definition. The approach used here assumes that the borderlines are drawn at the topographical bottlenecks so that the exposure of the coast from winds and waves from the open sea is minimized (see Chap. 3). The exposure and the section area affect the exchange of water between the coast and the sea and are included in regression models for the theoretical surface water and deep-water retention times, T_{SW} and T_{DW} (see Table 3.2). The morphometric parameters are also used to determine the coastal the form factor (Vd; see Chap. 3), which influences internal fluxes of TP.

The mean surface-water salinity influences aggregation of suspended particles and sedimentation. There is an empirically based sub-model to quantify this relationship between SPM, salinity and water clarity (see (5.1)). This regression may replace the dynamic SPM-model, but in all the following simulations, we have used the dynamic SPM-model as well as the dynamic TP-model. The median Secchi depth in the sea outside the coast (Sec_{sea} in m) is used to estimate the inflow of SPM from the sea to the studied coastal areas using this sub-model that relates SPM, salinity and Secchi depth. The inflow of SPM is calculated as, $Q_{SW} \cdot SPM_{sea}$ [$(m^3/month) \cdot (g/m^3) = g/month$].

Sub-models to calculate the emissions of SPM and TP from fish farming have been presented and discussed by Håkanson et al. (2004).

This modeling first calculates the TP-concentration in the water dynamically with the CoastMab-model (described in next section) and then uses regression models to predict three indicators for eutrophication:

1. The concentration of chlorophyll-a (Chl in μg/l). Empirical models to predict chlorophyll were discussed in Chap. 5. The model in Fig. 5.6 is based on empirical median TP-concentrations from the *growing season*. Chl is predicted here from the dynamically modeled TP-concentrations using the approach in Fig. 5.6 as the first step. For the prediction of *monthly* Chl from TN, empirical TN values are utilized, since TN is not modeled dynamically. Chl is calculated in the following manner from the TN-concentrations (and the same approach is used to relate dynamically modeled TP-concentrations to chlorophyll):

$$\mathrm{Chl} = ((\mathrm{SWT}+0.1)/20) \cdot 10^{\wedge}(2.115 \cdot \log(\mathrm{TN}) - 4.888) \qquad \text{(A.1)}$$

   Where ((SWT + 0.1)/20) is a dimensionless moderator based on the ratio between median monthly surface-water temperatures (SWT) divided by a reference surface water temperature 20○C for the growing season. The constant 0.1 is added since there is also primary production in the winter if SWT approaches zero. This expression will give a simple seasonal variability pattern to the chlorophyll values using the two regressions (Chl vs TN and Chl vs TP).
2. The Secchi depth (Sec in m), which is also used to predict SPM (mg/l); using the approach illustrated in Fig. 5.1; SPM is used in the dynamic model, where it influences the sedimentation of particulate phosphorus.
3. The oxygen saturation in the deep-water zone (O_2Sat in %; see Table A.2).

The data used for all these empirical models emanate from Wallin et al. (1992); Persson et al. (1994); Persson and Håkanson (1996) and Nordvarg (2001). Statistical methods (such as transformations and regressions) are discussed in Chap. 4 and will not be further elaborated in this section.

A.1.3 The CoastMab-Model for TP

The focus in this section is on the dynamic TP-model, but the dynamic SPM-model is also used to quantify sedimentation which is needed in the TP model, since sedimentation affects the oxygen conditions in the deep-water zone and thus also the diffusion of TP from sediments. The SPM-model also calculates the production of SPM in the coastal area (from Chl) and it also includes mineralization (i.e., bacterial decomposition of the organic fraction of SPM). The SPM-model has been presented before (Håkanson 2006) and the idea here is not repeat that information but to highlight the specific features and structures of the mass-balance model for TP.

The sub-model for phosphorus includes certain substance-specific parts:

1. The particulate fraction, i.e., the ratio between the particulate phase of phosphorus to the total concentration (PF = PP/TP). This is basically related to an equilibrium constant between dissolved (bioavailable) and particulate phosphorus (subject to gravitational sedimentation) and in this monthly modeling for coastal areas in the Baltic Proper, we set the PF to 0.56, as motivated by the data in Fig. A.2.
2. Diffusion, i.e., the diffusive transport of dissolved phosphorus from the sediments triggered mainly by low oxygen concentrations in the deep-water zone.

A.1.3.1 Basic Model Structure

This modeling uses ordinary differential equations and the temporal resolution is one month to reflect seasonal variations. In the basic model, there are four main compartments: surface water, deep water, areas where processes of fine sediment erosion and transport dominate the bottom dynamic conditions (ET-areas) and areas with continuous sedimentation of fine particles, the accumulation areas (A-areas). The inflow of TP to any given coastal area is handled by the following six fluxes:

1. Inflow of TP to coastal surface water from the sea (F_{inSW}).
2. Inflow of TP to the deep water from the sea (F_{inDW}).
3. Land uplift (F_{LU}), see Sect. A.2.
4. Emissions of TP from point sources (F_{PSSW}), in this case from fish cage farms. The sub-model for these emissions of TP is given in Håkanson (1999).
5. Tributary inflow (F_{inQ}).
6. Direct atmospheric fallout (F_{prec}).

The internal processes are: sedimentation from surface water to deep water (F_{SWDW}) and to areas of erosion and transportation (F_{SWET}), resuspension from ET-areas either back to surface water (F_{ETSW}) or to deep water (F_{ETDW}), sedimentation from

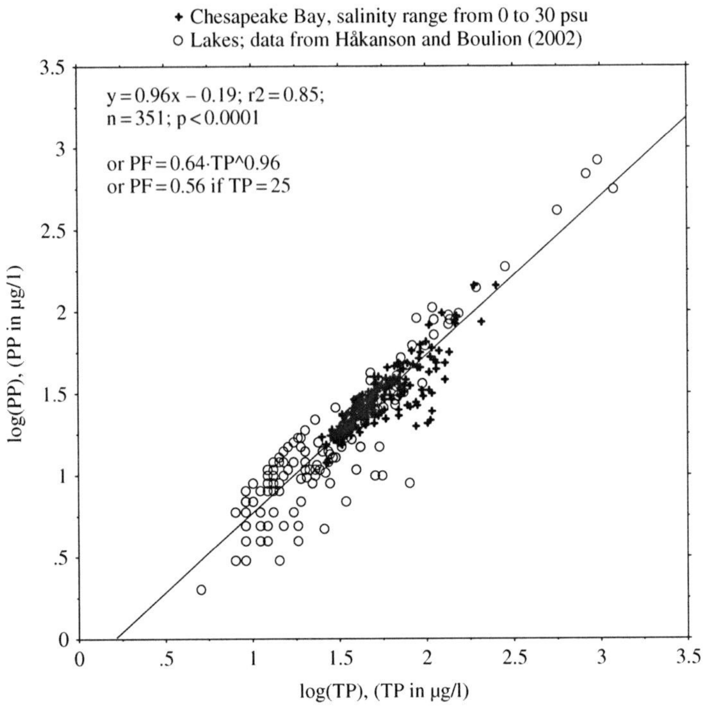

Fig. A.2 The relationship between empirical PP (particulate phosphorus; logarithmic values; PP in mg/m^3) and empirical TP (total-P; logarithmic values in mg/m^3). The figure also gives the regression line based on individual data (n = 351) for systems covering a salinity range from 0 to 30 psu. The data from Chesapeake Bay are median values from the surface-water layer from 1984 to 2006

deep water on accumulation areas (F_{DWA}), diffusion of phosphorus from accumulation area sediments (A-sediments) to deep water (F_{ADW}) and upward and downward mixing between the surface water and deep water compartments i.e., the transport from deep water to surface water (F_{DWSWx}) and from surface water to deep water (F_{SWDWx}). The transport from a coastal area is regulated by the outflow from surface water and deep water to the adjacent sea (F_{outSW} and F_{outDW}). The structure of the basic model is shown in Fig. A.3 and all equations are compiled in Table A.3.

When there is a partitioning of a flow from one compartment to two or more compartments, this is handled by a distribution (= partitioning = partition) coefficient. This could be a default value, a value derived from a simple equation or from an extensive sub-model. There are four such distribution coefficients (DCs) in the basic TP-model:

1. The DC regulating the amount in particulate and dissolved fraction (see Fig. A.2).
2. The DC regulating sedimentation either to areas of erosion and transport (ET-areas) above the theoretical wave base (F_{SWET}) or to the deep-water areas beneath the theoretical wave base (F_{SWDW}, see the ETA-diagram in Fig. 3.15).

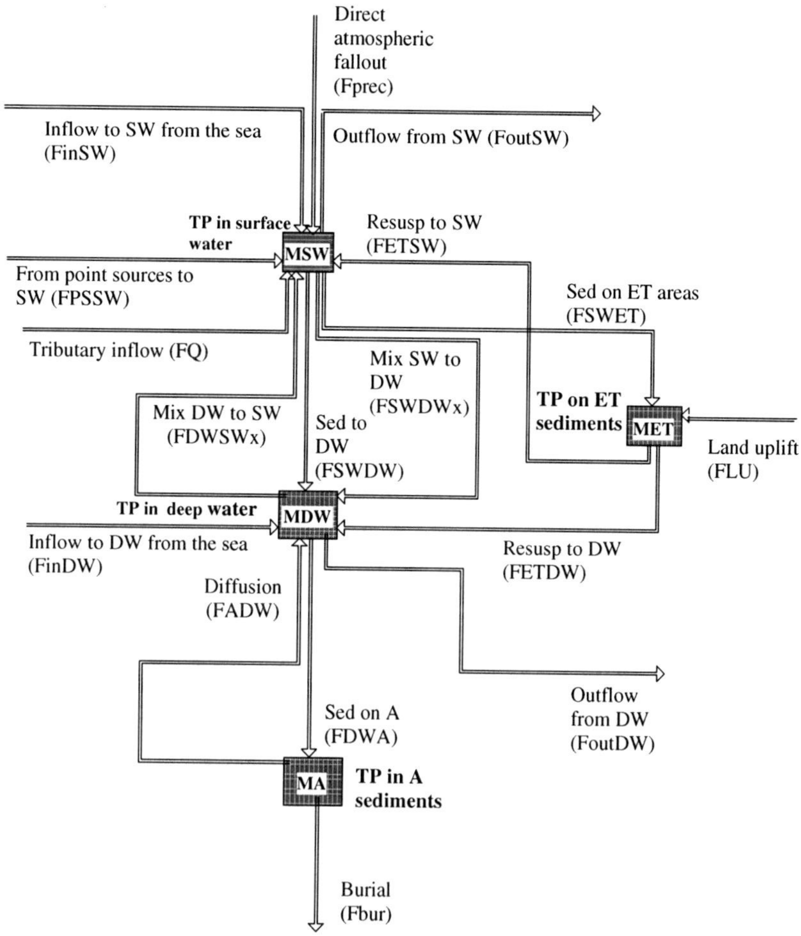

Fig. A.3 A general outline of the coastal model for phosphorus. Note that for simplicity point source emissions to the deep-water compartment have been omitted

3. The DC describing resuspension flux from ET-areas back either to the surface water (F_{ETSW}) or to the deep-water compartment (F_{ETDW}).
4. The DC describing how much of the TP in the water that has been resuspended (DC_{res}) and how much that has never been deposited and resuspended ($1\text{-}DC_{res}$).

A.1.3.2 Determination of the Different Compartments

From a mass-balance perspective, it is necessary that the four compartments (surface water, deep water, ET-areas and A-areas) included in the CoastMab-model are defined in a relevant manner.

Table A.3 A compilation of the differential equations for the basic dynamical coastal model for phosphorus, CoastWeb (excluding the differential equation for uptake and retention of phosphorus in biota, which is given in (A.33))

Surface water (SW):

$M_{SW}(t) = M_{SW}(t - dt) + (F_{inSW} + F_{prec} + F_{inQ} + F_{PS} + F_{farmSW} + F_{ETSW} + F_{DWSWx} - F_{outSW} - F_{SWDW} - F_{SWDWx} - F_{SWET})\cdot dt$

$F_{inSW} = 0.001\cdot C_{TPsea}\cdot V_{SW}/T_{SW}$ [data on C_{TPsea} in Table 1; V_{SW} = SW volume; T_{SW} = theoretical SW retention time; inflow to SW from the sea]

$F_{prec} = Prec\cdot Area\cdot C_{TPprec}\cdot 0.001\cdot 0.001/12$ [Prec = annual precipitation in mm; $C_{TPprec} = 5\ \mu g/l$; direct fallout from precipitation]

$F_{inQ} = 60\cdot 60\cdot 24\cdot 30\cdot C_{TPin}\cdot Q$ [C_{TPin} = conc. in river water; Q = mean monthly river water discharge; river inflow]

F_{PS} = [TP from other point sources = 0 in these simulations]

$F_{farmSW} = 1000\cdot DC_{farm}\cdot(Y_{em})\cdot 0.01\cdot((AFP\cdot FCR\cdot C_{TPfeed})-(AFP\cdot C_{TPfish}))$ [TP emissions from fish farm]

$F_{ETSW} = M_{ET}\cdot(1/T_{ET})\cdot(1-Vd/3)$ [Vd = the form factor = $3\cdot D_m/D_{max}$; resuspension ET to SW]

$F_{DWSWx} = M_{DW}\cdot R_{mix}\cdot V_{SW}/V_{DW}$ [V_{SW} = SW-volume; V_{DW} = DW-volume; mixing DW to SW]

$F_{outSW} = Q\cdot C_{TPSW}\cdot 0.001+M_{SW}\cdot(1/T_{SW})$ [outflow from SW]

$F_{SWDW} = M_{SW}(1-DF)\cdot(v_{SW}/D_{SW})\cdot(1-ET)\cdot(1\cdot(1-DC_{resSW})+Y_{res}\cdot DC_{resSW})$ [sedimentation SW to DW]

$F_{SWDWx} = M_{SW}\cdot R_{mix}$ [mixing SW to DW]

$F_{SWET} = M_{SW}\cdot(1-DF)\cdot(v_{SW}/D_{SW})\cdot ET\cdot(1\cdot(1-DC_{resSW})+Y_{res}\cdot DC_{resSW})$ [sedimentation SW to ET]

Deep water (DW):

$M_{DW}(t) = M_{DW}(t - dt) + (F_{SWDW} + F_{ETDW} + F_{ADW} + F_{inDW} + F_{SWDWx} + F_{farmDW} - F_{DWSWx} - F_{DWA} - F_{outDW})\cdot dt$

$F_{SWDW} = M_{SW}(1-DF)\cdot(v_{SW}/D_{SW})\cdot(1-ET)\cdot(1\cdot(1-DC_{resSW})+Y_{res}\cdot DC_{resSW})$ [sedimentation SW to DW]

$F_{ETDW} = M_{ET}\cdot(1/T_{ET})\cdot(Vd/3)$ [Vd = the form factor = $3\cdot D_m/D_{max}$; resuspension ET to DW]

$F_{ADW} = M_A\cdot R_{diff}$ [diffusion]

$F_{inDW} = 0.001\cdot(C_{TPsea}\cdot 1.25)\cdot Q_{DW}$ [data on C_{TPsea} in Table 1; C_{TP} set 25% in DW than in SW; inflow of TP to DW from the sea]

$F_{SWDWx} = M_{SW}\cdot R_{mix}$ [mixing SW to DW]

$F_{farmDW} = (1-DC_{farm})\cdot F_{farmSW}$ [$DC_{farm} = 0.5$; TP-emissions to DW from fish farm]

$F_{DWSWx} = M_{DW}\cdot R_{mix}\cdot V_{SW}/V_{DW}$ [V_{SW} = SW-volume; V_{DW} = DW-volume; mixing DW to SW]

$F_{DWA} = M_{DW}\cdot(v_{DW}/D_{DW})\cdot(1-DF)\cdot Y_T\cdot(1\cdot(1-DC_{resDW})+Y_{res}\cdot DC_{resDW})$ [sedimentation DW to A]

$F_{outDW} = M_{DW}\cdot(1/T_{DW})$ [outflow from DW to sea]

ET-areas (ET):

$M_{ET}(t) = M_{ET}(t - dt) + (F_{LU} + F_{SWET} - F_{ETDW} - F_{ETSW})\cdot dt$

F_{LU} = [TP from land uplift]

$F_{SWET} = M_{SW}\cdot(1-DF)\cdot(v_{SW}/D_{SW})\cdot ET\cdot(1\cdot(1-DC_{resSW})+Y_{res}\cdot DC_{resSW})$ [sedimentation SW to ET]

$F_{ETDW} = M_{ET}\cdot(1/T_{ET})\cdot(Vd/3)$ [Vd = the form factor = $3\cdot D_m/D_{max}$; resuspension ET to DW]

$F_{ETSW} = M_{ET}\cdot(1/T_{ET})\cdot(1-Vd/3)$ [Vd = the form factor = $3\cdot D_m/D_{max}$; resuspension ET to SW]

Table A.3 (continued)

A-areas (***A***):

$M_A(t) = M_A(t - dt) + (F_{DWA} - F_{bur} - F_{ADW})\cdot dt$

$F_{DWA} = M_{DW}\cdot(v_{DW}/D_{DW})\cdot(1-DF)\cdot Y_T\cdot(1\cdot(1-DC_{resDW})+Y_{res}\cdot DC_{resDW})$ [sedimentation DW to A]

F_{bur} = if $T_A > 48$ then $M_A\cdot(1/48)$ else $M_A\cdot(1/T_A)$ [T_A = the age of active A-sediments]

$F_{ADW} = M_A\cdot R_{diff}$ [diffusion; R_{diff} = the diffusion rate]

$DF = 1 - PF = 1 - 0.56 = 0.44$

$D_{DW} = (D_{max}-D_{wb})/2$ [D_{wb}]

$D_{SW} = D_{wb}/2$

$DC_{resSW} = F_{ETSW}/(F_{ETSW} + F_{prec} + F_{inSW} + F_{PS} + F_{inQ} + F_{farmSW})$

R_{mix} = if ABS(SWT-DWT) < 4°C then $R_{mix} = 1$ else R_{mix} = 1/ABS(SWT-DWT)

$R_{diff} = 0.0003/12\cdot Y_{O2}$; if $O_2Sat > 50\%$, $Y_{O2} = (2-1\cdot(O_2Sat/50-1))$ else $Y_{O2} = (2-(C_{TPsed}/1)\cdot 3000\cdot(O_2Sat/50-1))$

$Sec_{SW} = 10\text{^}(-(z+0.5)\cdot(\log(SPM_{SW})+0.3)/2+z)$; $z = (10\text{^}(0.15\cdot\log(1+Sal_{SW})+0.3)-1)$

$SPM_{sea} = 10\text{^}(-0.3-2\cdot(\log(Sec_{sea})-(10\text{^}(0.15\cdot\log(1+Sal_{sea})+0.3)-1))/((10\text{^}(0.15\cdot\log(1+Sal_{sea})+0.3)-1)+0.5))$

$V_{DW} = Area_A\cdot(D_{max}-D_{wb})\cdot Vd/3$

$V_{SW} = (V-V_{DW})$

$v_{def} = 6$ m/month

$v_{DW} = (v_{def})\cdot Y_{SPMDW}\cdot Y_{salDW}\cdot Y_{DR}\cdot Y_{DW}\cdot Y_{ZMT}\cdot((1 - DC_{resDW}) + Y_{res}\cdot DC_{resDW})$

$v_{SW} = (v_{def})\cdot Y_{SPMSW}\cdot Y_{salSW}\cdot Y_{DR}\cdot Y_{ZMT}\cdot((1 - DC_{resSW}) + Y_{res}\cdot DC_{resSW})$

Y_{ZMT} = If $Q > Q_{sea}$ then $Y_{ZMT} = (Sal_{sea}/Sal_{SW})\cdot(Q_{sea}+Q)/Q)$ else $Y_{ZMT} = (Sal_{sea}/Sal_{SW})\cdot(Q_{sea}+Q)/ Q_{sea})$ [Q-values in m^3/month; calculates sedimentation effects related to the "zone of maximum turbidity"]

$Y_{SPMSW} = (1 + 0.75\cdot(C_{SW}/50 - 1))$ [calculates how changes in SPM (C_{SW}) influences sedimentation]

$Y_{SPMDW} = (1 + 0.75\cdot(C_{DW}/50 - 1))$ [calculates how changes in SPM (C_{DW}) influences sedimentation]

$Y_{salSW} = (1 + 1\cdot(Sal_{SW}/1 - 1) = 1\cdot Sal_{SW}/1$ [calculates how changes in salinities > 1 psu influence sedimentation]

$Y_{salDW} = (1 + 1\cdot((Sal_{SW} + 3)/1-1))$ [calculates how changes in deep-water salinities influence sedimentation]

Y_{DR} = If DR < 0.25 then 1 else 0.25/DR [calculates how changes in DR/turbulence influence sedimentation]

$Y_{res} = (((T_{ET}/1) + 1)^{0.5}$[calculates how much faster resuspended sediments settle out]

$Y_{ET} = 0.99/ET$ [calculates how changes in ET among systems influence mineralization]

Y_{DW} = If $T_{DW} < 7$ (days), $Y_{DW} = 1$ else $Y_{DW} = (T_{DW}/7)^{0.5}$ [calculates how changes in deep-water turbulence influence deep-water sedimentation]

$d = 100\cdot 2.6/(100+(W_A+IG_A)\cdot(2.6-1))$; W_A and IG_A in % wet weight.

The water depth that separates the surface-water and the deep-water compartment could potentially be related to (i) water temperature conditions and the thermocline, (ii) vertical concentration gradients of dissolved or suspended particles, (iii) wind/wave influences and wave characteristics and (iv) sedimentological conditions associated with resuspension and internal loading (Håkanson et al. 2004). In this work, the separation is done by sedimentological criteria and the two volumes are separated by the theoretical wave base, D_{wb} (see Fig. 3.15). By definition, D_{wb} also determines the limit between ET- and A-areas. D_{wb} is calculated from the coastal area (note that the area should be in km^2 in (A.2), giving D_{wb} in m), which is related to the effective fetch and how winds and waves influence the bottom dynamic conditions:

$$D_{wb} = (45.7 \cdot \sqrt{\text{Area}})/(21.4 + \sqrt{\text{Area}}) \tag{A.2}$$

This algorithm is valid for closed lagoons and lakes. For coastal areas, the wind/wave impact from the sea lowers the theoretical wave base, as compared to closed lagoons. The larger the exposure (Ex = 100·At/Area), the larger the potential energy impact from the sea (Persson and Håkanson 1996) and the deeper the theoretical wave base (D_{wb}). This means that open coasts have larger areas above D_{wb} than enclosed coastal areas, if all else is constant. This is accounted for by the following approach using a dimensionless moderator, Y_{Ex}:

$$\begin{aligned} \text{If Ex} < 0.003 \text{ then } Y_{Ex1} &= 1 \text{ else } Y_{Ex1} = (\text{Ex}/0.003)^{0.25} \\ \text{If Ex} > 10 \text{ then } Y_{Ex2} &= 10 \text{ else } Y_{Ex2} = Y_{Ex1} \end{aligned} \tag{A.3}$$

The value for Y_{Ex2} is multiplied by the value for D_{wb}. This means that when the exposure varies between 0.003 and 10, D_{wb} is increased by the factor $(\text{Ex}/0.003)^{0.25}$. That is, if Ex = 0.01, the factor is 1.35 and D_{wb} likely at a water depth of $1.35 \cdot D_{wb}$ m rather than at D_{wb}.

There are several models to predict the ET-areas for coasts (Persson and Håkanson 1995), and some of those approaches require data on the filter factor (Ff) or the mean filter factor, i.e., how the conditions outside the defined coastal area (islands, etc.) work as an energy filter and reduce the impact of the waves from the sea (see Fig. 3.3). In this modeling, the filter factor has been omitted since it may be difficult to access reliable data on Ff. The accumulation areas are calculated from the depth of the theoretical wave base (A.2) as the area beneath the theoretical wave base. The fraction of ET-areas is used as a dimensionless distribution coefficient. It regulates the sedimentation of particulate TP either to deep-water areas or to ET-areas and hence also the amount of matter available for resuspension on ET-areas. For simplicity, this approach is used also when there is an ice cover, because the stratification is weak during the winter, primary production low and the error in predicting sedimentation of TP with this simplification small. ET generally varies from 0.15 (see Fig. 3.14), since there must always be a shallow shore zone where processes of erosion and transport dominate the bottom dynamic conditions, to 1 in large and shallow areas totally dominated by ET-areas, which is the situation in areas where $D_{wb} > D_{max}$. In this modeling, ET is, however, never permitted to become 1, since one can assume that in most coastal areas there are deep holes, sheltered areas or macrophyte beds which would function ecologically as A-areas. To estimate the fraction of ET-areas in such systems, the following expression is used to calculate a value for the theoretical wave base:

$$\text{if } D_{wb} > 0.95 \cdot D_{max} \text{ than } D_{wb} = 0.95 \cdot D_{max} \tag{A.4}$$

A.1.3.3 The Panel of Driving Variables

Table A.4 gives the panel of driving variables. These are variables needed to run the CoastMab-model for any given coastal area. No other parts of the model should be changed unless there are good reasons to do so.

A.1.3.4 Primary Inflows

Inflow of TP from the Sea

The inflow of TP to the surface water from the sea or adjacent coastal areas (F_{inSW}) is calculated from the surface water flow (Q_{SW}), which is calculated from the theoretical surface-water retention time (T_{SW}; see Table 3.2) and the concentration of TP outside the coast (C_{TPsea}; see Table A.1). This means that the inflow of TP from the sea is given by:

$$F_{inSW} = Q_{SW} \cdot C_{TPsea} = (V_{SW}/T_{SW}) \cdot C_{TPsea} \tag{A.5}$$

Where V_{SW} is the surface-water volume (in m^3).

Table A.4 Panel of driving variables

A. Parameters for morphometry and location:
1. Coastal area
2. Mean depth
3. Maximum depth
4. Section area
5. Latitude

B. Chemical variables:
6. Characteristic mean salinity outside and within the coastal area
7. Characteristic TP and SPM-concentrations in the sea outside the given coastal area (here the SPM-concentration in the sea outside the given coastal areas is predicted from the corresponding Secchi depth values)

C. Inflow variables:
8. Monthly emissions from point sources or tributaries
9. Land uplift

The deep-water inflow of TP from the sea (F_{inDW}) is quantified in the following way:

$$F_{inDW} = Q_{DW} \cdot (C_{TPsea} \cdot 1.25) = (V_{DW}/T_{DW}) \cdot (C_{TPsea} \cdot 1.25) \quad (A.6)$$

Q_{DW} = The deep-water flow (m^3/month), which is given by the ratio between the volume of the deep water (V_{DW} in m^3) and the theoretical deep-water retention time (T_{DW} in months; see Table 3.2), i.e., V_{DW}/T_{DW}.
C_{TPsea} = The TP-concentration in the surface water in the sea outside the coast. If empirical data are not at hand, the TP-concentration in the deep-water is set 25% higher than in the surface water (Håkanson 2006).

The exposure and the section area affect the exchange of water between the coast and the sea and are included in models for the theoretical surface water and deep-water retention times, T_{SW} and T_{DW} (Håkanson 2000; Håkanson et al. 2004; see Fig. A.4 and Table 3.2). Note that Fig. A.4 also gives brief overviews of calculation routines for coastal areas dominated by tides and by coastal currents.

Direct Atmospheric Fallout

The direct deposition (F_{prec}, g TP/month) is traditionally given by the annual precipitation multiplied by the TP-concentration in the rain and the coastal area and dimensional adjustments (Prec $0.001 \cdot \text{Area} \cdot C_{TPprec} \cdot 0.001 \cdot 1/12$ in (m/yr)$\cdot m^2 \cdot$(g TP/m^3)$\cdot$(1/month)). In all of the following calculations, we have set C_{TPprec} to 5 µg/l (see Håkanson 1999).

River Inflow

The inflow (F_{in} in g TP per month) to a coastal estuary from rivers is generally calculated from water discharge (Q) times the TP-concentration in the tributary (C_{in}), i.e.:

$$F_{in} = Q \cdot C_{in} \quad (A.7)$$

In this modeling, the inflow has been calculated from a river model for the four estuaries included in this study (see Table A.1; using the catchment model from Håkanson 2006) when empirical data have not been available.

A.1.3.5 Internal Processes

Sedimentation

Sedimentation of particulate TP depends on:

1. A default settling velocity, v_{def}. Here, 72 m/yr is used as a general default value for the complex mixture of substances making up SPM and the carrier particles

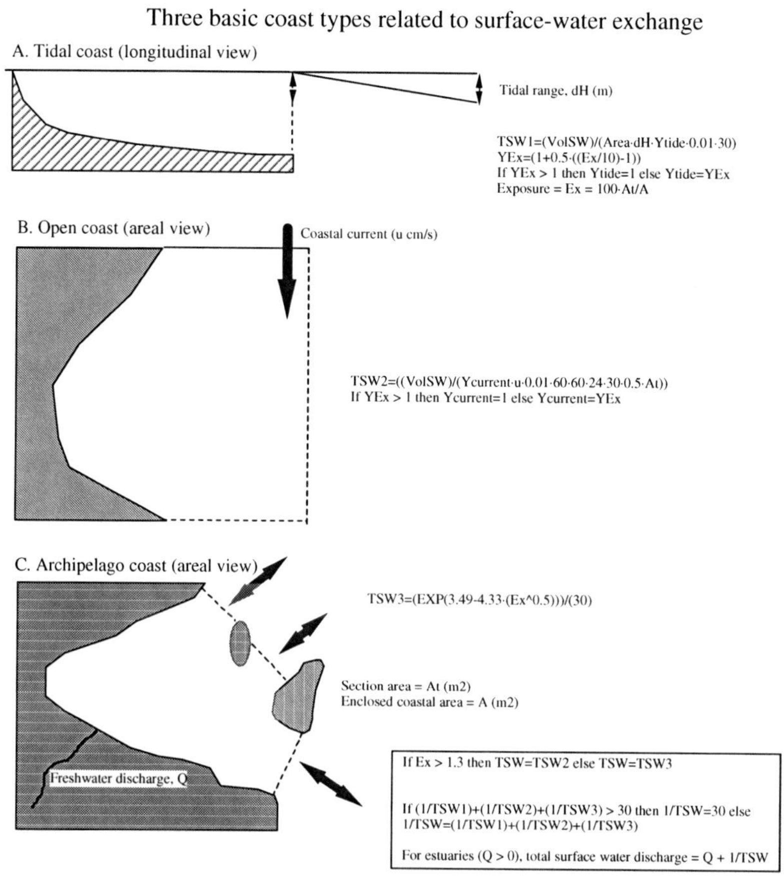

Fig. A.4 Illustration of the three main coast types (tidal coasts, open coasts and archipelagos) and how the theoretical surface water exchange (T_{SW}) may be estimated for each type (based on Håkanson 2000)

of particulate TP in coastal waters (from Håkanson 2006). Note that this default settling velocity is related to conditions in very turbid freshwater systems (SPM = 50 mg/l). The default settling velocity is changed into a rate (1/month) by division with the mean depth of the surface-water areas (D_{SW}) for sedimentation in these areas and by the mean depth of the deep-water areas (D_{DW}) for sedimentation in deep-water areas.

2. The SPM-concentration will also influence the settling velocity; the greater the aggregation of suspended particles, the bigger the flocs and the faster the settling velocity (Kranck 1973, 1979; Lick et al. 1992). This is expressed by a dimensionless moderator (Y_{SPM}; see (A.12)).
3. The higher the salinity, the greater the aggregation, the bigger the flocs and the faster the settling velocity (Kranck 1973, 1979). This is expressed by a dimensionless moderator for salinity (Y_{sal}) operating on the default settling velocity

(see (A.13)). The effect of salinity is of special importance in estuaries where fresh and salt water meet and the effect called "zone of maximum turbidity" occurs (Y_{ZMT}). This effect is also accounted for in this modeling.

4. The turbulence, which is very important for the fall velocity of suspended particles (Burban et al. 1989, 1990). Generally, there is more turbulence, which keeps the particles suspended, and hence causes lower settling rates, in the surface water than in the calmer deep water. The turbulence in the surface-water compartment is also generally greater in large and shallow systems (with high dynamic ratios, DR) compared to small and deep coasts. In this modeling, two dimensionless moderators (Y_{DR} and Y_{DW}; see (A.14) and (A.20)) related to the theoretical deep-water retention time and the dynamic ratio are used to quantify how turbulence is likely to influence the settling velocity in the surface-water and deep-water compartments.
5. The fraction of resuspended matter (DC_{res}). The resuspended particles have already been aggregated and they have also generally been influenced by benthic activities, which will create a "gluing effect", and they have a comparatively short distance to fall after being resuspended (Håkanson and Jansson 1983). The longer the particles have stayed on the bottoms, the larger the potential gluing effect and the faster the settling velocity if the particles are resuspended. The age of the particles on ET-areas is calculated by (A.18).

Sedimentation from surface water to ET-areas (F_{SWET}) and to the deep water (F_{SWDW}) is given by (A.8) and (A.9).

$$F_{SWET} = M_{SW} \cdot PF \cdot R_{SW} \cdot ET \tag{A.8}$$

$$F_{SWDW} = M_{SW} \cdot PF \cdot R_{SW} \cdot (1 - ET) \tag{A.9}$$

M_{SW} = the total mass of TP in the surface-water compartment (g);
PF = the particulate fraction of TP (PF = PP/TP = 0.56, see Fig. A.2);
R_{SW} = the sedimentation rate in the surface water (1/month);
ET = the fraction of ET-areas (ET = $Area_{ET}$/Area).

Sedimentation from deep-water areas on A-areas is given by:

$$F_{DW} = M_{DW} \cdot PF \cdot R_{DW} \tag{A.10}$$

The basic sedimentation rates for surface water and deep-water areas may be written as $R_{SW} = v_{SW}/D_{SW}$ and $R_{DW} = v_{DW}/D_{DW}$. The mean depths of the surface and deep-water areas, D_{SW} and D_{DW}. The monthly settling velocity for the surface-water compartment (v_{SW}) is calculated from the default settling velocity of v_{def} (m/yr) accordingly:

$$v_{SW} = (v_{def}/12) \cdot Y_{SPMSW} \cdot Y_{salSW} \cdot Y_{DR} \cdot Y_{ZMT} \cdot ((1 - DC_{resSW}) + Y_{res} \cdot DC_{resSW}) \tag{A.11}$$

v_{def} = 72 m/yr;
Y_{SPMSW} = the dimensionless moderator expressing how changes in SPM in the surface water influence v_{SW};

Y_{salSW} = the dimensionless moderator expressing how changes in salinity in the surface water influence v_{SW};
Y_{DR} = the dimensionless moderator expressing how changes in dynamic ratio (turbulence) influence v_{SW};
Y_{ZMT} = the dimensionless moderator expressing increased sedimentation in the zone of maximum turbidity;
Y_{res} = the dimensionless moderator expressing how changes in the age of the resuspendable ET-sediments influence the settling velocity;
DC_{resSW} = the resuspended fraction in the surface-water compartment.

Y_{SPMSW} is given by:

$$Y_{SPMSW} = (1 + 0.75 \cdot (SPM_{SW}/50 - 1)) \quad \text{(A.12)}$$

This dimensionless moderator quantifies how changes in SPM_{SW} (in mg/l) influence the fall velocity of the suspended particles. The amplitude value is calibrated in such a manner that a change in SPM_{SW} by a factor of 10, e.g., from 2 mg/l (which is a typical value for low-productive coastal systems) to 20 mg/l (which is typical for highly productive systems), will cause a change in the settling velocity by a factor of 2. The borderline value for the moderator is 50 mg/l, since it is unlikely that coastal areas will have higher mean monthly SPM-values than that. In this modeling, SPM has a default settling velocity of 72 m/yr in systems with SPM-values of 50 mg/l, and in systems with lower SPM-concentrations the fall velocity is lower, as expressed by (A.12).

The dimensionless moderator for salinity (Y_{sal} or Y_{salSW} or Y_{salDW}) is given by:

$$\text{If salinity} < 1 \text{ then } Y_{sal} = 1 \text{ else } Y_{sal} = (1+1) \cdot (Sal/1 - 1) = 1 \cdot Sal/1 = Sal \quad \text{(A.13)}$$

The norm-value of the moderator is 1 and the amplitude value is 1. This means that if the salinity (in psu) changes from 5 to 10, the moderator (Y_{sal}) changes from 5 to 10 and the settling velocity increases by a factor of 2.

In estuaries, where salt water meets fresh water, there is increased flocculation and sedimentation of suspended particles. This is quantified by the following algorithm:

$$\begin{aligned} &\text{If } Q > Q_{sea} (\text{values in } m^3/\text{month}) \text{ then} Y_{ZMT} = ((Sal_{sea}/Sal_{SW}) \cdot (Q_{sea} + Q)/Q) \\ &\text{else} Y_{ZMT} = ((Sal_{sea}/Sal_{SW}) \cdot (Q_{sea} + Q)/Q_{sea}) \quad \text{(A.14)} \end{aligned}$$

Where Y_{ZMT} is a dimensionless moderator quantifying how the default fall velocity v_{def} increases in estuaries depending on the fresh-water inflow (Q), the salt-water inflow (Q_{sea}), the salinity in the coastal area (Sal_{SW}, the surface water salinity) and the mean salinity in the sea outside the given coastal area (Sal_{sea}). The greater the difference in salinity between the sea and the coast, the higher the ratio Sal_{sea}/Sal_{SW} and the higher the flocculation and sedimentation. The higher the fresh-water inflow relative to the salt-water inflow, the higher the effects related to the "zone of maximum turbidity", ZMT, and the higher the value of Y_{ZMT}. $(Q_{sea}+Q)/Q$ or

$(Q_{sea}+Q)/Q_{sea}$ attain values between 1 and 2 and Sal_{sea}/Sal_{SW} may attain different values depending on the prevailing conditions; if Sal_{sea} is 10 and Sal_{SW} 5 and if the $(Q_{sea}+Q)/Q$ ratio is 2, then Y_{ZMT} is 4 and the fall velocity is 4 times higher than the default value (6 m/month).

The dimensionless moderator for the dynamic ratio (the potential turbulence in the system), Y_{DR}, is given by:

$$\text{If DR} < 0.25 \text{ then } Y_{DR} = 1 \text{ else } Y_{DR} = 0.25/\text{DR} \quad (A.15)$$

Systems with a DR-value of 0.25 (see Fig. 3.14 and Håkanson and Jansson 1983) are likely to have a minimum of ET-areas (15% of the area) and the higher the DR-value, the larger the area relative to the mean depth and the higher the potential turbulence and the lower the settling velocity.

The resuspended fraction of TP in the surface-water compartment is calculated by means of the distribution coefficient (DC_{resSW}), which is defined by the ratio between resuspension from ET-areas to surface water (SW) relative to all fluxes (except mixing) to the surface-water compartment:

$$DC_{resSW} = F_{ETSW}/(F_{ETSW} + F_{prec} + F_{inSW} + F_{PS} + F_{inQ} + F_{farmSW}) \quad (A.16)$$

F_{ETSW} = resuspension from ET-areas to surface-water areas (g TP/month);
F_{prec} = inflow of TP from direct precipitation (g TP/month);
F_{inSW} = inflow of TP from the sea (g TP/month);
F_{PS} = inflow of TP from point source emissions (g TP/month);
F_{farmSW} = inflow of TP from fish cage farm(s) in the coastal area (g TP/month).

DC_{resSW} is calculated automatically in the model.

The dimensionless moderator expressing how much faster resuspended particles settle compared to primary particles is given by:

$$Y_{res} = ((T_{ET}/1) + 1)^{0.5} \quad (A.17)$$

Where T_{ET} is the mean retention time (the mean age) of the particles on ET-areas in months estimated from:

$$\text{If } Sal_{SW} = Sal_{Sea} \text{ then } T_{ET} = 1 \text{ else } T_{ET} = (1 + 1\cdot((Sal_{Sea}/(Sal_{SW} + 1))^{0.5} - 1)) \quad (A.18)$$

If the surface-water salinity in the coastal area (Sal_{SW}) is the same as the surface-water salinity in the sea outside the coast (Sal_{Sea}), one should expect a dynamic water exchange between the coast and the sea and the mean age of the ET-sediments is 1 month. If, e.g, the salinity in the sea is 30 psu, and the surface-water salinity in the coast is 6 psu, there must be a more limited water exchange between the coast and the sea, and the age of the ET-sediments is longer (2.5 months). If the particles have stayed on the ET-areas long, they are likely to be aggregated and will settle out faster when resuspended. This is given by $Y_{res} = \sqrt{(2.5 + 1)} = 1.87$ for a coastal area with T_{ET} = 2.5, which means that resuspended particles settle 2.5 times faster than the primary materials.

The corresponding equation for the settling velocity in deep-water areas (v_{DW}) is given by:

$$v_{DW} = (v_{def}/12)\cdot Y_{SPMDW}\cdot Y_{salDW}\cdot Y_{DR}\cdot Y_{DW}\cdot Y_{ZMT}\cdot((1-DC_{resDW}) + Y_{res}\cdot DC_{resDW}) \quad (A.19)$$

v_{def} = 72 m/yr;
Y_{SPMDW} = the dimensionless moderator expressing how changes in SPM in the deep water influence v_{DW};
Y_{salDW} = the dimensionless moderator expressing how changes in salinity (in psu) in the deep water influence v_{DW}; this value can be measured but in the following simulations, we assume that Sal_{DW} is 3 psu higher than Sal_{SW}; this is motivated by the salinity distributions for the Baltic Sea (Håkanson 1991);
Y_{DR} = the dimensionless moderator expressing how variations among systems in the dynamic ratio and turbulence (DR) are likely to influence the settling velocity for SPM;
Y_{DW} = the dimensionless moderator expressing how hydrodynamic properties, as expressed by the deep-water retention time, T_{DW} influence the turbulence in the deep-water zone and v_{DW};
Y_{ZMT} = the dimensionless moderator expressing increased sedimentation in the zone of maximum turbidity;
DC_{resDW} = the resuspended fraction in the deep water;
Y_{res} = the dimensionless moderator expressing how changes in the age of the ET-sediments influence v_{DW}.

Y_{SPMDW} is given in the same manner as Y_{SPMSW} as:

$$Y_{SPMDW} = (1+0.75\cdot(SPM_{DW}/50-1)) \quad (A.20)$$

The dimensionless moderator for salinity (Y_{sal} or Y_{salSW} or Y_{salDW}) is the same as in (A.13); the moderator for Y_{ZMT} is the same as in (A.14); and the moderator for Y_{DR} the same as in (A.15).

Y_{DW} is calculated from:

$$\text{If } T_{DW} < 7\text{days then } Y_{DW} = 1 \text{ else } Y_{DW} = (T_{DW}/7)^{0.5} \quad (A.21)$$

Where T_{DW} is the theoretical deep-water retention time in days. If T_{DW} is 1 week or shorter, Y_{TDW} is 1; if T_{DW} is 120 days (the maximum value in Baltic Sea coastal areas, see Table 3.2), then Y_{DW} is 4.1 and the settling velocity is 4.1 higher than in a more turbulent situation when T_{DW} is 1 week.

The resuspended fraction of SPM in the deep-water compartment is given by:

$$DC_{resDW} = F_{ETDW}/(F_{ETDW} + F_{inDW} + F_{SWDW} + F_{ADW} + F_{farmDW}) \quad (A.22)$$

F_{ETDW} = resuspension from ET-areas to deep-water areas (g TP/month);
F_{inDW} = inflow of TP to the deep-water compartment from the sea (g TP/month);

F_{SWDW} = sedimentation, i.e., transport from surface water to deep-water areas (g SPM/month);
F_{ADW} = diffusive transport of TP from A-sediments (g TP/month);
F_{farmSW} = inflow of TP to the deep water from fish cage farm(s) in the coastal area (g TP/month).

DC_{resDW} is, just as DC_{resSW}, calculated automatically in the model.

Resuspension

By definition, the materials settling on ET-areas will not stay permanently where they were deposited but will be resuspended by wind/wave activity. If the age of the material (T_{ET}) is set to be long, e.g., 10 years, these areas will function as accumulation areas; if, on the other hand, the age is set to 1 week or less, they will act more as erosion areas. In this modeling, it is assumed that the mean age of these deposits are estimated by (A.18). Resuspension back into surface water, F_{ETSW}, i.e., mostly wind/wave-driven advective fluxes, is given by:

$$F_{ETSW} = (M_{ET} \cdot (1 - Vd/3))/T_{ET} \tag{A.23}$$

Resuspension to deep-water areas, F_{ETDW}, by:

$$F_{ETDW} = (M_{ET} \cdot (Vd/3))/T_{ET} \tag{A.24}$$

M_{ET} = the total amount of resuspendable matter on ET-areas (g);
Vd = the form factor; note that Vd/3 is used as a distribution coefficient to regulate how much of the resuspended material from ET-areas that will go the surface water or to the deep-water compartment. If the coast is U-shaped, Vd is about 3 (i.e., $D_{max} \approx D_m$) and all resuspended matter from ET-areas will flow to the deep-water areas. If, on the other hand, the coast is shallow and Vd is small, most resuspended matter will flow to the surface-water compartment (see Fig. 5.24);
T_{ET} = the age of TP on ET-areas.

Diffusion

Diffusion of phosphorus from A-sediments back to the deep-water layer (F_{ADW} in g TP per month) is given by:

$$F_{AW} = M_A \cdot R_{diff} \tag{A.25}$$

R_{diff} is the diffusion rate (1/month), which depends on the oxygen conditions in the sediments. The default value is 0.0003 (1/year; Håkanson 1999). In this modeling, the default value will be influenced by the estimated value of the oxygen saturation in the deep-water zone (O_2Sat), which is calculated from the empirical model given in Table A.2.

$$R_{diff} = (0.0003/12) \cdot Y_{O2} \quad (A.26)$$

Where Y_{O2} is a dimensionless moderator expressing how changes in O_2Sat are likely to influence the diffusion of phosphorus from A-sediments. Y_{O2} is given by:

$$\text{If } O_2Sat > 50(\%) \text{ then } Y_{O2} = (2 - 1 \cdot (O_2Sat/50 - 1))$$
$$\text{else } Y_{O2} = (2 - Amp \cdot (O_2Sat/50 - 1)) \quad (A.27)$$

Where the amplitude value (Amp) is given by:

$$Amp = 3000 \cdot (C_{TPsed}/1) \quad (A.28)$$

This means that if O_2Sat = 100%, Y_{O2} = 1 and the diffusive flux is small (0.0003/12); if O_2Sat = 50%, Y_{O2f} = 2; if O_2Sat = 20, Y_{O2} = 1802 if the TP-concentration in the A-sediments is 1 mg/g dw and Y_{O2} = 3602 if the TP-concentration in the A-sediments is 2 mg/g dw. In such situations, diffusion is likely the most dominating TP flux in the system. The amplitude values have been derived from many iterative calibration steps.

Burial

If the sediments are oxic (i.e., when the bioturbation factor, BF, is likely high), the age of the A-sediments (T_A) and hence burial (F_{bur}) will be influenced by the biological mixing by zoobenthos (Håkanson and Jansson 1983).

$$\text{If } O_2Sat < 20(\%) \text{ then } BF = 1 \text{ else } BF = (1 + D_{AS})^{0.3} \quad (A.29)$$

If O_2Sat is lower than 20%, zoobenthos are likely to die and bioturbation halted. D_{AS} is the depth of the bioactive A-sediment layer. The default value for D_{AS} is set to 10 cm (Håkanson and Jansson 1983). This means that $BF = (1 + D_{AS})^{0.3} =$ 2.05 and the sediments 2.05 times older than calculated from the ratio between the depth of the active A-sediments (D_{AS} in cm) and sedimentation (Sed in cm/yr). Sedimentation, in turn, is calculated from the dynamical SPM-model.

Burial (F_{bur}) is then given by:

$$\text{If } T_A > 48(\text{ months}) \text{ then } F_{bur} = M_A \cdot 1/48 \text{ else } F_{bur} = M_A/T_A \quad (A.30)$$

Where M_A is the total amount of TP in A-sediments (g) and T_A is the age of the active A-sediments. Settling the boundary condition to T_A = 48 months means that we assume that all mobile phosphorus in the active A-sediments that can be lost from the sediments due to diffusion will do so in this time span (the rest of the phosphorus will be in more stable forms). T_A depends on bioturbation and is given by:

$$T_A = 12 \cdot BF \cdot D_{AS}/Sed \quad (A.31)$$

Sedimentation (Sed in cm/yr) is calculated from sedimentation on accumulation areas (Sed_A in g dw/m^2·day) by the following dimensional adjustments:

$$Sed = 12 \cdot (Sed_A \cdot 10^{-4}/AreaA)/(100 - W_A) \cdot (1/d) \quad (A.32)$$

The default value for the water content (W_A) and organic content (= loss on ignition, IG_A) of A-sediments from the bioactive layer (0–10 cm) is set to 70% ww and 7.5% dw, respectively (Håkanson et al. 1984). The bulk density (d in g ww/cm^3) is calculated from W_A and IG_A by a standard formula (Håkanson and Jansson 1983; see Table A.3).

Uptake and Retention in Biota

To calculate the TP-uptake and retention in biota, this modeling uses a similar approach as presented by Håkanson and Boulion (2002). This means that the uptake and retention in biota is given by:

$$M_{BLTP}(t) = M_{BLTP}(t-dt) + (F_{TPbioup} - F_{TPbioret})\cdot dt \qquad (A.33)$$

$M_{BLTP}(t)$ = the mass (amount) of TP in biota (g);
$F_{TPbioup} = M_{SWTP}\cdot Y_{SWT}\cdot(30/T_{BL})\cdot(1-PF_{SW})$; the biouptake of TP in biota (g/month);
$F_{TPbioret} = M_{BLTP}\cdot 30/T_{BL}$; the retention (= outflow) of TP from biota with long turnover times (g/month);
$Y_{SWT} = (SWT/SWT_{MV})$; the dimensionless moderator regulating the temperature dependent biouptake of TP. SWT = the surface water temperature (°C); SWT_{MV} = the mean annual surface water temperature (°C; e.g., 11.85°C in the Himmerfjärden Bay);
$T_{BL} = (100\cdot 11 + 10\cdot 300 + 1\cdot 450)/111$; the average turnover time for the functional groups included in biota with long turnover times (these are the organisms which are not included in the TP in water; predatory zooplankton, T = 11 days; prey fish, T = 300 days and predatory fish, T = 450 days); or if the calculation is done for biota with short turnover times, the corresponding T_{BS} – values are phytoplankton = 3.2 days, bacterioplankton = 2.8 days and herbivorous zooplankton = 6 days (from Håkanson and Boulion 2002);
$PF_{SW} = Y_{PF} + (M_{BLTP}/(M_{SWTP} + M_{BLTP}))$; the particulate fraction of phosphorus in surface water (dim. less);
PF_{def} = if $Ex < 0.01$ and $ET > 0.56$ then $PF_{def} = (ET + 0.4\cdot(Ex/0.01-1))$ else $PF_{def} = DC_{resSW}\cdot(Y_{SWT})^{0.5}$; PF_{def} = a default PF-value;
Ex = the exposure (= 100·At/Area);
ET = the fraction of ET-areas (A_{wb}/Area; A_{wb} is the area above the wave base; Area is the costal area).

Outflow, Mixing and Stratification

If the water depth that separates the surface-water and the deep-water compartments is defined in a relevant way, this will also imply that outflow from these compartments (F_{outSW} and F_{outDW}) can be calculated in a simple and mechanistic manner. The outflow from surface water is given by:

$$F_{outSW} = Q \cdot C_{TPSW} + M_{SW}/T_{SW} \tag{A.34}$$

Q= the water discharge from the tributary in m^3 per month;
C_{TPSW} = the TP-concentration in the surface water in g/m^3 or mg/l;
M_{SW} = the mass or amount of TP in the surface-water compartment (g);
T_{SW} = the theoretical surface-water retention time (1/month).

The outflow from the deep water is given by:

$$F_{outDW} = M_{DW}/T_{DW} \tag{A.35}$$

M_{DW} = the mass or amount of TP in the deep-water compartment (g);
T_{DW} = the theoretical deep-water retention time (1/month).

The following sub-model for mixing gives the monthly mixing rate (R_{mix}; 1/month) as a function of the absolute difference between mean monthly surface and deep-water temperatures.

$$\text{If ABS(SWT} - \text{DWT)} < 4(^\circ\text{C})\text{then } R_{mix} = 1 \text{ else } R_{mix} = 1/\text{ABS(SWT} - \text{DWT)} \tag{A.36}$$

That is, if the absolute difference between surface-water (SWT) and deep-water temperatures (DWT) is smaller than 4°C, the system is not stratified and R_{mix} is set to 1. It has also been tested if the mixing rate, R_{mix}, during homothermal conditions should be larger than one for very large and shallow systems with high dynamic ratios (DR) and lower for small and deep coasts, but that does not seem to improve model predictions (see Håkanson and Eklund 2007c). Complete mixing, i.e., setting the mixing rate to one (rather than to a higher number) generally seems to be sufficient. In calculating mixing, the same amount of water should be transported from surface water to deep water and from deep water to surface water. The downward flux, i.e., mixing from surface water to deep water, is given by:

$$F_{SWDWx} = M_{SW} \cdot R_{mix} = (M_{SW}/V_{SW}) \cdot (V_{SW}/T_{SW}) = C_{SW} \cdot Q_{mix} \tag{A.37}$$

Where the concentration of TP in surface water (C_{SW}) is equal to M_{SW}/V_{SW} and the water flux related to mixing (Q_{mix}) is equal to V_{SW}/T_{SW}, where T_{SW} is the theoretical surface-water retention time. This means that mixing from deep water to surface water will cause a transport given by:

$$F_{DWSWx} = M_{DW} \cdot R_{mix} \cdot (V_{SW}/V_{DW}) \tag{A.38}$$

Where $M_{DW} = V_{DW} \cdot C_{DW}$. If the theoretical wave base (D_{wb}) is very close to the maximum depth (D_{max}), and hence the deep-water volume is very small and the ratio V_{SW}/V_{DW} very large, the flux from deep water to surface water can be so large that it will become difficult to get stable solutions using, e.g., Euler's or Runge–Kutta's calculation routines. This means that the following boundary condition is used for mixing:

$$\text{If } R_{mix}\cdot(V_{SW}/V_{DW}) > 100 \text{ then } F_{DWSWx} = M_{DW}\cdot 100$$
$$\text{else } F_{DWSWx} = M_{DW}\cdot R_{mix}\cdot(V_{SW}/V_{DW}) \quad \text{(A.39)}$$

The coastal water is not likely stratified if the dynamic ratio (DR = $\sqrt{}$(Area$\cdot 10^{-6}$) /D_m) is higher than 3.8 (Håkanson and Jansson 1983). Then, the system is probably not dimictic but polymictic. A boundary condition for this is added to the model: If DR > 3.8 then $R_{mix} = 1$.

A.1.4 Results

This section will first give results from the calibrations of the model; the following parts will present the validations, calculations of fluxes and sensitivity analyses.

A.1.4.1 Calibrations

There is only one substance-specific part in the dynamic model, which has been subject for calibrations, the amplitude value in the algorithm for diffusion (A.28). This has been done for several coastal areas and Fig. A.5 gives results from one of those (Laitsalmi, Finland, which has been randomly selected for this illustration). The calibrations have focused on the dynamically modeled TP-concentrations in water (surface water plus deep water) and on the TP-concentrations in A-sediments. There are empirical data on the mean TP-concentrations in coastal water and two empirical reference values for TP-concentrations in A-sediments, an expected low reference value of 0.75 mg/g dw and a high reference value of 2.5 mg/g dw (Håkanson 1999).

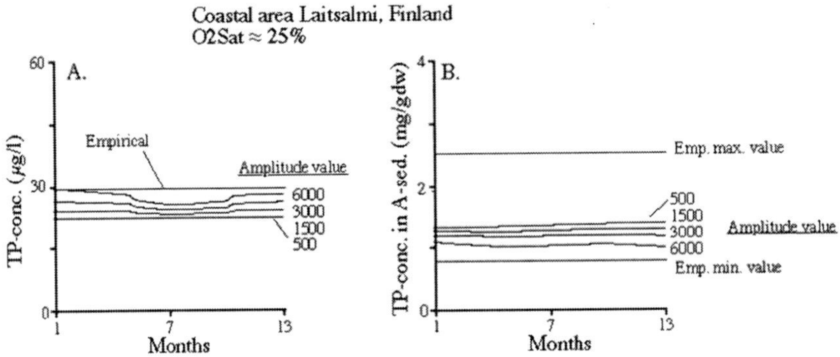

Fig. A.5 Calibrations in the Laitsalmi area, Finland (see Table A.1) (A) for modeled values of TP-concentrations in water and (B) for modeled TP-concentration in A-sediments (0–10 cm). The amplitude value in the algorithm for diffusion has been changed (between 500 and 6000) and the consequences simulated for TP-concentrations in water and sediments. The modeled output is tested against empirical data

The modeled values will be compared to these reference values. From Fig. A.5, one can see that the value used for the amplitude value, which regulates diffusion of phosphorus from A-sediments, clearly influence the predictions in this area (and in most coastal areas). If the amplitude value is set too high, diffusion will be too high, giving too high TP-concentrations in water and too low TP-concentrations in A-sediments. In all following simulations, we have used a default amplitude value of 3000.

A.1.4.2 Validations – Blind Tests

Note that in the following blind tests, there has been no tuning of the model constants. Only the obligatory driving variables (from the panel of driving variables, see Table A.4) have been changed for each coastal area. The results of the validations will be presented in the following way. First, comparisons between empirical data, uncertainties in empirical data and model-predicted values will be given for each area followed by more detailed information for two of the areas. The criteria to evaluate the model results are: "excellent fit" – modeled value are within an interval of one standard deviation of the empirical mean value ($\pm$ 1SD $\approx$ 70% confidence interval), "good fit" – modeled value are within an interval of two standard deviations of the empirical mean value ($\pm$ 2SD $\approx$ 95% confidence interval), "acceptable fit" – modeled value within three standard deviations of the empirical mean, and "poor fit" for modeled values outside three standard deviations of the empirical mean value. The predicted TP-concentrations in the A-sediments are compared to expected empirical low and high reference values. Characteristic uncertainties in data for empirical water variables from Baltic coastal areas are discussed in Chap. 4 and used in the following tests. The basic question is: how well does the model predict considering all the uncertainties in the empirical data used to run the model and the assumptions behind the given algorithms?

Table A.5 shows the result with respect to the different criteria used to evaluate the model behaviour for each studied coast. Note that the TP-concentrations in water and in A-sediments have been modeled dynamically, whereas the other variables have been calculated from empirical regressions.

The overall conclusion is that the dynamic model seems to work very well. For 14 of the 21 studied areas there is an excellent fit between modeled and empirical TP-concentrations in water. For one area only there is an "acceptable fit" while for the remaining areas is there a "good fit". The predicted TP-concentrations in A-sediments are within the empirical reference interval for all areas. For the variables predicted by regression models, the results are generally "excellent" or "good", only for one area is there a "poor fit".

For the coastal areas Ronneby (an estuary in southern Sweden) and Gävle (an estuary in northern Sweden), the results are illustrated in Figs. A.6 and A.7. Figure A.6A gives the mean empirical TP-concentrations in water (surface plus deep water; called TP emp. MV, curve 1), the uncertainty in the empirical value

Table A.5 A comparison of empirical data, uncertainties in empirical data and model-predicted values for the studied coastal areas. The evaluated variables are the dynamically modeled TP-concentrations in water and A-sediments, chlorophyll-a, sedimentation in surface and deep water, Secchi depth and oxygen saturation in deep water

Coastal area	TP-conc. (water) (µg/l)	TP-conc. (A-sed.) (mg/g dw)	Chl conc. (fr. TN emp) (µg/l)	Chl conc. (fr. TP mod) (µg/l)	Sed. SW (g/m^2 day)	Sed. DW (g/m^2day)	Secchi (m)	O_2sat (%)
Sweden (South)								
Matvik	+++	I	+++	++	++	+++	+++	+++
Boköfjärd	++	I	+++	+	+++	+++	++	+++
Tärnö	+++	I	+++	+++	+++	+++	+++	+++
Guavik	+++	I	+++	+++	+++	++	+++	+++
Järnavik	+++	I	+++	++	+++	+++	+++	+++
Spjutsö	++	I	++	–	+++	+++	++	+++
Ronneby	+++	I	+++	+++	++	++	+++	+++
Sweden (East)								
Lilla Rimmö	+++	I	+++	+++	+++	+++	+++	++
Eknön	+++	I	+++	++	+++	++	++	+++
Lagnöstr.	++	I	++	++	++	+++	++	+++
Gräsmarö	+++	I	+++	+++	+	+++	++	+++
Ålön	++	I	+++	+++	+++	+++	+++	+++
Sweden (Bothnian Sea)								
Gävle	++	I	++	+++	+++(a)	(a)	++	+++
Gårdsfjärden	+++	I	+++	+++	0	0	+	+++
Finland (Åland)								
Svibyviken	+++	I	0	0	0	0	+++	0

Table A.5 (continued)

Coastal area	TP-conc. (water) (μg/l)	TP-conc. (A-sed.) (mg/g dw)	Chl conc. (fr. TN emp) (μg/l)	Chl conc. (fr. TP mod) (μg/l)	Sed. SW (g/m^2 day)	Sed. DW (g/m^2day)	Secchi (m)	O_2sat (%)
Finland (South West)								
Särkensalmi	+++	I	++	++	+++	++	–	+++
Käldö	+++	I	+++	+++	++	+++	++	++
Haverö	+++	I	+++	+++	+++	+++	+	+++
Hämmärönsalmi	+++	I	+++	+++	+++	+++	+++	+++
Laitsalmi	++	I	+++	+++	++	++	++	+++
Kaukolanlahti	+	I	–	–	++	+++	–	–

+++ Excellent fit – modeled values within MV emp ±1SD (70% conf.interval).
++ Good fit – modeled values within MV emp ±2SD (95% conf.interval).
+ Acceptable fit – modeled values within MV emp ±3SD (99% conf.interval).
– Poor fit – modeled values outside 3SD.
I modeled values within the expected min–max interval for emp. data.
0 No emp. data available or emp. data analysed by means of different sampling or analysis method.

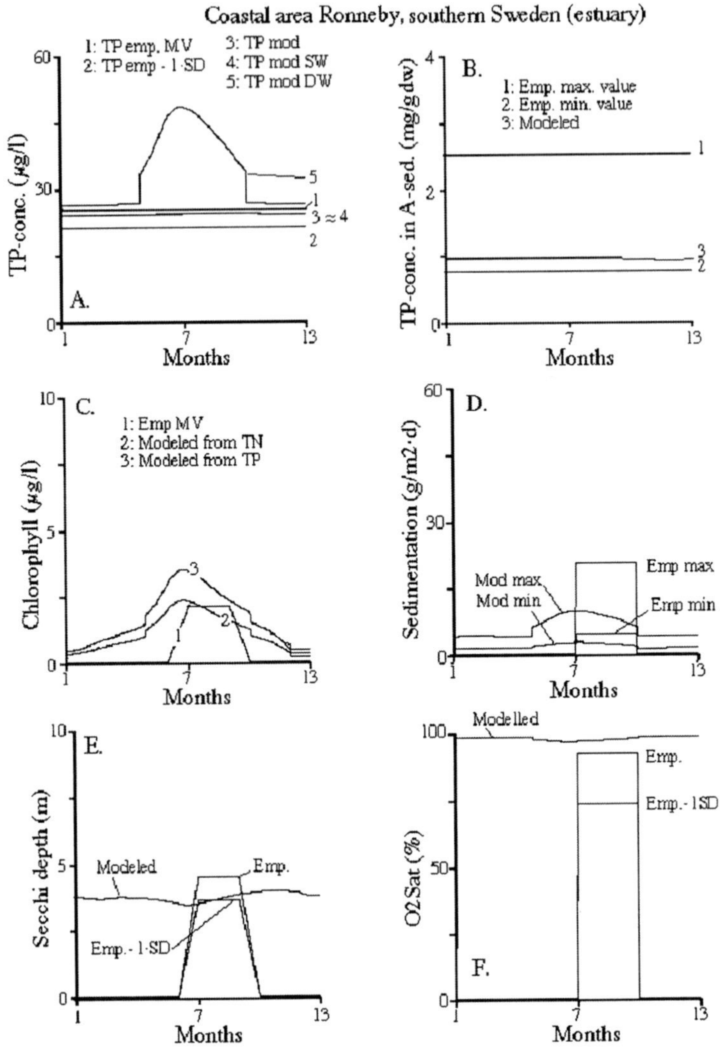

Fig. A.6 Validations in the Ronneby coastal area (see Table A.1) (**A**) gives the empirical mean value (curve 1) and the uncertainty in the empirical mean value (minus 2·SD, curve 2) and modeled values for TP-concentrations in water (curve 3), in surface water (curve 4) and in deep water (curve 5), (**B**) gives modeled TP-concentrations in A-sediments (0–10 cm) versus expected maximum and minimum TP-concentrations in A-sediments, (**C**) gives empirical data (curve 1) versus modeled chlorophyll values calculated from empirical TN-concentrations (curve 2) and dynamically modeled TP-concentrations (curve 3), (**D**) gives empirical and modeled minimum and maximum values for sedimentation, (**E**) gives modeled Secchi depths versus empirical data and uncertainty in empirical data (MV minus 1·SD) and (**F**) gives modeled values of O_2Sat and empirical data and uncertainty in empirical data (MV minus 1·SD)

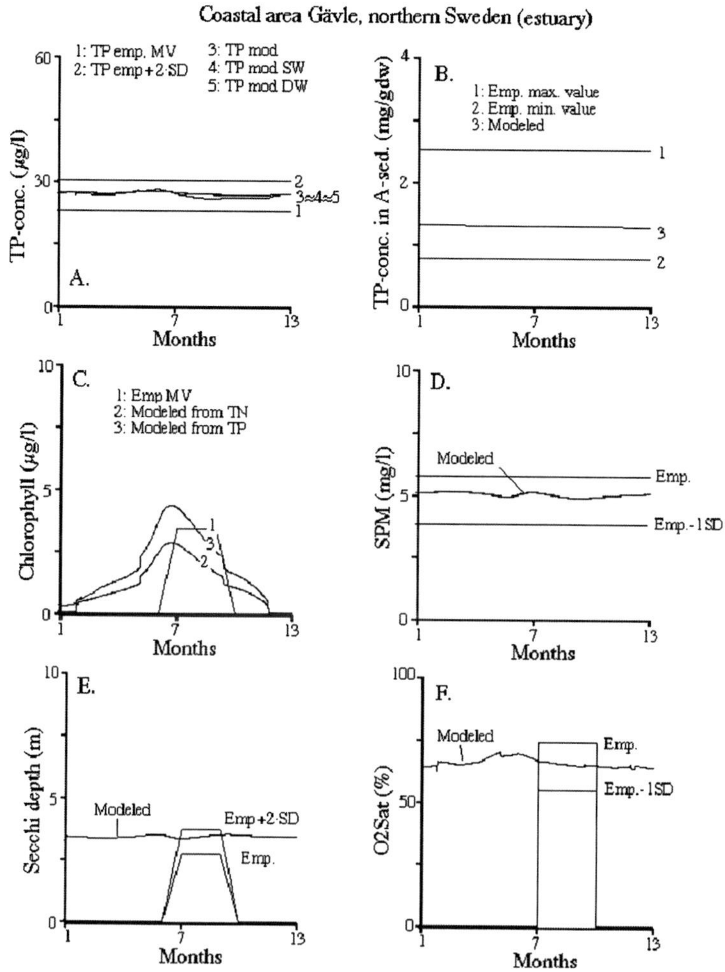

Fig. A.7 Validations in the Gävle coastal area (see Table A.1) (**A**) gives the empirical mean value (curve 1) and the uncertainty in the empirical mean value (minus 2·SD, curve 2) and modeled values for TP-concentrations in coastal water (curve 3), in surface water (curve 4) and in deep water (curve 5), (**B**) gives modeled TP-concentrations in A-sediments (0–10 cm) versus expected maximum and minimum TP-concentrations in A-sediments, (**C**) gives empirical data (curve 1) versus modeled chlorophyll values calculated from empirical TN-concentrations (curve 2) and dynamically modeled TP-concentrations (curve 3), (**D**) gives modeled values of suspended particulate matter and empirical data and uncertainty in empirical data (MV minus 1·SD). (**E**) gives modeled Secchi depths versus empirical data and uncertainty in empirical data (MV plus 2·SD) and (**F**) gives modeled values of O_2Sat and empirical data and uncertainty in empirical data (MV minus 1·SD)

(curve 2; MV minus 2 standard deviations; called TP emp-2·SD), modeled mean values (curve 3; TP mod; these data should be compared to curve 1), modeled TP-concentrations in the surface water (curve 4; TP mod SW) and modeled TP-concentrations in the deep water (curve 5; TP mod DW).

The modeled values (curve 3) are close to the empirical mean value (curve 1) and within the uncertainty of the empirical mean value (curve 2). The concentration of TP in A-sediments are predicted within the expected interval (Fig. A.6B).

Chlorophyll-a is predicted better from TN than from TP (Fig. A.6C), but still there is an excellent fit for chlorophyll-a in both cases. Secchi depth and oxygen saturation in the deep-water zone are predicted within one standard deviation of the empirical mean value (Fig. A.6E and F) and the modeled values for sedimentation in surface and deep-water areas are within the 95% confidence limits for the empirical data (Fig. A.6D). For the studied area Gävle, the modeled values of TP-concentrations in water (curve 3 in Fig. A.7A) are higher than, but close to, the empirical mean value (curve 1) and well within the 95% uncertainty bands (± 2 standard deviations) of the empirical mean value (curve 2). The predicted TP-concentration in the A-sediments is within the expected range (Fig. A.7). There is a better prediction of chlorophyll-a from TP than from TN (Fig. A.7C). For this coastal area, there are no empirical data on sedimentation, but data on the SPM-concentration, which is predicted very well (Fig. A.7D). Also the Secchi depth and oxygen saturation in the deep-water zone are predicted close to the empirical mean values and within the uncertainty band of the empirical mean value (Fig. A.7E and F).

As an illustration of the predictions of TP-concentrations in water and chlorophyll-a concentrations, the error functions for these variables are shown in Figs. A.8 and A.9.

For TP, the mean error is 0.028, the median error 0.005 and the standard deviation 0.18, which is just a little higher than the coefficient of variation (CV) related to the uncertainty in the empirical data, 0.16. This means that one cannot expect to obtain better results since the reliability in the empirical data will set the

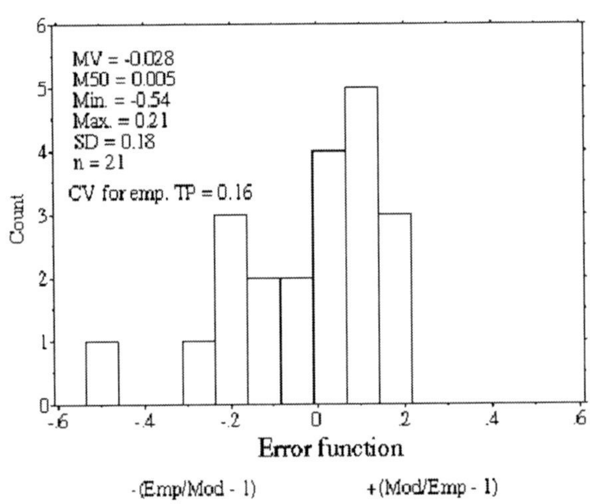

Fig. A.8 Results (error function and statistics) when modeled mean TP-concentrations in water are compared to empirical values for the 21 coastal areas

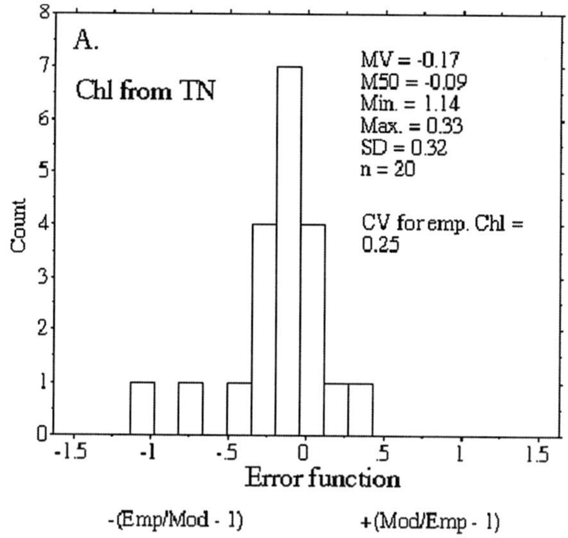

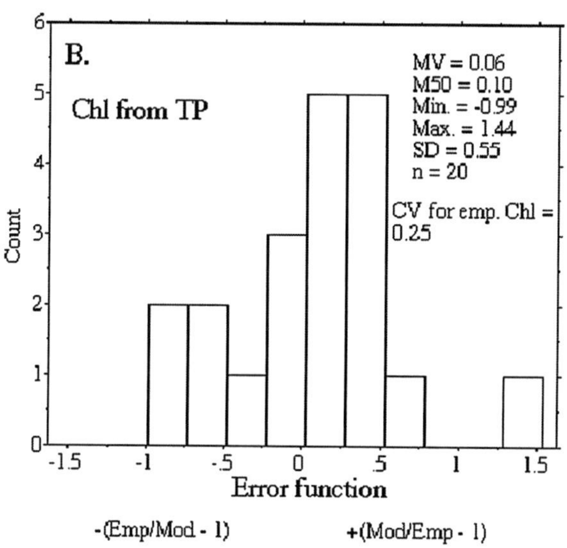

Fig. A.9 Results (error function and statistics) when (**A**) modeled chlorophyll-a concentrations calculated from empirical TN-concentrations are compared to empirical values and (**B**) when modeled chlorophyll values calculated from modeled TP-concentrations are compared to empirical values for the 21 coastal areas

limit to the predictive success. Figure A.9A shows the error function for chlorophyll predicted from empirical TN-concentrations using the regression model and Fig. A.9B gives the corresponding error function for chlorophyll, as predicted using dynamically modeled TP-concentrations and the regression model to relate modeled TP-concentrations to chlorophyll.

One can note that the key part in this modeling concerns the dynamic TP-model, but it is interesting to see that also chlorophyll can be predicted well from modeled TP-concentrations (mean error = 0.06, standard deviation for the error = 0.55); the uncertainty (CV) in the empirical chlorophyll values is 0.55.

A.1.4.3 Calculation and Ranking of Fluxes

It is important to identify and differentiate between small and large fluxes of nutrients so that realistic expectations can be obtained for various remedial measures intended to reduce coastal eutrophication. Figure A.10 compares fluxes in three coastal areas, one with direct contact with the sea, Järnavik, on the Swedish south coast, one area situated in the Finnish archipelago, Hämmärönsalmi, and one area from northern Sweden, Gävle. The aim of this test is to see which are the dominating fluxes in the three cases.

From Fig. A.10, one can note that:

- There are no major differences between the three coastal areas. The surface-water fluxes of TP dominate in all.

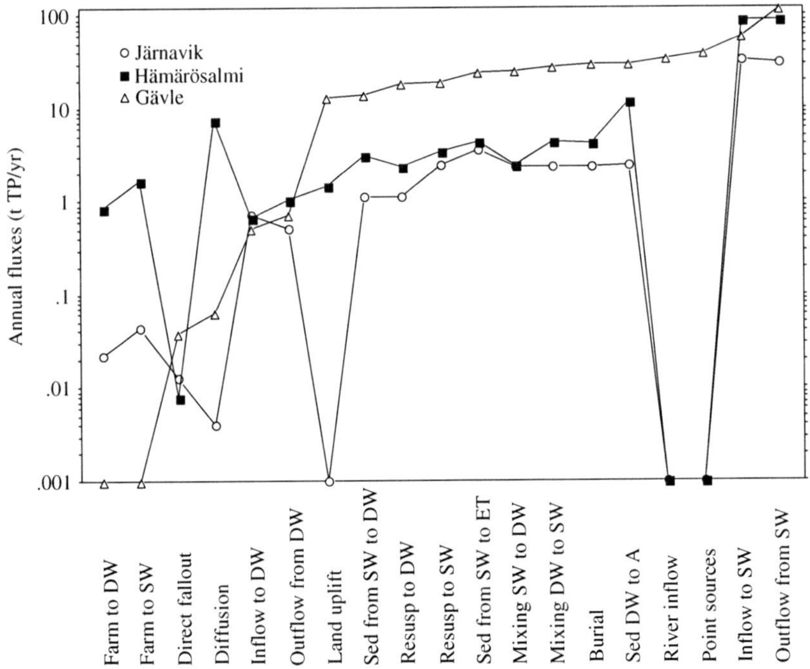

Fig. A.10 Simulations giving a ranking of all the TP-fluxes (transport processes) for area Järnavik, Sweden, which has direct contact with the Sea, area Hämärösalmi, which is deep in the Finnish archipelago, and area Gävle which has an outside archipelago

- There is a logical difference in the ranking of the fluxes in the three cases, especially concerning land uplift, which is more important in the area Hämmärönsalmi, which is part of a relatively shallow archipelago system. Land uplift is zero in the coastal area Järnavik.
- The coastal area Gävle is exposed to the largest loading from anthropogenic point sources.

These results show that the conditions in the sea or the adjacent coastal water play an important role for the transport of TP to and from these areas, thus also controlling the TP-concentrations in these coasts to a large extent. This is also a logical consequence of the fact that the characteristic theoretical surface-water retention times in these coastal areas are generally short (in the order of days, or in exceptional cases, of weeks).

Figure A.11 gives a compilation of seasonal TP-concentrations on a monthly basis for the 21 coastal areas included in this study. It is interesting to note:

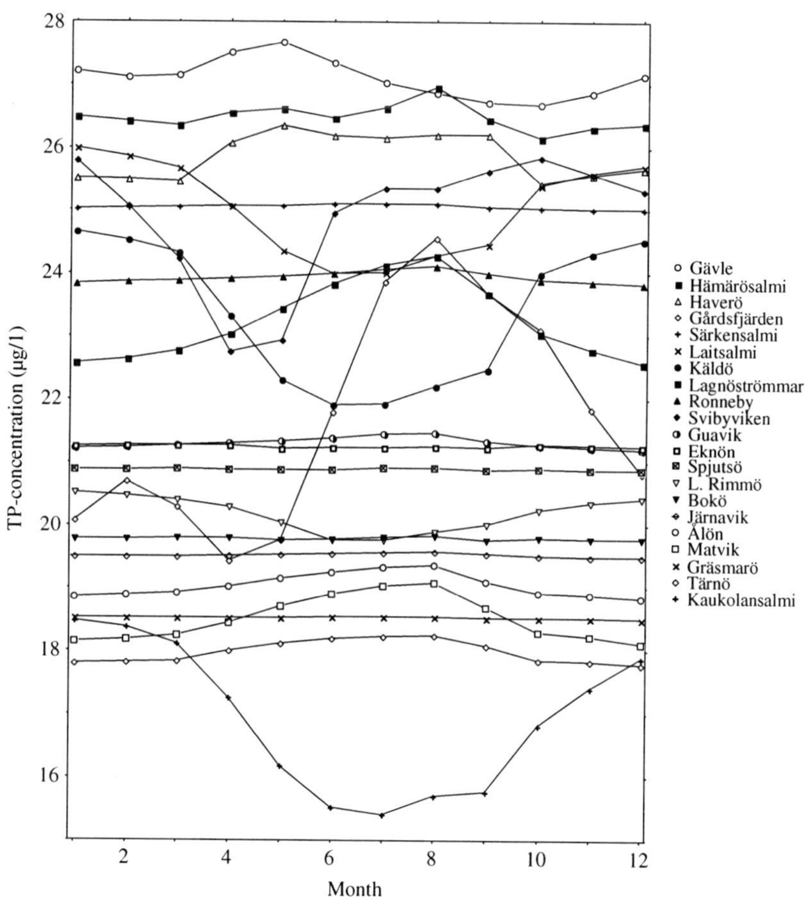

Fig. A.11 Simulations to show how TP-concentrations vary seasonally (monthly) in the studied coastal areas

- That area-characteristic TP-values vary from about 16 to 28 μg/l.
- That there are coastal areas without any clear seasonal TP-patterns, areas with higher TP-concentrations in the summer and fall and areas with lower values in the summer, depending on the characteristics of the coast.

A.1.4.4 Sensitivity Analyses

The idea here is to get a quantification of the role of a given flux for the value of the target variable (the TP-concentration in water), while all else is constant. 100 runs have been simulated and normal frequency distributions have been applied for the uncertainty in the fluxes. A uniform uncertainty (CV) of 0.5 is being used for all fluxes. Figure A.12 shows results for one coastal area (Ronneby, S. Sweden, an estuary). This figure also ranks the importance of the various uncertainties for the predictions of TP in this area. From these presuppositions, one can note that the three most important uncertainties concern the TP-fluxes to (F_{inSW}) and out of the surface water (F_{outSW}) and tributary inflow (F_{in}) to the surface water. All other uncertainties in the TP-fluxes are of less importance and the uncertainty in land uplift (F_{LU}) is of no significance since land uplift is zero in this region.

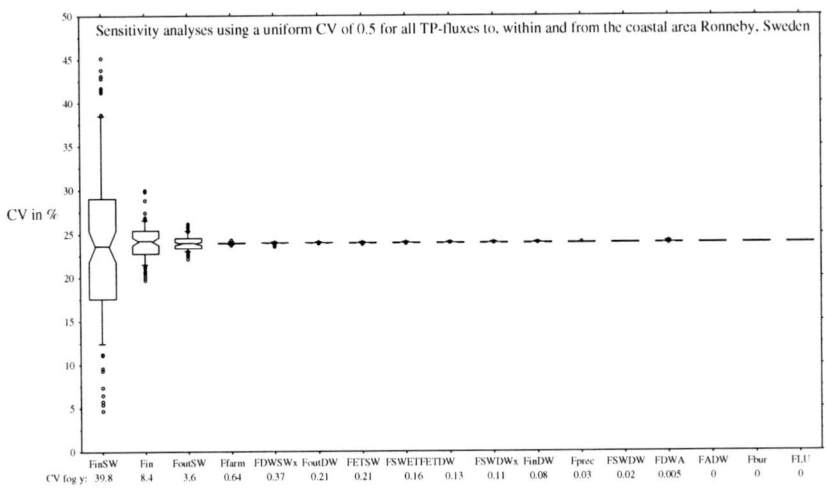

Fig. A.12 Sensitivity tests where all fluxes in the model are accounted for, one by one, and all else kept constant. A uniform uncertainty for all the fluxes has been used (a CV of 0.5 and a normal frequency distribution around the mean value). The figure also ranks the importance of the fluxes in relation to the prediction of the target variable, TP-concentration in water in the Ronneby coastal area, under these presuppositions. The figure gives the box-and-whisker plots (median, 25 and 75 quartiles, 10 and 90 percentiles and outliers) as well as the CV for the y-variable (related to the in the fluxes in July)

A.1.5 Comments

This section has discussed a general dynamic mass-balance model for phosphorus in coastal areas handling all important fluxes of phosphorus to, from and within coastal areas, as defined according to the topographical bottleneck method. This type of modeling makes it possible to perform different simulations by adding, changing or omitting fluxes, evaluate ecosystem responses, and thereby predict effects of different approaches to reduce nutrient input to the studied area. This allows for a good estimation of what can be expected in terms of improved environmental conditions as a result of different remedial strategies. Many of the structures in the model are general and have also been used with similar success for other types of aquatic systems (lakes and rivers) and for other substances (mainly SPM and radionuclides; Håkanson 2000, 2006).

The results have shown that it was possible to predict the TP-concentration in different coastal areas very well. The associated bioindicators (chlorophyll-a, Secchi depth and oxygen saturation in the deep-water zone) are also predicted well in most cases. A general conclusion is that the TP-concentration in the outside sea controls the conditions in a given coastal area to a large extent even if the coast is rather enclosed.

It should be stressed that although the critical model tests include data from 21 coastal areas, all of these are from the Baltic Sea. This means that for the future, it would be very interesting to also test the model for coastal areas from other parts of the world. The model is based on general mechanistic principles at the ecosystem scale, but further tests may reveal shortcomings so that the boundary conditions of the model may have to be improved.

When using the model, no tuning should be performed. The model should be adjusted to a new area only by changing the obligatory driving variables. Since the utilized driving variables emanate from standard monitoring programs, or can be calculated from bathymetric maps, the model could have great practical utility in coastal management. The driving variables include coastal area, section area (between the defined coastal area and the adjacent sea), mean and maximum depths, latitude (to predict water temperatures, stratification and mixing), salinity and TP-concentration in the sea. Many of the model structures are general and could be used for areas other than those included in this study, e.g., for open coasts, estuaries or tidal coasts and also for other substances than phosphorus.

The model has also been scrutinized with sensitivity tests, which show that the most important factor regulating model predictions is generally the TP-concentration in the sea outside the coast.

This modeling is not meant to describe conditions or handle changes at individual monitoring sites, but to evaluate the general monthly or seasonal conditions at the ecosystem-scale (for defined entire coastal areas). Working at this scale allows important simplifications to be made, as compared to modeling on finer spatial and temporal scales. Finally, it may be said that simplifications are always needed in modeling, and the main challenge is to find the simplest and mechanistically best model structure yielding the highest possible predictive power in blind tests using the smallest number of driving variables.

A.2 Nutrient Input from Land Uplift

The amount of suspended particulate matter (SPM) always depends on two main causes: Allochthonous inflow and autochthonous production. In the northern part of the Baltic Sea, however, there is also another source, land uplift (see Fig. 1.1 and Voipio 1981). Thousand-year-old sediments influence the Baltic ecosystem today.

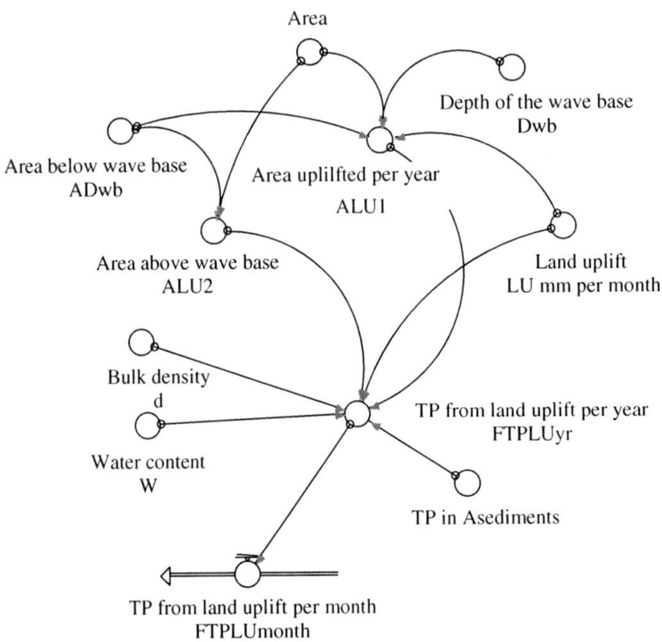

Equations using data from the Gulf of Riga
FTPLUmonth = (10^6)·FTPLUyr/12
Area = 16700·10^6 (m2)
ALU2 = Area-ADwb (m2)
ADwb = 3500·10^6 (m2)
ALU1 = (Area-ADwb)·0.001·LU·12/Dwb (m2)
d = 100·2.6/(100+(W+IG*(1-W/100))·(2.6-1)) (g ww/cm3)
Dwb = (If (45.7·(Area·10^(-6))^0.5)/(21.4+(Area·10^(-6))^0.5) > 0.95·Dmax then 0.95·Dmax else (45.7·(Area·10^(-6))^0.5/(21.4+(Area·10^(-6))^0.5))) (m)
LU = 0.55/12 (mm/month)
FTPLUyr = 12·(ALU2+ALU1)·0.001·LU·((1-(W-15)/100))·(d+0.2)·((TP)/2)·1000·10^-6 (g TP/year)
TP = MTPA/((10^3)·VolAsed·d·(1-W/100)) (mg TP/g dry weight)
W = (if DR > 6 then 65 else if DR > 0.5 then 75 else if DR > 0.045 then 85 else 95) (% ww)

Fig. A.13 The land uplift sub-model

When the old bottom areas rise after being depressed by the glacial ice, they will eventually reach the depth above which the waves can exert a direct influence on and resuspend the sediments (the theoretical wave base). The land uplift in the Baltic Sea (measured in relation to the sea surface) varies from about 9 mm/yr in the northern part of the Bothnian Bay to about 0 for the southern part of the Baltic Sea (Fig. 1.1).

The transport of SPM, TN or TP from land uplift (F_{LU}) may be estimated using the method illustrated in Fig. A.13 based on the hypsographic curve and the depth of the theoretical wave base depth (D_{wb}). Instead of moving the wave base downward when the effective fetch increases, D_{wb} is moved upward due to land uplift (see Fig. A.14). When there is land uplift, the new supply of matter eroded from the glacial clays exposed to winds and waves does not emanate just from the newly raised areas but also from increased erosion of previously raised areas.

Order-of-magnitude calculations on the role of land uplift for the overall budgets for nitrogen and phosphorus given in Table 4.1 for the Baltic Proper. About 80% of the materials settling beneath the theoretical wave base of 44 m in this part of the Baltic Sea emanate from land uplift (from Jonsson et al. 1990). The contribution of TN and TP from land uplift has been estimated from empirical values of the TP- and TN-concentrations in the sediments which will come above the wave base due to land uplift (about 1 and 3 mg/g ww for P and N, respectively), and an estimated average sedimentation of 0.15 cm/yr beneath the wave base. Much of that information comes from Jonsson (1992) but also from unpublished reports from Uppsala

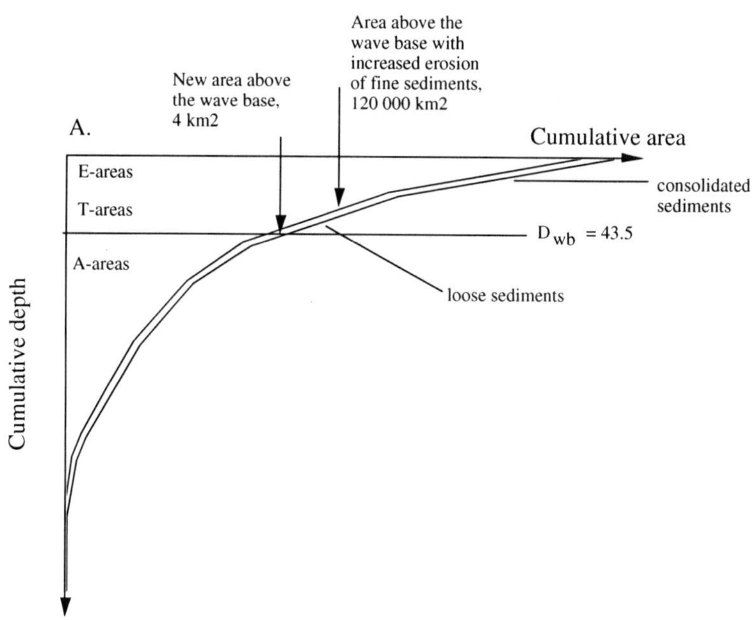

Fig. A.14 Illustration of how land uplift influences the area exposed above the theoretical wave base

University. The area above the theoretical wave base is 120,000 km^2 in the Baltic Proper (including the Gulf of Finland and the Gulf of Riga).

Finally, it may be mentioned that the materials (nutrients, clay particles, organic matter, iron, manganese, etc.) added to the Baltic Sea ecosystem from the ongoing land uplift may influence the system in many ways not accounted for in these simulations for the individual coastal areas. One should expect that the sub-basins with the highest land uplift (the Bothnian Sea and the Bothnian Bay) would be influenced the most. One such also expect that there may be several compensatory effects related to the eutrophication so that a higher loading of nutrients from land uplift may be counteracted by increased concentrations of SPM (clay particles), which would cause increased flocculation, aggregation and sedimentation (and higher settling velocities). It has been beyond the scope of this book, which has a focus on general tools and criteria for coastal management, to discuss the specifics related to land uplift in the Baltic Sea more thoroughly.

Index

Printed in the United States
203475BV00001B/18/P